AMEDEO AVOGADRO

CHEMISTS AND CHEMISTRY

A series of books devoted to the examination of the history and development of chemistry from its early emergence as a separate discipline to the present day. The series will describe the personalities, processes, theoretical and technical advances which have shaped our current understanding of chemical science.

MARIO MORSELLI

AMEDEO AVOGADRO

A Scientific Biography

D. REIDEL PUBLISHING COMPANY
A MEMBER OF THE KLUWER ACADEMIC PUBLISHERS GROUP
DORDRECHT / BOSTON / LANCASTER

Library of Congress Cataloging in Publication Data

CIP

Morselli, Mario, 1922-
Amedeo Avogadro, a scientific biography.

(Chemists and chemistry)
Bibliography: p.
Includes indexes.
1. Avogadro, Amedeo, 1776-1856. 2. Chemists–Italy–Biography.
I. Title. II. Series.
QD22.A8M67 1984 540'.92'4 [B] 83-24564
ISBN 90-277-1624-2

Published by D. Reidel Publishing Company.
P.O. Box 17, 3300 AA Dordrecht, Holland.

Sold and distributed in the U.S.A. and Canada
by Kluwer Academic Publishers,
190 Old Derby Street, Hingham, MA 02043, U.S.A.

In all other countries, sold and distributed
by Kluwer Academic Publishers Group,
P.O. Box 322, 3300 AH Dordrecht, Holland.

Printed in The Netherlands

CONTENTS

LIST OF ILLUSTRATIONS

ACKNOWLEDGEMENTS

First of all, my boundless gratitude goes to those dear friends and relatives, on both sides of the Atlantic, who for over a decade were able to endure my toil with Amedeo Avogadro.

I am most indebted to those who through these years advised me or read through several versions of the manuscript. I would like to recall Mary Twitchell, Richard Kremer, Sara Schechner, and especially Barbara Reeves at Harvard. Their advice, skill and patience were invaluable. I would like to express my appreciation to Frank James for his final reading of the manuscript.

Above all, I feel most grateful to John H. Brooke, Robert Fox, Noel G. Coley, William Brock, and Jed Buchwald, whose scholarship helped me through successive drafts of this work. Without their experience and their encouragement, I would never have succeeded in publishing it.

For assistance with the primary sources I wish to thank the Director and the staff of the Biblioteca dell'Accademia delle Scienze of Turin, also the Directors and staff of the Biblioteche Civiche of Turin, Biella, and Vercelli; Giovanni Zaccone who helped me by researching the Archivio di Stato of Turin.

The Special Collections Libraries at Cornell, Dartmouth, and Boston Public Library were of great assistance, and I benefited repeatedly from the vast Italian history collection in the Widener Library at Harvard, especially the H. Nelson Gay Italian Risorgimento Collection.

The staff of the Bailey Library of the University of Vermont and especially John Buechler, Director of its Special Collections, generously offered their time and precious facilities year after year.

Throughout the entire ordeal, two persons kept me going. They had faith in me. They helped me morally and materially. For all this, I will never be able to express in words what I feel for them.

Plates 1, 4, 5, 7, and 8 were taken personally by the author. Plates 2, 3, and 6 are reproduced from the volume *Immagini di Torino nei secoli*, Tipografia Torinese Editrice, 1973. I am grateful to Prof. Ada Peyrot, one of the coeditors, for her gracious consent to photograph some of the pages.

South Burlington, Vermont,
North Danville, Vermont,
1968–1982

M.A.M.

INTRODUCTION

> Quantum refert in quae tempora vel optimi cujusque virtus incidet.
>
> (From an architrave on the Palatine.)

The history of the development of scientific thought offers several examples of unconventional, perplexing and untimely conjectures advanced by obscure men. The molecular hypothesis set forth by Amedeo Avogadro in 1811 is one of these. It was little noticed when proposed and it soon fell into oblivion.

Although numerous attempts have been made in the past one hundred years to identify the main reasons for the neglect and the indifference which surrounded the molecular hypothesis, very little effort has been devoted to showing how historical circumstances and Avogadro's own personality hindered the acceptance of the concepts he had proposed.

After half a century of intense scientific activity, Avogadro passed away as little known in Italy as abroad. He had few close friends and his personal life seemed centered on his large family and his work. Outside this limited circle, he avoided social engagements and intellectual contacts. This attitude may have been encouraged by the political situation in Piedmont, his native country, which remained essentially hostile to intellectual activity until 1840, by which time Avogadro was sixty-four. The only respite in Piedmont had come during the short-lived Napoleonic regime; and quite remarkably, the molecular hypothesis was conceived and published at the very end of this period.

Avogadro's name is universally recognized, although his work is often misrepresented. Avogadro is described in most textbooks as a chemist, yet he never undertook chemical manipulations. Naturally he is known as a scientist, but his formal education was in jurisprudence, and he was a lawyer and public administrator for many years. He was interested in most fields of physics (except optics), and he approached chemistry from a strictly physical and mathematical point of view. In an age of skilful experimentalists, he remained a theoretician all his life. His experimental works were few and insignificant; and he often used the empirical results of others in order to demonstrate the validity of his own speculations.

Other aspects of his personality also set Avogadro apart. In the scientific community of the first half of the nineteeth century, Avogadro emerges as an independent mind; he had no mentor. While he drew most of his concepts of heat and molecular forces from the French School, at the same time he proved his intellectual emancipation from Laplace and Berthollet by being the first among the followers of their program to implicitly accept the Daltonian theory of atoms. In spite of some marginal reservations, he understood the importance of atomic theory and incorporated the core of it in his gas hypothesis.

Avogadro's writings evince such personal qualities as insight, imagination, lack of intellectual prejudice, and a highly developed ability to correlate heterogeneous notions. These attributes distinguish his approach from the empirical one adopted by most of his contemporaries.

By 1809 Avogadro was focused on his attempt to reconcile the chemical behavior of gases with the "theory of definite proportions" (as Avogadro himself qualified the atomic theory in his 1814 mémoire), seeing the need for an original path to the problem. Gay-Lussac had already come very close to the conclusions eventually reached by Avogadro, and, as Ampère's 1814 essay demonstrated, the stage was set to harmonize Gay-Lussac's generalization and Dalton's findings.

As a conceptual bridge between atomists and volumists, the molecular hypothesis was generally ignored. Those who were aware of it failed to recognize its significance. With hindsight, they cannot be faulted for their attitude. While physics advanced remarkably in the first quarter of the nineteenth century, chemists were still struggling in their endeavor to give a quantitative structure to their science, and some of them were too busy opening new paths – eight previously unknown elements were isolated and identified in the four years immediately preceding 1811 – to pay much attention to yet another generalization put forward by a scarcely known provincial thinker.

A contrast to Avogadro is provided by another physicist, a generation older, who came from a provincial town in neighboring Lombardy: Alessandro Volta (1745–1827) taught in the local secondary schools before his appointment at the University of Pavia in 1778, two years after Avogadro's birth. He spent most of his life within 100 miles of Avogadro's residence. Volta owed his far-reaching reputation to two circumstances, both absent in Avogadro's case: the broad experimental basis of his memorable researches, and the continuous and close contacts he maintained with some of the best known, influential natural philosophers and scholars of his time. In the course of his frequent journeys he met Voltaire in Geneva; he was befriended by Lavoisier, Berthollet, and Laplace, among others; he discussed his findings with Franklin and Priestley, set forth his observations before the Royal Society in London and regularly corresponded with fellow physicists from Germany, Holland, and Switzerland. Upon his return from his first trip to Paris at the age of thirty-seven, and well before the discovery of the 'artificial electric organ', Volta was widely recognized for his merits as a natural philosopher.

With the exception of the first fifteen years of his career, Avogadro worked almost entirely outside the main currents of European scientific thought. He did not travel abroad and corresponded little; and his writings lacked the brilliancy and appeal of those reporting experimental discoveries. His image was too indistinct to assure him even a *succès d'estime*. Apparently he never tried to alter his reserved nature, and there was still less he could do to change the time of appearance of his gas hypothesis. Its acceptance in its current form was delayed until both physics and chemistry reached their maturity at the end of the nineteenth century.

This book contains a concise description of the historical milieu in which Avogadro lived and worked, followed by a thorough discussion of the molecular hypothesis. In my critical examination of his major legacy to science, special consideration has been given to its vicissitudes in the fifty years that preceded the Karlsruhe Congress and in the forty years that followed it. This should be regarded not as an attempt to rewrite the history of the physical sciences in the nineteenth century, but as an examination of the various understandings of the molecular hypothesis by those who investigated the constitution of the matter, from Ampère to Clausius and from Berzelius to J. J. Thomson.

One of the aims of this study is to offer to the English-speaking world a detailed examination of what Avogadro did in addition to proposing the gas hypothesis. The reader then will be better prepared to judge the meaning and the value of the minor writings. But the main purpose of the book is to place Avogadro and his work fully within the context of his times.

CHAPTER ONE

AMEDEO AVOGADRO, HIS COUNTRY AND HIS TIMES

Generally recognized as an Italian, though the definition is correct only in a strict geographic sense, Amedeo Avogadro was a citizen of the Kingdom of Sardinia for sixty-eight years and, during the more crucial period of his cultural formation and for the elaboration of the molecular hypothesis, a subject of Napoleonic France.

During his lifetime, the Sardinian Kingdom represented a mosaic of regions having very little in common except the bond of the ruling dynasty, the House of Savoy. It covered an area of approximately 25,000 square miles which straddled the Western Alps east to the Apennines, from Lake Geneva to the river Magra on the Tuscan border, and included Savoy, Piedmont, Liguria, the County of Nice and Sardinia. Within these borders lived a great variety of peoples: in fact, historical and ethnic features kept the Piedmontese apart from the Sardinians and separated the Ligurians from the Savoyards as if these populations were living at the opposite corners of Europe.

Piedmont, where Avogadro's ancestors had dwelt for over five hundred years, stands in a unique position among the Italian regions. Eventually the prime mover in shaping the political unification of Italy, Piedmont remained for centuries strikingly distant from the cultural and historical centers of the peninsula. Its territory, from the Middle Ages through the seventeenth century, was shared by the House of Savoy and the Marquisates of Saluzzo and Monferrato, with the struggle for regional supremacy coming to an end only in 1713 when, after the peace of Utrecht, Victor Amadeus II of Savoy was given the royal crown and recognition as the only ruler of Piedmont.

Although the political influence and the military power of these three principalities were felt to some extent by the other Italian States from the second half of the fourteenth century until the end of the reign of Amadeus VIII, first Duke of Savoy in 1439,[1] the ineptitude which characterized the policy of Amadeus' successors (with the remarkable exception of Emmanuel Philibert who ruled from 1553 to 1580), had much to do with the almost total absence of the northwestern regions of the peninsula from the Italian scene during the *Quattrocento* and *Cinquecento*.

The complex political, social, and economic conditions which made possible such an unique occurrence as the Italian Renaissance were not present

Plate 1. Bust of Amedeo Avogadro by Pietro Canonica, erected on the 24 September 1911 in Turin by the International Committee to celebrate the centenary of Avogadro's formulation of the molecular hypothesis.

in Piedmont, which remained untouched by the great cultural movement involving all the other regions in those years. One may then understand why no major manifestation of art – with the exceptions of works by the painter Gaudenzio Ferrari in Novara and Vercelli, some examples of Lombard Gothic architecture in the last two towns and the cathedral of San Giovanni in Turin designed by the Florentine Pontelli – occurred in the entire region. For the Dukes of Savoy, it has been said, the humanities were an exotic fruit, an ornament. Only a few scholars lived in the larger centers, and their work was not appreciated. In this sense, one can agree with the view that the spirit of the Renaissance scarcely penetrated Piedmont or was just smuggled in.[2]

Several reasons lie behind this situation. Besides the intellectual mediocrity of several representatives of the Savoyard dynasty, among the rulers of the various states of the peninsula during that period, they were the only ones of non-Italian origin. Even after the conquest of their cisalpine territories and the removal of their seat of government from Chambery to Turin in 1563, the Savoy kept substantially foreign traditions and customs. They never adopted the Italian vernacular used at the time at, for example, the courts of the Gonzaga in Mantua, the Bentivoglio in Bologna, and the Malatesta in Rimini. It should be added that, from the *Quattrocento* through to the first half of the nineteenth century, the House of Savoy remained one of the more absolute European monarchies. Their traditional sense of discipline and strong belief in a simple, orderly life eventually influenced the behavior and even the way of life of their subjects, especially the aristocracy.

While the Savoy could not rival the Medici or the Visconti in personal resources, they were not less wealthy than the Gonzaga or the Montefeltro; yet from their almost puritanical attitude towards anything which might seem superfluous or sinful, one can infer that nothing even comparable to the treasures of the ducal palaces of Urbino and Mantua would be found in their residences on either side of the Alps.

Along similar lines it is possible then to understand the position of the Piedmontese aristocrats towards the cultural experience of the Renaissance, quite in contrast with the scholarly interests and the mercenarism displayed by many members of the patrician families of Venice, Rome, and Florence. Generation after generation, the Piedmontese nobility, through close contact with their monarchs, acquired a pride, a reticence, a sense of thriftiness, a disdain for extravagance, and a gallantry which found no parallel among the Italian upper classes.[3]

Only in the second half of the seventeenth century may a reversal of this attitude be observed, as clearly indicated by the remarkable stimulus given to architecture by Charles Emmanuel II, his son and grandson, who called to Turin the Modenese Guarini and the Sicilian Juvara, both prominent architects. As a result of the extensive work of these exponents of the Italian Baroque, Turin acquired some of its most distinguished buildings, which represent the city's oldest monuments (other than Roman ruins) and its most noteworthy art.[4] Both the Palazzo Carignano and what became the palace of the Royal Academy of Sciences were due to Guarini, who pushed the Baroque style to its greatest extremes. To Juvara was given the design of the magnificent votive church of Superga, of the royal castle of Stupinigi, the so-called Versailles of Piedmont, and Palazzo Madama, all built in the first two decades of the eighteenth century. At the same time Turin also acquired a modern urban plan characterized by wide, regular, right-angled streets which makes it rather unusual among the Italian cities.

Bypassed by the Renaissance and by the humanistic experience, the Piedmontese entered the modern age with a frail foundation for the development of a truly sophisticated cultural tradition.

Probably more than in other Italian regions, the scarce diffusion or the absence for all practical purposes of a common official language hampered the subjects of the Savoyard Kingdom. Italian emerged slowly as a *lingua franca* for that small fraction of the population which was literate. For those of more advanced education, Latin, even in the early nineteenth century, was often more familiar than Italian.[5] At home, the Piedmontese of the middle classes, and more so the aristocracy, used either French corrupted by local expressions or a vernacular. And most of the time, everybody, old and young, rich and poor, spoke their local dialects of which the Kingdom of Sardinia, no less than the rest of Italy, was rich: at least six of them in Liguria and Piedmont plus the Sardinian and the *patois* of Savoy and of the Vare region beyond the Alps. One could hear Italian only in the schools, the courts, political and scientific gatherings, and read it in the journals.[6] A great deal of correspondence, private or otherwise, was conducted in French and in this language were many of the books read by the educated. On the other hand, a good deal of theatre and poetry were performed and written in dialect. A further element typical of Piedmont was, through the first half of the nineteenth century, the attraction of the mostly francophone upper classes towards the French-speaking centers of culture and business, such as Lyon, Dijon, Paris, Geneva and Neuchatel. One could fairly say that Turin was culturally closer to these cities than to Naples, Rome, Venice, or Florence.

A series of historical circumstances, enhanced by the peculiar strategic location of the country on the western threshold of the peninsula and by the political ambition of its rulers, whose martial spirit endured even in peacetime, did not help to modify the prevailing trend of indifference towards any cultural initiative. For generation after generation, after 1580, the main effort of the ruling dynasty was aimed rather at insuring the survival and independence of their country by skillfully playing the game of diplomacy among the Great Powers. To this purpose, the resources of the small state were mainly devoted to raising and maintaining the army at its maximum strength.

During the seventeenth century the Dukes of Savoy allied themselves alternatively with the Spaniards against the French and then with the French against Philip of Spain. Thereafter, with the Austrians as allies, Victor Amadeus II fought Louis XIV of France and defeated him near Turin in 1706. As a consequence of a series of victories over the French during the War of the Spanish Succession, the Treaty of Utrecht granted Victor Amadeus the crown of Sicily and the Marquisate of Monferrato. Six years later, in 1720, when Sicily was returned to the Bourbons he received Sardinia in exchange and the new Kingdom of Sardinia was thus established.[7]

The centuries-long dream of kingship of the House of Savoy had finally become a reality. However, in the second half of the eighteenth century the increasing military might of Prussia altered the entire balance of power on the continent. Among the repercussions of this shift in the European political equilibrium was the steady decline in importance of the Savoyard Kingdom as a potential political and military ally. This oddly happened while the King was the most brilliant leader of the dynasty in the century, the son of Victor Amadeus II, Charles Emmanuel III, who ruled from 1730 to 1773. In 1756, when the former enemies of the Austrian wars allied themselves – with the exception of Britain – to fight Prussia in the Seven Years' War, no power asked for the support of the Kingdom of Sardinia.

The man who succeeded Charles Emmanuel III proved to be a weak and mediocre ruler in every sense. Victor Amadeus III, in whose reign Amedeo Avogadro was born and spent the first twenty years of his life, became king when a middle-aged man in 1773. Instead of concentrating on broad and badly needed programs for the improvement of the general conditions of his kingdom, he ruled as if it were his own family, trying to take care of and provide for even the most minute detail of not only the government's day by day administration, but also the private lives of the individual citizens. The whole country eventually suffered the consequences of the policy adopted by this paternalistic and narrow-minded autocrat. While the main resources of Piedmont, whose economy had been bled by endless wars, remained its agriculture, industry in the last decades of the eighteenth century was so scanty and backward that almost all manufactured goods came from abroad with obvious consequences for the coffers of the State.

Though Victor Amadeus III deeply admired Frederick the Great, he surrounded himself with mediocre men in resplendent uniforms, whereas the sophisticated Prussian king preferred the company of intellectuals. Married to a devout and bigoted Spanish princess, the Savoyard ruler behaved, at least formally, as a very strict Catholic. Not only did he authorize the reinstatement of the Inquisition in his States, which meant the reappearance in Piedmont of torture wheels, but he compelled his subjects to receive Communion at Easter and to abstain from meat on Fridays.

In the general gloom of the Kingdom of Sardinia there were still a few beacons of erudition. Beside the University of Turin founded in 1404, thus a relatively recent institution by Italian standards, another institution had emerged, the Academy of Sciences. This had been formed in 1759 as a private society by Count Angelo Saluzzo, Gian Francesco Cigna, and Joseph Louis Lagrange. It was looked upon with mistrust in those early years and surprisingly it was the then Prince of Piedmont, later Victor Amadeus III, who asked and obtained in 1762 from his father some tolerance for the new 'association of learned people'. After he became king, Victor Amadeus III granted royal patents in 1783 to the Academy of Sciences, which by then had lost its most eminent member with the departure of Lagrange for Berlin.

In 1776 other small and informal circles of intellectuals were established in Turin. The *Accademia Sanpaolina* met in the house of the Counts of San Paolo, while Count Prospero Balbo founded the *Accademia dei Crescenti*. There was also a *Patria Società Letteraria* which published a periodical, *Ozii letterarii*. Intellectual curiosity and the desire to exchange ideas brought together philologists, writers, poets, mathematicians, and physicists. These gatherings offered them the opportunity, quite rare in those years, of meeting foreign scholars and through them of knowing what took place beyond the borders of the Sardinian States.

Around 1780, during Avogadro's childhood, other groups of men, mostly members of the middle class, but also a few aristocrats, all moved by common leanings, began to associate not only in Turin but throughout Piedmont. Their discussions, not always limited to poetry and literature, often involved the political, social, and economic doctrines then emerging in France.[8] Despite all the precautionary censorship of Victor Amadeus, the ideas expressed by Diderot, Voltaire, Condillac, and Rousseau eventually penetrated his Kingdom and reached Turin. The writings of these men became the subjects more commonly debated and greatly influenced the attitude of the members of such circles.

To a few of them, who later on would become the hard core of the Jacobinism in Piedmont, a freely elected republican regime appeared more and more clearly as the only acceptable solution to the social and political problems of their country. To others, Jansenism represented the righteous alternative to the abuses and corruption of the Roman Church. Not surprisingly, most of these men, aware of the dangers involved in their ideas and afraid of the police, met surreptitiously, sometimes in Masonic lodges which were widely diffused.[9]

Needless to say, of paramount importance was the control of ideas and, accordingly, a tight censorship was enforced on every cultural manifestation. After 1776 all books printed inside and outside Piedmont needed to be approved by a board of reviewers. Piedmontese and Sardinian citizens who published abroad without the *imprimatur* of the board were heavily penalized. In order to prevent the 'contamination' of youth, the King even denied his subjects permission to attend courses at the nearby University of Pavia because of the political ideas expressed by some faculty members.[10]

Under these conditions, rather than surrender to the general dullness and conformity pervading the entire country, several people left Piedmont in the decades immediately preceding the French Revolution. Among these were Carlo Denina, professor at the University of Turin and distinguished Hellenist, who went into exile because his political views were unorthodox;[11] Giuseppe Baretti, a noted literary critic, had preceded him; and Count Vittorio Alfieri, Italy's most famous tragedian, spent only a few years of his adult life in Piedmont, not finding there the freedom he needed for his work.[12] In those years the Kingdom of Sardinia lost two other citizens, Joseph Lagrange and Claude Berthollet, who later rose to become one of Europe's greatest mathematicians and France's most renowned chemist, respectively.

In the last quarter of the eighteenth century, during Amedeo Avogadro's youth, the cultural isolation of Piedmont grew even more extensive. To its rulers, the most serious threat was France. Victor Amadeus III and his advisors did not fail to grasp the importance of preventing the libertine customs – so widely accepted at the Court of Versailles – and the more subtle but not less pernicious philosophical doctrines then discussed beyond the Alps from reaching their subjects. Thus, at that time, although French still remained the language spoken by the King and the aristocracy in Turin, no effort was spared to prevent the circulation of ideas between the two countries; personal contacts and travel abroad were also highly discouraged.

While this was taking place, the masses, often without outside instigation, had begun to move. The first riots took place in Piedmont in 1776, the main complaint then being the unjust distribution of taxes. Next in turn came Sardinia, where the economic conditions of the laborers were frightful; and Savoy, the transalpine region, whose communities had been more exposed to the political ideas coming from France.

In 1792, under the renewed pressure of the French loyalists who had taken refuge in Piedmont, Victor Amadeus III began to assemble troops on the western borders of his Kingdom.[13] The features of the new French regime were so antithetical to his social and political views that he had in fact little choice other than to join the coalition then forming against France. The Sardinian army, although fighting hungry and ill-equipped opponents, was quickly defeated by a superior strategy near Mondovi and shortly afterwards Victor Amadeus III asked for an armistice which was signed on 27 April 1796. As a result, the French occupied most of northern and central Italy for the following three years. Later in 1796, upon the death of Victor Amadeus III, his son Charles Emmanuel IV became King.

By 1798, the political situation quickly deteriorated; on the fourth of July, an understanding was reached in Milan between the Sardinian plenipotentiary and the French General Brune, according to which the French troops – to 'safeguard' Piedmont's security – would occupy Turin's citadel for two months. Then the pace of the events quickened: in the latter part of September and again later on, severe fighting took place between the French troops and the people of Turin. By the end of November, Barthélemy Joubert, the new French Commander-in-Chief, following orders from Paris, proclaimed the total occupation of Piedmont and incorporated all the Sardinian forces into the French army. A few days later, after being subjected to further humiliation by the French, Charles Emmanuel left Turin for exile in Sardinia.[14]

In those final days of the eighteenth century, among the Piedmontese Jacobins and the more impassioned supporters of the republican regime, a movement had been taking shape whose final aim was political union with France. Although they represented a minority whose feelings were scarcely identifiable with those of the population, a few of the idealists among them saw this project as the only way for their small country to closely participate in the most extraordinary political and social upheaval of the time.

Beyond the Alps, the government in Paris not only looked with favor on this movement, but actually encouraged it; when in February 1799 the Senate of Turin approved the formal union with France, the *Directoire* immediately sent Joseph Musset as plenipotentiary commissioner to Piedmont. Without wasting time he dismissed the provisional government, established new civil and criminal courts, and reorganized the administration of the region into four departments, according to the new French system. Shortly afterwards he abolished the *Camera dei Conti* and, rather ironically, the same Senate which had just voted for the union with France.

In the summer of 1799, during Bonaparte's campaign in Egypt, a combined Austrian-Russian army, following several victories in northern Italy, entered Piedmont and occupied Turin. For one year Vienna administered Piedmont more as a conquered than an allied country. What was left of the Piedmontese army was compelled to swear allegiance to the Austrian Emperor.

After the return of Napoleon from Egypt and his successful *coup d'état* on 9 November 1799, a new invasion of Italy began in the spring of the following year, this time directly through Piedmont. Surprised by the unexpected maneuvering, the Austrians were compelled to evacuate the whole region in June 1800, after a battle near Marengo, whose outcome determined the political structure of the Italian peninsula for the following fourteen years. For a while, after Marengo, Bonaparte avoided any definite decision on the future of Piedmont. Then, on 21 April 1801, it was declared a military division at the disposal of the Consulate. Its political incorporation as a department of the French Republic came only on 21 September 1802. At that time, still in exile, Charles Emmanuel IV abdicated in favor of his brother Victor Emmanuel I.[15]

If Piedmont, in the eighteenth century, had kings of uneven talent, the state administration was staffed by a body of civil servants, from both the aristocracy and the emerging middle class, which was unparalleled in Italy for its honesty and reliability.

The family of Amedeo Avogadro had for generations symbolized the qualities that made the Piedmontese civil service unique. The Avogadros are so intertwined with the history of Piedmont that it is difficult to glance through local annals without finding their name.

Their origin can be traced back in the thirteenth century to the alpine valleys north of Vercelli and west of the Sesia river.[16] In the chronicles of many communities of this region, the public activities of the Avogadros, mainly lawyers and churchmen, were recorded. As a result of the branching of the family through the centuries and of the titles conferred on them, at least a dozen different patronymics are reported, Amedeo Avogadro's branch, Quaregna, dating from the middle thirteenth century.[17]

In 1232 the Avogadros di Casalgualone are mentioned on the occasion of a truce signed between Vercelli and Novara at the time of the renewal of the Lombard League against Frederick II. The two communes were often in conflict during the feudal wars in northern Italy, and Ardizzone Avogadro is recorded as a member of the military expedition led by the Vercellese people against Novara in 1218; his family was then leading the Guelph faction in Vercelli. Among the fiercest defenders of the Church's rights against the Emperor, Frederick II, is recalled Martino Avogadro di Quaregna, Bishop of Vercelli. Another member of the family, Guglielmo, appears in 1274 as one of the knights fighting on the side of the Milanese Ottone Visconti against the Ghibellines led by the powerful Torielli, rulers of Novara. In 1307, Giacomo and Pietro Avogadro di Quaregna were among those involved in the capture near Biella of Fra' Dolcino, leader of a group of heretics.

Simone Avogadro di Collobiano, lord of Vercelli, although a Guelph himself, welcomed Henry VII, the German Emperor, when he crossed the Alps as King of the Romans and entered Vercelli in 1310.[18] Simone Avogadro remained until his death a leading figure in the Guelph movement. The Guelphs, for most of the first half of the fourteenth century, had to struggle not only with the Emperor, Henry VII, but with the most powerful of his Ghibelline party supporters in Italy. Several times in this period, the Avogadros were compelled to leave Vercelli and their castles only to return again as soon as the events turned in their favor. This led eventually to the end of Vercelli as a free commune and to its transfer in 1335 to the domain of the Visconti, which lasted until 1427, when the entire district from the Alps to the Po river was acquired by Amadeus VIII of Savoy.

Twenty-three years earlier, in August 1404, in a small village near the Piccolo San Bernardo pass in the Alps, representatives of the Avogadro family swore their allegiance to Amadeus VIII, then still Count of Savoy, who had begun to expand his possession over eastern Piedmont.[19] This was the beginning of their relationship with the dynasty that would eventually last for over five centuries. The Avogadros served first their dukes, and later their kings as members of both military and civil administration: they were generals, magistrates, diplomats, and members of Parliament.

Alberto Avogadro, whose Latin verses were good enough to attract the attention of Cosimo de Medici in the sixteenth century, spent most of his life outside Piedmont. In Novara also the Avogadros remained prominent; the chronicles mention that when in 1630 the great plague which killed millions in northern Italy reached Novara, Fabrizio Avogadro was one of the few eminent citizens who did not leave.[20] Of more recent interest is Augusto Avogadro di Collobiano, a career diplomat, who was appointed in 1838 as the first representative of the Kingdom of Sardinia in Washington, during the administration of Martin Van Buren. Likely related to him was the Lieutenant Avogadro di Collobiano, one of the officers leading the troops which charged inside the courtyard of the University of Turin to control the rioting students during the constitutional revolt in January 1821. At that time an Abbé Avogadro di Valdengo was Rector of the *Collegio delle Province* in Turin and a member of the *Magistratura della Riforma* whose function was to supervise the educational system of the Kingdom.

As earlier noted, the cradle of the Avogadros was in the northern districts of the province of Vercelli, in the alpine foothills around the Sesia river basin. Here one can still find the small communities whose names appear as patronymics of the various branches of the family. And in this region, a few miles from Biella, lies the tiny commune of Quaregna, where Filippo Avogadro, Amedeo's father, was born in 1734.[21] After studying at the same *Collegio delle Province* where Amedeo eventually tutored, Filippo earned a degree in law from the University of Turin in 1756. He then began a long career in the judicial system of the Sardinian Kingdom. In 1786 Victor Amadeus III appointed him a member of the Senate, the highest judicial body of the State, and then Advocate General in 1777. Ten years later the King promoted him to President of a section of the Senate and shortly afterwards created him Count of Quaregna. In 1795 he was named Regent of the *Gran Cancelleria* and Chairman of the *Magistratura della Riforma.*[22]

When the French occupied Piedmont in 1799, Filippo Avogadro received the appointment of President of the Senate of Turin. A few months later, this body was dismissed by the *Directoire*. Such a measure may in part explain why there is no record of sanctions against him when the Austro–Russians occupied Piedmont after June 1799.

The attitude of the Piedmontese nobility after the departure of the King and throughout the French occupation, is, to say the least, perplexing. Those who had the opportunity to do so left for exile rather than stay. However, a portion of the aristocracy, despite their early links with the dynasty, took a position not unlike that of the middle class – professionals, educators, tradesmen – and welcomed the new political philosophy inspired by the events of 1789 and brought to Piedmont by *l'Armée d'Italie.*[23] This conduct was often rewarded by the French authorities who, in Piedmont, as well as in other districts, appointed to important administrative positions those among the upper classes who had remained in their country. In the case of Filippo Avogadro, his acceptance of a high office may be interpreted either as a demonstration of public spirit, or as an act of realism, or as the result of an opportunistic inclination, when this represented a typical behavior of many well-to-do. In any event, the Avogadros di Quaregna took a more lenient attitude towards the French than many of their peers did.

They may have been influenced also by the policy adopted by General J. F. Menou. A resolute foe of the Jacobins, a veteran of the Egyptian campaign, appointed *Administrateur Général* of Piedmont after 1802, Menou partially succeeded through a series of measures in regaining the trust and the support of the Piedmontese patriciate. Thus, in April 1805, when Napoleon passed through Turin on his way to receive the crown of King of Italy in Milan, about forty members of the aristocracy accepted the invitation to pay homage to their new ruler at the Stupinigi castle.[24]

Filippo Avogadro was forty-two and near the apex of his career in the judicial system of the Kingdom of Sardinia at the time of Amedeo's birth on 9 August 1776. The family was living in the western section of Turin, near Porta Susa, and he was baptized in the Church of Carmine, the local parish.

Little is known about his primary education beside the fact that, like most members of his social class, he received it at home, quite likely under the supervision of priests. His secondary education was obtained in the schools of Turin. After passing his baccalaureate in 1792, he entered the Faculty of Law of the University of Turin, where he spent the next three years, receiving the degree in jurisprudence in 1795 and the title of Doctor in Canon Law the following year. His brother Felice followed the same curriculum and eventually became a judge in Biella. Two other brothers, Lorenzo and Luigi, chose military careers.[25] Upon his graduation in March 1796, Amedeo went to work in the office of the *Avvocato dei Poveri* in Turin and then he moved to the *Avvocatura Generale,* the same body that his father had administered from 1777 to 1787.

He remained there until April 1801 when he took the position of *Secrétaire* in the Department of Eridanus, recently created by the French Government, having Turin as its capital. This appointment may have been the result also of the prominent position that his father Filippo held under the French occupation as President of the Piedmontese Senate before its dissolution in 1799, as well as a membership on the government commission established after the battle of Marengo and the subsequent reoccupation of Piedmont by the French; this commission had broad political and administrative functions all over the region.

When in 1801 France decided to annex Piedmont and a group of prominent citizens was chosen to go to Paris to express the gratitude of the region to the First Consul, Filippo Avogadro was not among them. And when General Menou in 1802 was appointed head of the French administration for the Cisalpine Departments, the local government commission ceased to exist. Emerging personal rivalries and, after the Empire was proclaimed in 1804, possible misgivings among the French towards the men who, already prominent under the Savoyard Dynasty, had accepted positions during the First Republic, meant for some of them the gradual removal from any further assignment. In the case of Filippo Avogadro, his age, too, may have been an important element. In 1807, at seventy-three, he retired from all public activities and died five years later.

From 1800 through 1814 Turin was just another departmental seat of the First Republic and later on of the Empire. Although the Governor-General of Piedmont, the Roman Prince Camillo Borghese, brother-in-law of Napoleon, was living there with his retinue, the town could certainly not match the brilliant social and cultural atmosphere of Milan, which had been raised by Napoleon in 1802 to capital of the Italian Republic and afterwards of the Kingdom of Italy. The importance of Turin, whose population in the meantime had dwindled to 66,000 (the same as in the middle of the eighteenth century), remained mostly limited to its strategic location on the main road connecting eastern France with the Po valley and the Illyrian Provinces, although a commercial bank and an exchange were established there in 1802. French administrators and troops were all over the region, their language became the only official one, and the circulation of French newspapers and books increased. As a result, the influence of the French culture, at least on the upper classes of the Piedmontese population, became more and more extensive in those years.

The inception of the scientific education of Amedeo Avogadro may be tentatively set around 1798. Since he worked on a regular basis first at the *Avvocatura Generale*, then as *secrétaire du department* until 1806 and after that as *répétiteur* at the College of the Provinces in Turin, as well as auditing Vassalli Eandi's physics courses at the University, one may assume that the young Avogadro found time and the state of mind for reading extensively in science only in his leisure.

In 1804, while still at the prefecture, Avogadro became a corresponding member of the Academy of Sciences of Turin. Quite likely the appointment was the result of the favorable review by Vassalli Eandi of two essays on

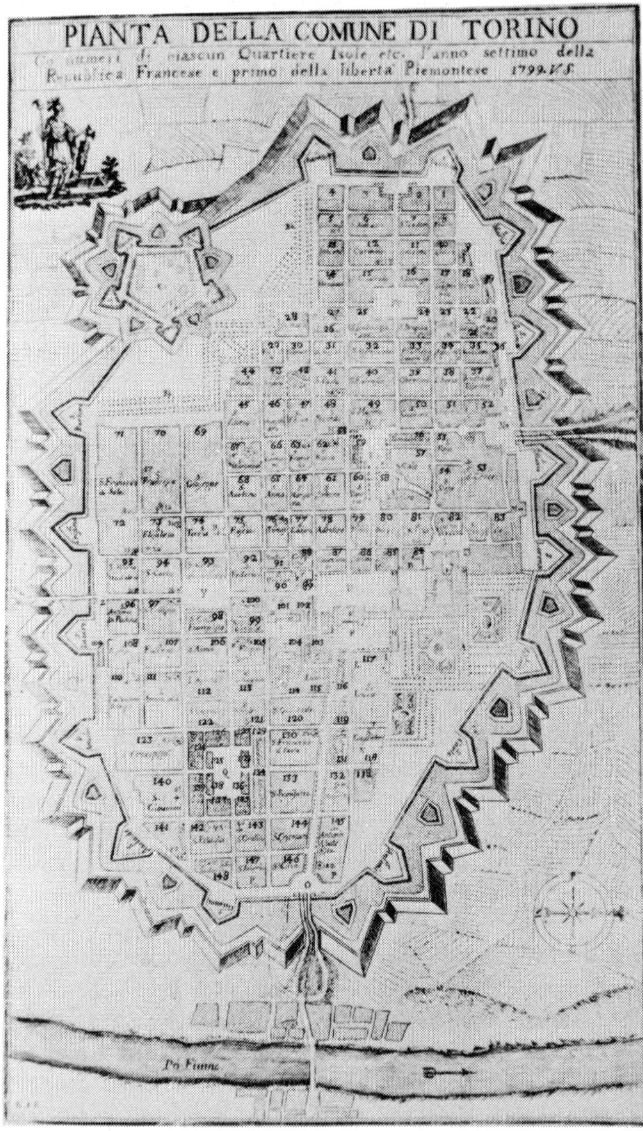

Plate 2. Map of the Commune of Turin, printed in 1799, the first year of its occupation by the French Republican Armies. In that year Avogadro was working as a young lawyer in the office of the *Avvocato dei Poveri* (legal counsel for the poor). Note the grid layout of the streets of Turin, quite unique for the Italian cities at the time. Avogadro's house was located in the Carmine (the 12th quarter in the map), East of the Cittadella. The Collegio delle Province had its seat in Piazza Carlina (no. 9 on the map). One can estimate that the total area of Turin at the time was 4.5 sq km.

electricity he had submitted to the Academy in that year and in 1803, although only a summary of them appeared in the *Memorie della Reale Accademia delle Scienze di Torino.*[26]

Father Antonio Vassalli Eandi, together with Francesco Cigna and Giuseppe Saluzzo belonged to the same generation of Piedmontese natural philosophers which emerged in the second half of the eighteenth century. Some of them reached considerable distinction and exerted a significant influence on Avogadro's early electrical investigations. They had all studied under Giambattista Beccaria (1716–81), professor of physics at the University of Turin after 1748. A Franklin supporter, well-known by Lavoisier, he maintained a correspondence with Joseph Priestley, and became the only Piedmontese member of the Royal Society. Beccaria's most renowned pupil, Joseph Louis Lagrange, left Turin for Berlin in 1766, a few years after founding with Cigna and Saluzzo what became the Academy of Sciences of Turin.[27]

At the turn of the century, Father Vassalli Eandi held the chair of physics previously occupied by Beccaria and by his uncle G. A. Eandi.[28] Those few years between 1800 and 1806, immediately preceding his appointment at the College of the Provinces, should be considered critical in shaping Avogadro's approach to and appreciation of physical sciences. Certainly, the ground on which Vassalli Eandi's lectures fell was very fertile: a brilliant intelligence aided by an intense eagerness and a deep sense of curiosity, a sharp intuition, the constant adherence to a strict logic, all these elements, which characterized his intellectual life, were already present in Avogadro at the beginning of his scientific involvement; time would only sharpen them.

With the first two essays on electricity, by 1804, Avogadro made his debut in the small scientific world of the Piedmontese capital. In the following two years his interest in the electrical behavior of bodies continued and he sent to Jean Claude de La Méthérie in Paris the results of his thoughts on the subject.[29] They were based upon a series of observations conducted almost forty years earlier by Symmer and later on by Cigna, Beccaria, and Volta aimed at explaining the unusual behavior of a layer of insulated material placed between bodies charged with opposite electricities. The facts observed had thus far eluded an acceptable explanation. Without recourse to mathematics, Avogadro analyzed the electric state of a solid insulator, such as two charged and then grounded glass plates when separated from each other, and postulated them as formed in turn by a very large number of extremely thin plates, each one displaying the same electrical behavior as the whole insulator. In the subsequent elaboration of his speculations, Avogadro, after discarding the two-fluid theory of electricity, argued that the electric charge essentially consisted in the displacement of a fluid from a layer of molecules of the insulator to the next layer. These ideas were published in 1806 and 1807 in the de La Méthérie's *Journal de Physique, de Chimie, d'Histoire Naturelle et des Arts* and, although neglected at the time, they represent the main contribution of their author to the theory of dielectrics, as Avogadro himself stated in an 1842 note, after the publication by Michael Faraday of the XIth Series of his *Experimental Researches in Electricity.*[30]

The span of time immediately preceding Amedeo Avogadro's application for a teaching position in Vercelli coincided with his gradual parting from well-defined training and a career that had continued for almost a decade and that would impress a distinct mark on his intellectual life. Since he had been born into a family of jurists, it was to be expected that within this tradition he would have devoted the rest of his life to the legal profession. When later in his life, the Chair he was holding at the University of Turin was abolished, he made good use of his legal background as a member of the *Camera dei Conti*, the body responsible for reviewing all the legal actions concerning the *Demanio* or royal estates; it had also both civil and criminal jurisdiction over all crimes and wrongdoings affecting the royal estates.

It is not easy to explain the determining influences for such a major shift in intellectual interests in a personality as steadfast as that of Avogadro. Even in the early part of the nineteenth century – when separations between fields of human endeavor were not so sharp that people could not enjoy a broad range of concerns – one cannot find many examples of individuals who underwent extreme variations in their intellectual pursuits.

In the physical sciences, men who eventually achieved success and fame at that time were either substantially self-educated or trained in a parallel discipline, quite often medicine; only a minority had formal and thorough training in their main field of activity. However, Avogadro represents the peculiar case of a man whose later informal acquisition of a vast amount of knowledge in totally unrelated branches of learning was superimposed upon a long and profound basic education. It is fair to assume, however, that the young Avogadro had a special inclination toward mathematical and physical studies even before his graduation from law school in 1795, and that he enhanced these natural leanings by reading assiduously and associating with people also involved in scientific work.

An avid and proficient reader of the scientific literature, Avogadro conceivably realized the gulf between his cultural milieu and that of his fellow scientists beyond the Alps. This contrast between the lonely, scarcely known individual and those who practiced science in his time emerges strikingly when one takes into consideration the cultural atmosphere of other European countries in the early part of the nineteenth century. In France, the new educational institutions established during the Revolution found strong support in the Napoleonic regime, which provided the most propitious conditions for the flourishing of a brilliant generation of scientists. Quite significant was the emergence of patronage, especially in Paris.[31] Long years of close association and a combination of friendship, sharing of common views, deference and gratitude contributed to the establishment of very firm ties between pupils and their teachers. With the passing of time these relations, in many instances, continued and extended to the next generation of young scientists; in turn, a few chosen among them became the leaders of new schools. As a typical manifestion of the system, a certain degree of loyalty and coherence with the ideas upheld by the patrons was expected; for those who took an independent position and for the outsiders, the chances of success and even of recognition were often very limited, regardless of their intrinsic merits and their achievements.

Although Avogadro worked for over a quarter of a century within the French scientific tradition, he derived no benefit from its favorable circumstances.[32] He remained isolated; he did not belong to any group or academic *clique*.

He should also be set aside from Davy, Ampère, and Dalton, because he had already a broad cultural preparation when he turned his interest from law to physics. One may, in fact, detect the unmistakable impression that his juridical studies left on both his mental process and style of writing. The latter is too often extremely involute; the former leans mostly on sheer speculative power leading to a wide range of generalizations to which he tried subsequently to adapt the empirical data. Needless to say, with this approach he moved against the grain of orthodox scientific logic and drew either the indifference or the criticisms of his fellows.

A comparison with the two men whose speculations were instrumental for the development of the molecular hypothesis is revealing: Avogadro stands closer to John Dalton, an independent mind, who brought forward bold new ideas, and despite the dissensions and criticisms, never compromised or deviated from his tenets as long as he lived.[33] In contrast to these two is the figure of Joseph Louis Gay-Lussac who, in reporting in 1809 the results of his investigations on the regularity of combining gas volumes, found himself in conflict with the opinion of Claude Berthollet;[34] his attempt to reconcile his position with that of his celebrated mentor simply caused more ambiguities.[35] A few years later, Gay-Lussac himself admitted that, on such crucial issues as the oxygen theory of acids and the elementary nature of chlorine, under the influence of Berthollet, he could not make a concession.[36]

Avogadro's social status and his family's long cultural tradition set him quite apart from Dalton, Dumas, or Faraday, who all came from humble backgrounds and had to overcome the inadequacies of their modest upbringing. He did not suffer, moreover, from the intellectual stubborness which typifies the attitude of many individuals lacking an extensive formal education.

Yet, when in 1806, the turning point came in Avogadro's life and he left the practice of law and an administrative position with the French Government in Piedmont in order to devote himself to teaching and to the study of natural philosophy, it is beyond doubt that he had to face not only the limitations of his scientific training, but also the added drawback of being isolated from Europe's cultural centers. There is scant evidence to substantiate the assertion that he had many friends outside Piedmont.[37] His personal contacts with transalpine physicists and chemists were limited indeed: no profuse correspondence existed with them, although it is known that much later he entertained a relationship with Auguste de La Rive in Geneva and with Faraday who appreciated the value of some of his works.[38] He never journeyed abroad like Berzelius, Davy, Berthollet, Gay-Lussac or, among the Italians, Volta.

In the years immediately preceding his appointment at the College in Vercelli in 1809, another road was open to Avogadro. A person more ambitioous and eager for honors would probably have left Piedmont and its narrow provincial atmosphere. The circumstances were favorable: Paris, the capital of the Empire of which Piedmont was then an integral part, attracted young intellectuals from all over Europe. Those were the years when the *élite* of French science gathered between Arcueil and Auteuil.

How agreeable would the noisy, brilliant and often frivolous world of the French capital have been to an individual so disinclined to social engagement and flattery as Amedeo Avogadro?[39] From what is known of him, the answer to this is not difficult. Very fond of his large family, Avogadro actually led a much more retiring life than his contemporary John Dalton, who certainly had not the reputation of being a particularly convivial man.

In contrast with Avogadro's fate is that of two among his compatriots who were destined to achieve great fame during their lifetimes. Both emigrated rather than wither in the intellectual mediocrity of the Savoyard capital; Claude Louis Berthollet, born in Savoy, received a degree in medicine at the University of Turin and at twenty-four, four years before Avogadro's birth, left for Paris, where he studied with Macquer and became an intimate friend of Lavoisier. By then a French citizen, he survived the Revolution; later on, although one of the most influential scientific advisers of Napoleon, who made him Count of the Empire, he went through the Restoration and was appointed Peer of France by Louis XVIII. Regardless of his merits as a scientist, Berthollet still remains a prominent example of a survivor, at least from the political point of view.

The other, Joseph Louis Lagrange, spent the first thirty years of his life in his native Turin. Early recognized as the greatest mathematician of his time, he left for Berlin in 1766 and stayed there for twenty-one years as the director of the science section of the Prussian Royal Academy. He then moved to Paris, invited by Louis XVI; his fame was so outstanding that, despite his friendly relations with the Bourbons, he, too, went unmolested through the French Revolution. He never returned to Piedmont.[40]

In 1806, at the time that his first essay on electricity was published in France, Avogadro, then thirty, left his administrative position with the French government to begin his teaching career. At first, he was *répétiteur* at the *Collegio delle Province* (College of the Provinces), a boarding institution connected with the University of Turin, having at the time about 100 students chosen among those of high academic standing. It had been established in 1730 by Victor Amadeus II for those who wished to attend the University but lived outside Turin, and it was for a few generations considered the seedbed of the Savoyard ruling class.[41]

In 1809 came the appointment to professor of 'positive philosophy' (or mathematics and physics) at the former Royal College of Vercelli. The decree of the appointment was issued by the Imperial University of Turin which acted as a ruling body for the educational system of the Cisalpine Departments.

Vercelli, where Avogadro would spend eleven years, is 35 miles east of Turin; with a population at the time of 16,000, it had been made seat of the Department of the Sesia and, within the French administrative system, a seat of a court of appeals and entitled to a *lycée*. However, no such a school had been established in Vercelli even after the Imperial decrees of 17 March 1808 and 15 November 1811, when the number of the lyceums in the Empire was raised to 150. Rather, the school where Avogadro taught was the result of the merging of the *Regie Scuole* and of the *Collegio Dal Pozzo*, the latter formerly a boarding institution founded in 1573 and which after 1727 sent its few students to attend classes at the *Regie Scuole*. These royal schools had been established by Victor Amadeus II in 1727 when reorganizing the educational system of Piedmont; their curriculum consisted mainly of grammar, humanities, philosophy, and rhetoric, and their diploma of proficiency was necessary in order to be accepted by the Universities in the Kingdom.[42] There is no record of how many students were attending the College of Vercelli while Avogadro taught there. The overall impression emerges that, despite the impressive name, it was a small school with modest programs.

In 1809, de La Méthérie published in the *Journal de Physique* a six-page note by Avogadro, shortly before his appointment to the College of Vercelli, stating that oxygen is not 'the principle of acidity' and introducing a series of considerations on the concept of alkalinity and acidity, to which he remained faithful for the rest of his life.[43]

It is quite possible, as will be discussed later in greater detail, that by the fall of 1809, when he moved to Vercelli, Avogadro was already gathering the material which would help him to formulate the ideas he would submit to the attention of the scientific world in his most celebrated work, the 'Essai d'une manière de déterminer les masses relatives des molécules élémentaires des corps . . . '. This appeared in the *Journal de Physique* for July 1811, in the same issue as a long article by Berzelius,[44] and was followed three years later by 'Mémoire sur les masses relatives . . . ' again in the *Journal de Physique*, which represents a major corollary to the concepts expressed earlier.

In the 1811 essay Avogadro explained the simple combining ratios of gases found by Gay-Lussac by asserting that under the same physical conditions, equal volumes of gases contained the same number of particles. This idea had been considered by others at the time, including Dalton, but then rejected, because it implied *ipso facto* the concept of a divisibility of elementary particles which appeared nonsensical. Resorting to what has been called an auxiliary or an *ad hoc* hypothesis, Avogadro postulated the existence of two orders of particles at the elementary level.[45] This assumption applied to both solid and gaseous elements, to the former when in the form of vapors. From the equal volumes–equal numbers generalization Avogadro derived the tool that in his view would offer chemists an unerring way to determine the relative weights of the elements and their compounds, regardless of their physical state, and their chemical composition as well.

In the period of time between the appearance of Avogadro's first fundamental paper in 1811 and that of the second in 1814, the fate of Napoleonic Europe was sealed, and the contours and political structure of the Italian peninsula for the next half century were established. Victor Emmanuel I returned from exile in Sardinia to Turin in May 1814. Through the final act of the Congress of Vienna, Piedmont and Sardinia were restored to him, and he also received Genoa and the entire Ligurian region, which was incorporated into the Kingdom of Sardinia after eight centuries as an independent republic.

The signing of the Paris Treaties in 1814 and 1815 and the conclusion of the Congress of Vienna represented for the Italian States the beginning of a period of harsh reaction and, at the same time, of deep social, political, and intellectual turmoil. The restoration of Victor Emmanuel brought back the aristocracy once more as the ruling class. After fourteen years of exile or of seclusion their social and political vision had remained unchanged. Through their influence on the sovereigns, and their positions in the diplomatic, high bureaucratic, and armed services, they would prevail for another third of a century. As for those members of the middle class and of the aristocracy who had been involved with the French administration or had served in Napoleon's army, several were ousted from their positions or seriously hindered in their careers. The return of the Savoyard rulers coincided with a reinstatement of the old codes and most of the pre-Napoleonic jurisdictions. Past privileges were reinstated. For example, whereas the Protestants and Jews were left very little freedom of worship, the Jesuits were permitted to return and were welcomed at Court.

For Amedeo Avogadro, the nine years between the publication of the 'Essai d'une manière' and his appointment at the University of Turin in 1820 represented a period of fecund scientific activity, during which he published twelve essays and articles. The social disorders and political confusion of the last years of the French occupation and the restoration of the House of Savoy probably encouraged Avogadro's natural reserve. He seems to have lacked any personal ambition beyond the scrupulous execution of his own work and his professorial duties. Unlike other aristocratic supporters of the French, he lived through the entire period from Marengo through the restoration without being affected, at least professionally, by the events which rocked his native region for more than fourteen years. Although he had served in the French administration, he did not suffer retribution in Vercelli for that earlier association. He apparently made little impact on Vercelli either, since almost no mention was made of him in the detailed chronicles of the town.[46] The young Count of Quaregna was recalled only as a professor of physics in a brief description of the local school system. Ignazio Avogadro della Motta, another member of the family and Mayor of Vercelli in those years, received much more attention.

In January 1815, Amedeo Avogadro, then thirty-eight, married Felicita

Mazzia in Biella, then a small town north of Vercelli.[47] As is often the case with men of his disposition, Avogadro found more satisfaction and enjoyment in the serene atmosphere of his family – they had seven children – than in the pursuit of social and professional success.[48]

After the 'Mémoire sur les masses relatives . . . ' was published in 1814, Avogadro turned his attention for the following six years, while still teaching in Vercelli, to the study of the physical behavior of gases and vapors. This was the beginning of twenty years of work on the problem of specific heats of bodies in their three states, and on many attempts to correlate them with other physical and chemical characteristics: affinity for caloric, refractive power, affinity for oxygen, electropositivity, and so on.

How the scientific community of the time reacted to this series of speculations which contained calculations based upon others' experimental data (and almost never on his own), *ad hoc* assumptions, and often daring conclusions, will be examined later.

The first two major essays on the specific heats of gases of 1816 and 1817 appeared in the *Biblioteca Italiana* of Milan, a brand new journal with very limited diffusion beyond the Alps; a well-known article in the *Annales de Chimie* in early 1813 by the French, François Delaroche and Jacques Berard, apparently prompted Avogadro's interest in this field of research, in which he incorporated his famous gas generalization and which he approached consistently through the caloric theory of heat even after its decline had begun in the 1830s.[49]

Including a short note in October 1818, five articles were then published by 1820 in the *Giornale di Fisica, Chimica e Storia Naturale* (formerly *Giornale di Brugnatelli*) of Pavia. Here, rather than with specific heats, he dealt with the elastic force of steam and with the correlation between heat and the expansion of mercury, a topic to which he returned in 1833.

In the last year of his stay in Vercelli, the Royal Academy of Sciences of Turin made him a permanent member as a result of his contributions.[50] He was then forty-three and was still teaching in a secondary institution. Only a year later he received his appointment to the newly established Chair of Mathematical Physics (*Fisica Sublime*) at the University of Turin. Thus, in the fall of 1820, Avogadro returned to live in his native town which he was never to leave again, except for the summer vacations he used to spend with his family in their large country house in Quaregna.

The University of Turin,[51] founded in 1404, was reorganized in the early

Plate 3. Courtyard of the University of Turin, built in 1713 by order of the Duke Victor Amadeus III. (Water color by L. Premazzi, 1841, in the Royal Library of Turin.)

part of the eighteenth century by Victor Amadeus II, first King of Sardinia, who donated his personal library of 10,000 volumes to provide the university with its first library. Ten years later, in 1730, the University had a faculty of twenty-one and a student population of about 2,000.[52]

In those years the academic program at Turin included philosophy, law, theology, medicine, and surgery. Physical and natural sciences seemed neglected in general; an exception was physics in the second half of the eighteenth century, when the chair was held by G. B. Beccaria.

Typical of the mental attitude and of the prevailing intellectual leaning of Piedmontese society is the fact that during the seven years of secondary schooling required to enter the University, young Piedmontese devoted most of their time to the study of Latin.[53] While a remarkable proficiency was reached in this language, little attention was paid to either modern or medieval history. Next to Latin, the more important subjects appeared to be religious instruction, rhetoric, and philosophy, with particular emphasis on the Scholastic and Patristic doctrines. Thus, the main purpose of the whole educational system of the Sardinian Kingdom remained that of molding the future ruling class of the country into essentially devout people of stern ethical discipline. There was no room for discussion or dissipation of any kind. As a result of all this there emerged in Piedmont in the last decades of the eighteenth century a superficially moralistic and at the same time very dull, bored society, where everything appeared preordained from the top and everyone from his early years knew what to do according to his social rank. It was scarcely a place for independent and rebellious spirits.

For the eighteen years which followed the armistice of Cherasco in April 1796 a state of almost complete upheaval prevailed over the entire educational system of Piedmont. The series of reforms and counter-reforms instituted during the French occupation and the short-lived Austrian–Russian occupation in 1799 had only deleterious consequences.

The University of Turin, which had closed after the events of 1797, opened again in January 1798, and under a decree of reorganization promoted by the Jacobins Carlo Botta, and Carlo Cavalli, several chairs were abolished, including theology and canon law, in the latter of which Amedeo Avogadro had taken his degree only two years earlier, and chairs of chemistry and obstetrics were created. The faculty were required to take an oath of allegiance to the French Republic. Following the Austro-Russian seizure of Piedmont in 1799, the University was closed again for almost a year, and new reforms were prepared to offset those of the Jacobins. These, however, were interrupted by the second French occupation, which came after the battle of Marengo in 1800. Still another academic reorganization was then studied and eventually adopted, making the entire Piedmontese educational system conform to the one already established in France.

The University of Turin, which after 1808 became one of the Imperial Universities created by Napoleon to administer education all over his domains, drew a definite financial benefit from the new situation.[54] In spite of these new developments, however, the intellectual life in Piedmont did not change considerably even during the Napoleonic era, most cultural endeavors being devoted solely to the base flattery of the great Corsican, his family, and his triumphs.

With the fall of Napoleon and the Savoyard restoration, everything on the surface reverted to prerevolutionary times. In its formal customs, such as wearing pigtails, the speech mannerism, and the use of snuffboxes, Piedmont seemed to have returned to the life of 1798. Even the almanac of Court and the privileges of that time were restored. The *Magistratura della Riforma* was reinstated. In 1814, at least nine professors were removed from their teaching positions for their political leanings and stands during the Napoleonic regime. Among them were Vassalli Eandi and Giobert.[55] The attitude of the restored King Victor Emmanuel I towards education in general could probably be identified with that of his brother Charles Felix who considered all professors as 'loathsome' and 'all those who have studied at the University as corrupt. . . .'[56]

Fortunately, Count Prospero Balbo, a moderate, who had been Rector of the University during the Napoleonic era, was appointed in 1818 head of the *Magistratura della Riforma.*[57] He reestablished a number of important chairs at the University: mineralogy, zoology, and eastern languages. In 1819, he was able to obtain from the French government the payment of the past interests on the University's old endowments. In the meanwhile, faculty and students at Turin were subjected to the strictest surveillance by the police, since the Court always looked with great suspicion at any movement which might have liberal implications.

It is possible that through Count Balbo's support – he had been President of the Turin Academy of Sciences when Avogadro was made an ordinary member – the scientific merits of a little known professor at the College of Vercelli were finally recognized. And in 1820, Avogadro received the appointment to the newly created chair of *fisica sublime.*[58]

A few months after this event, Avogadro read an extensive memoir to the Turin Academy of Sciences. The 160-page long 'Nouvelles considérations sur la théorie des proportions déterminées dans les combinaisons et sur la détermination des masses des molécules des corps' was the most important chemical essay published by the author of the molecular hypothesis.[59] In it, ten years after the 'Essai d'une manière', he reasserted in the clearest and most direct way the equal volumes–equal numbers generalization, stressing its superiority over the use of chemical equivalents as a precise method to determine the weights of chemical elements and compounds. Avogadro had by then reached the conclusion that only by adopting a 'nombre proportionel', obtainable through the link he had established between Gay-Lussac's gas volumes law and the law of definite proportions, would it be possible to avoid the uncertainties and the variability, observed in other approaches to the problem of combining weights, which still seriously hampered chemistry. As for the recent laws proposed by Dulong–Petit and by Mitscherlich on atomic heats and isomorphism, Avogadro criticized them as too limited and premature.

Historically, the 'Nouvelles considérations' are significant because this essay contained Avogadro's reactions to two articles which had appeared since 1811 and postulated concepts quite relevant to the molecular hypothesis: Ampère's letter to Berthollet of 1814 and William Prout's still anonymous note of 1815. He took exception to both of them, although for different reasons, which will be discussed below.

In the 'Nouvelles considérations', Avogadro made an extensive review of a large number of the inorganic compounds known at the time, in order to see how his interpretation would modify, if necessary, the conclusions on the combining ratios of the elements reached by the experimentalists, especially Berzelius. Again in 1821, in the Academy's *Memorie*, Avogadro published another long essay, this time devoted to organic compounds.[60] Two years earlier, in his famous 'Essai sur les proportions chimiques' Berzelius had suggested that, unlike inorganic combinations, organic compounds which usually contain three or four elements might have these combined in any proportion. Now, on the basis of arguments relying more on his unitarian view of nature than on any experimental evidence, Avogadro in his 'Mémoire sur la manière . . . ' disagreed with the Swede and supported the view that the law of multiple proportions applied also to organic chemistry. A series of elaborate equations using compositions found by organic analysis then led him to write empirical formulas for a number of compounds ranging from ethanol (which is correct) to sugars. These two essays, plus those of 1811 and 1814, are the only ones devoted exclusively to chemistry, a subject that Avogadro, as is apparent, approached only from a theoretical viewpoint. In an age when this science was already flourishing with the work of excellent experimentalists, it is not surprising that such theoretical contributions as those of 1821 from Turin left very little trace.[61]

Less than a month after Avogadro had read the "Nouvelles considérations' to the Academy, a series of events took place in Piedmont which not only shook the foundations of the country's entire political system, but also had a direct, long-lasting effect on Avogadro's life and work.

Despite the reactionary tide which had flooded the Sardinian Kingdom after the Savoyard restoration, too many new social, cultural, and economic interests had been aroused by the French Revolution and the Napoleonic occupation to have them suddenly suffocated and completely forgotten when the old ruling classes returned to power.

After 1815 a small group of people, mostly young, in the University, and both outside and inside the civil and military administrations, had kept alive the hope for a less absolutistic form of government. A new constitution, granting more freedom to the individuals and extending civil rights to the bourgeoisie and to less wealthy citizens, became part of the programs secretly discussed by these men. A few of them, like the young Cesare Balbo, son of Prospero Balbo, were members of the Piedmontese aristocracy who had served under Napoleon in different capacities. Other aristocrats, like Santorre di Santarosa and Giacomo Provana di Collegno, found some encouragement for their ideas from another young man who was no less than a member of the royal family and presumed heir to the throne of Sardinia, Charles Albert, Prince of Carignano.[62] Timid and insecure, Charles Albert, after being reared and educated in Napoleonic France, found himself in those years deeply divided between the liberal ideas assimilated during his youth and the reactionary influence of the court of Turin.[63] His conduct with the young Piedmontese liberals after 1820 was at least ambiguous and would eventually lead to their doom.

In 1821, just months after a short-lived constitution had been granted by the Bourbons in Southern Italy, Santarosa and his friends thought the time for action in Piedmont had come. Quite likely, they either misinterpreted or were deceived by the position Charles Albert had taken. The insurrection began in January, with the occupation of the University of Turin by a large number of students, reacting to the arrest of some of their colleagues. Despite the attempt by Prospero Balbo to find a peaceful solution to the crisis, the Governor of Turin ordered the troops to dislodge the insurgents. A long, uneven fight followed; but the bloody repression failed to quell the riots, which soon engulfed Turin and other towns in the provinces. Shortly after Alessandria, Piedmont's main military base, had fallen into the hands of the

Constitutionalists on March 10, Victor Emanuel I abdicated in favor of his brother Charles Felix, who was out of the country at the time, and named Charles Albert as regent. The King had earlier promised a bicameral system of government similar to that of the British. By then, the capital and other cities had abandoned the royalist cause and the young regent was reluctantly forced to announce that a constitution would be granted to the Kingdom.[64] But on March 21, one week after granting the constitution, Charles Albert betrayed the liberal cause and surreptitiously left Turin for Novara to join the troops which had remained loyal to the old regime. Charles Felix sought the aid of Austrian troops and, together with their own divisions, thoroughly defeated the insurrectionists in mid April.

Upon the return of Charles Felix to Turin in September 1821, the reaction that followed was long and severe: thousands arrested, seventy directly involved in the plot sentenced to death; more than 300 people were dismissed from their positions. The institutions of higher education, for their participation in the revolt, did not escape the reactionary wave: both the universities of Genoa and Turin were closed for a full year; several faculty members were suspended or forced to retire.[65] The students who were not arrested had to comply with extremely rigid disciplinary rules, including the notorious *admittatur.*[66] Typically, as a result of the renewed influence of the clergy on the court, special emphasis was given to the compliance with religious observances. Those who wished to remain at the University had to confess monthly, attend spiritual exercises, and be present at their parish functions, even during vacation.

These sanctions, which were chiefly implemented in 1822, were followed by others: several chairs at the University of Turin were abolished by royal decree, including, on 24 July 1822, that of mathematical physics, which had been assigned to Amedeo Avogadro only twenty-four months earlier. As a consequence of what seems to be a purely arbitrary decision, this man, who had not been involved in any conspiracy, was for twelve years denied the right to carry on his mission as a scientist and a teacher. After his dismissal, however, the Royal Government continued to pay him the meager salary of 600 Piedmontese Lire per year, as the decree said, 'until he (Avogadro) may be assigned to another occupation . . . '. In 1824 they raised it to 900 Lire as a form of pension for his previous teaching position in Vercelli.[67] It was certainly a difficult existence with a large family to support.

The reasons behind the removal of Avogadro from his academic position remain obscure. He was not personally involved in politics although, according to a confidential police report, he was regarded as 'not a staunch supporter of the Throne and of his Majesty . . . '.[68] A possible explanation may be offered by the fact that Giovanni Plana, the renowned astronomer at Turin and a very close friend of Avogadro, was suspended temporarily from his position in 1822, as well as the professor of chemistry, Giovanni Giobert. Giobert had strongly compromised himself since 1798 by taking a definite republican and francophile stand. Avogadro, who also accepted a number of positions under the French occupation, likely had foes among the courtiers and members of the Sardinian Government. Resentment against those who had collaborated with the Napoleonic regime was still strong following the Savoyard Restoration, and the 1821 revolt by the constitutionalists had disclosed that many of their leaders had grown and matured under that regime. Avogadro's close friendship with Count Prospero Balbo, who also lost his position as *Magistrato della Riforma* following the 1821 events, may also shed some light on his dismissal from the University; sanctions applied at the time to other Faculty members at Turin were certainly less serious.[69]

Whatever the motivations behind the removal from his office, the fact remains that even during those twelve years he continued to work and to publish the result of his studies. A dozen articles appeared between 1823 and 1834, when his 'forced retirement' came to an end. Two of them, just notes, appeared in France in Ferussac's *Bulletin des Sciences*, the others entirely in Italy, and they dealt mostly with the specific heats of solids, liquids, and gases, and again with his attempt to correlate chemical, electro-chemical, and thermal properties of bodies. Most remarkable, however, was a memoir that he read a few months before his dismissal at the Academy of Turin, in which he reported a series of experiments conducted with a very sensitive multiplier built by his colleague, Vittorio Michelotti, in the physical cabinet of Turin by introducing some improvements to the Schweigger and the Poggendorf instruments. By using it, Avogadro arrived at some new conclusions about the electrochemical series of metals that he found at variance with that determined by Berzelius.[70]

The 1822 memoir, published in 1823, is important not so much for its content, but because it represents one of the few experimental works that Avogadro, assisted or unassisted, conducted during his scientific life. The others worthy of recall, besides an investigation of the vapor pressure of mercury in 1833, reported the results of observations conducted in electro-chemistry with Giuseppe Domenico Botto, who taught experimental physics at Turin, and on the specific heats of a number of solid and liquid elements and their compounds.[71] The two articles on the latter subject attracted the unfavorable reaction of both Berzelius and Regnault. How the scientific reputation of Avogadro suffered from their remarks is difficult to assess, but certainly they contributed to his characterization as a poor experimenter.[72]

It was only in 1832, one year after Charles Albert's accession to the throne of Sardinia, that the government decided to reinstate the chair of *fisica sublime* at Turin. The former holder of that chair found himself once again set aside, and this time *la raison d'ètat* was not the only factor involved. The choice had fallen on one of the prominent European mathematicians of the time, Augustin Cauchy, who had resigned his chair at the *École Polytechnique* rather than take an oath of loyalty to the Orleanist 'usurper', Louis Philippe, in 1830. Charles Albert who, like other European sovereigns, scarcely concealed his dislike for the new French King, took this occasion also to express his feelings by appointing Cauchy, then in exile, to the faculty position at Turin. He taught there, however, for only two academic years, resigning to become tutor to the grandson of the deposed King Charles X in Prague.

The vacant chair was now assigned again to Amedeo Avogadro, who resumed his teaching in November 1834. However, for reasons which are not clear, he continued to be called professor emeritus rather than professor, to which he was entitled.[73]

After 1832 the number of Avogadro's publications declined, which would seem to indicate that he devoted most of his time to gathering material for the preparation of his most extensive work, the four-volume *Fisica dei corpi ponderabili*, the publication of which by the Royal Printing House took four years beginning in 1837. As the title indicates, this treatise – the largest in Italy up to that time on theoretical physics – was mainly devoted to the study of the constitution of heavy ordinary bodies, more specifically of bodies formed by tightly joined although not touching molecules having a sensible mass. The nature of molecules, the nature of molecular forces, and how physical states could be interpreted in terms of these forces were the main topics examined by Avogadro in his *Fisica*. Electricity and magnetism, considered imponderable, were not discussed, and chemistry was touched on only peripherally, while crystallography and changes of physical state took up a great deal of space, together with the author's favorite themes, specific heats, and densities and their significance in terms of 'atomic masses'.[74] Avogadro's inquiry included, among other subjects, molecular forces in gases, intermolecular distances, and their effect on density of gaseous substances; all this prompted him to a restatement of his 1811 assumption that, since at the same temperature and volume the molecules are equidistant, they are also in equal number, regardless of their nature. It cannot be denied that the clearest explanation of the arguments which led to his 'fundamental principle' – as he described it – definitely appeared on this occasion.[75]

Pervading the treatise is the adherence of its author to the caloric theory of heat, which was by then well into decline.[76] However, it should be mentioned that on more than one occasion[77] Avogadro was able to recognize that, as Ampère had already proposed in 1832 and again in 1835, another 'intervening fluid', the ether, could be invoked and that vibrations could take place at the level of the 'partial molecules' composing the 'integral molecules'. In other words, a possible alternative – the vibrations of particles transmitted by vibrations of the ether – was being offered to the caloric theory of heat, but always in a very tentative form.[78]

When, in September 1840, the second national congress of Italian scientists met in Turin, Avogadro read three papers in the meetings of the mathematical and physical section. In one of them he discussed conclusions which he had reached on the specific heats of compound gases and their relationship to the specific heats of their component gases; this topic had also formed the subject of a short note published in the *Bibliothèque Universelle* in the same year.[79] On this occasion he restated before his audience the 1811 gas hypothesis; unfortunately, he also discussed again his notion of 'fractionary' atoms, in the same context as he had earlier used 'partial molecules'. In another session of the 1840 congress, Avogadro went back to his distinction between 'apparent' and 'real' neutrality of a substance and defined the 'neutralizing power' of the element from its 'atomic mass', which he now considered the keystone of the molecular theory. Except for a few favorable comments by Auguste de la Rive in the discussion of the specific heats of solids, which formed the subject of a third paper, there is no record in the minutes of the congress of comments to Avogadro's contributions.[80]

In the early 1840s Avogadro directed his attention to two totally unrelated fields of research. A first study resulted in a highly elaborated mathematical analysis of the dielectric and was published in 1844, while a second article which appeared in the same year represented Avogadro's major contribution to a subject which from time to time still attracted the attention of natural philosophers all over the world, a clearer method of classifying and designating chemical compounds.[81] In his 'Saggio di teoria matematica' written in the first part of 1842, Avogadro referred at first to his two memoirs of 1806 and 1807 in which he had attempted to explain the modifications occurring in the molecules of an insulator placed between charged plates.

After 1838, Faraday's *Experimental Researches* had thrown new light on these facts, as Avogadro pointed out in his essay. As a matter of fact, Avogadro gladly recognized that in order to explain the force of induction, Faraday had advanced considerations quite similar to his own, or had assumed the existence of a modified state of the dielectric. A reprint of the 'Saggio di teoria matematica'[82] was also forwarded to Faraday who apparently found it 'too mathematical' and passed it to William Thomson for review.[83] The purpose of Avogadro's study was to determine by mathematical means the law of the distribution of electricity on different points of two insulated spheres touching each other. He recognized the extremely complex nature of the task before him and conceded that his was just an 'essay of calculation', because a solution to such an intricate problem could only be suggested, but not fully given for the time being.

In the 'Proposizione di un nuovo sistema di nomenclatura chimica' written over a period of several years, but not submitted for publication until early 1843,[84] Avogadro argued that recent developments in chemistry had made apparent the inadequacy of the then prevailing system of nomenclature. What he suggested was the elaboration of a generally acceptable system to designate the known compounds and those yet to be found. In analogy with zoology and botany, Avogadro suggested for each compound the use of a generic name followed by a specific one. Further, in order to overcome the lengthy expressions used for complex compounds, he recommended the adoption of shorter names or syllables and numbers in a given sequence so as to avoid errors and confusion. Avogadro devoted a great deal of time and effort to this study another article followed the first with more than 100 pages of the *Memoire* in which he extended his proposal to the organic compounds as well. It is worth mentioning that on this occasion he completely dropped the term 'molecule' and, like most of his contemporaries, found himself debating whether the proportions of elements entering chemical compounds should be referred to as numbers of atoms or as equivalents. He recognized that there was still much uncertainty among chemists about the 'true atom' of some elements.[85]

In the last years of his scientific carreer, which extended beyond his retirement from active teaching in 1850 at the age of seventy-four, Avogadro turned to the question of atomic volumes. Between 1846 and 1852 he published four extensive articles dealing with this subject in the Memoirs of the Academy of Turin, some of which also appeared in abstract in the *Annales de Chimie* and in the *Bibliothèque Universelle* and were reviewed by Berzelius in the *Jahresberichte* of these years (1845–49). Here his purpose was to attempt to establish a link between the densities of liquid and solid elements (and later compounds) and their molecular masses, as he had already proved to be possible for gaseous substances. And here he embarked on a series of long speculations which incorporated very daring and certainly arbitrary assumptions. Among them, probably the most vulnerable to criticism, was once more the suggestion that for several elements, not only the halogens, sulfur and phosphorus, a 'division' or a 'reunion' of 'chemical atoms' may take place within their 'integral molecules'; only for heavy elements, such as a gold and platinum, were the integral molecule and the 'chemical atom' identical, he claimed.[86] In the case of other metals such as manganese and zinc, his calculations, based upon their densities and their position in the electrochemical series, indicated that their integral molecule contained two 'chemical atoms' and, therefore, that their atomic weight should be doubled, and likewise with other metals. This 'adjustment' of the atomic weights to fit physical considerations did not represent by any means an attitude peculiar to Avogadro: Dulong and Petit had done the same after 1819, and Mitscherlich, who had postulated that crystalline form was independent of the 'chemical nature of the atom', with his celebrated observations of 1820 and 1821, induced Berzelius himself to modify several of his earlier atomic weights. It is fair to say, however, that when Avogadro's speculations on atomic volumes appeared a quarter of a century later, the prevailing scientific mood had become rather different; furthermore, they had no experimental support; and finally, they suggested an almost arbitrary fractioning of atoms, whenever this was necessary to reconcile the theory with the actual densities of the elements.[87]

To argue that this was not one of the major obstacles to the understanding and, therefore, to the acceptance of Avogadro's 1811 gas hypothesis would be quite difficult. At the same time, the approach followed late in his scientific life seems to confirm that Avogadro's understanding of the terms integral molecule, simple atom, and so forth as he used them at the time, is not encompassed by our current conceptions of such terms as 'molecule' and 'atom'. And this is quite understandable in view of what has just been said in relation to the 'chasm' between the physical and chemical interpretation given to the elementary components of matter.[88]

An 1853 *Memoire della Reale Accademia delle Scienze* of Turin was the last of Avogadro's many contributions to this journal; it was a seventy-page analysis of Regnault's recent series of observations on the behavior of gases subjected to different degrees of compression; Amedeo Avogadro was then seventy-seven. He died in Turin in the summer of 1856, a few months after the

Plate 4. Entrance to the summer house of the Avogadro family in Quaregna. Note the Count's crown above the iron gate.

Plate 5. Mausoleum of the Avogadro family in Quaregna. Amedeo Avogadro, his wife Felicita and their children are buried here. The family coat of arms is above the entrance to the mausoleum.

Congress of Paris in which Cavour participated as Prime Minister of Sardinia, concluded the Crimean War. Both these events meant the rise of the Savoyard Kingdom as a major European power, just on the eve of the political unification of Italy, which followed in 1861. Certainly Avogadro's life spanned the most eventful years of his native region, from the semi-obscurity of Victor Amadeus III through the Napoleonic Empire, and the first sparks of the Risorgimento to the emergence of Camillo Cavour as the leading figure of the national resurgence.

Shortly after his death the editors of *Nuovo Cimento* wrote a short obituary of Amedeo Avogadro, where they recalled his retiring disposition and the simplicity of his life.[89] In the history of science, it was said, his studies on atomic volumes and their relationship with the chemical affinities and the electrochemical series, and the electrochemical researches conducted with Michelotti on the chemical theory of the voltaic pile would be remembered. However, no mention at all appeared on this occasion of the 1811 gas generalization, whose full significance would remain unrecognized in Italy as well as in the rest of the scientific world until the late 1860s.[90] One of the editors of *Nuovo Cimento*, Raffaele Piria, teaching chemistry at the University of Turin at the time of Avogadro's death, was in 1857 apparently unaware of the notion of polyatomic molecules postulated by Avogadro almost half a century earlier.[91] When Eugenio Sismonda, Secretary of the Physical and Mathematical Sciences Class of the Academy of Sciences of Turin gave the 1856 annual report, he did not fail to commemorate the death of his colleague and to give a full list of Avogadro's articles published in *Memorie della Reale Accademia delle Scienze di Torino*; he mentioned that some of his scientific writings had also appeared in *Journal de Physique de La Méthérie* after 1806.[92] That was all. Yet it would be unfair to blame only his contemporaries for Avogadro's neglect. As will be examined later, so many factors of such a wide range delayed the understanding and the approval of his fundamental idea as to make this extremely improbable, at least in his lifetime. Only when a 'crisis situation' intervened in the second half of the century, did anyone realize that the molecular hypothesis – the key to the puzzle – was already available.

NOTES

1. Amadeus VIII, first Duke of Savoy, annexed most of Piedmont to his dominion. After forty-nine years of reign he abdicated and in 1439 was elected Pope at the Council of Basel. He was a schismatic Pope, and a few years later he renounced in favor of the legitimate Nicholas V.

2. See G. Vinay, *L'Umanesimo subalpino nel secolo XV* (Turin, 1935). Also D. Hay, *The Italian Renaissance* (Cambridge, 1961).
3. They became known in the sixteenth century as the 'Macedonians of Italy'. See V. Gioberti, *Del primato morale e civile degli Italiani*, vol. 1 (Turin: UTET) repr., p. 137.
4. M. S. Briggs, *Baroque Architecture* (London, 1913), p. 97.
5. Until 1852 Latin was the medium of instruction in most subjects at the Universities of Turin and Genoa.
6. Nassau William Sr. in *Journals kept in France and Italy from 1848 to 1852* (London, 1871), p. 291, quotes the Marchioness Arconati as saying in Turin: 'A speech in Genoese or Piedmontese would be unintelligible to two thirds of the Assembly (the Chamber of Deputies). Except the Savoyards who sometimes use French, the Deputies all speak Italian. But this to them is a dead language in which they have never been accustomed even to converse.'
7. The feeling of being just a pawn among the great powers for centuries had ultimately a very negative effect on the formation of a unified national spirit among the Italians in general; a spirit which had already become part of the common heritage of the French and British people.
8. Other wide-ranging problems involving the juridical and economic policies of Piedmont were debated in the last two decades of the eighteenth century. They dealt with the agriculture's leasing contracts; unemployment, banking and currency reforms, equal distribution of taxes, etc.
9. According to E. Passerin, *L'Italia nell'etá delle Riforme, Storia d'Italia*, vol. 3 (Turin, 1959), pp. 43 ff: Freemasonry was extensively diffused in the Sardinian States in the eighteenth century. Even members of the Court and the King himself joined in 1780, although Pope Benedict XIV had formally condemned these associations in 1751. Quite likely, the aristocracy did not belong to the more politically progressive lodges which came into existence in the last decade of the century.
10. Pavia, in Lombardy, was under Austrian occupation. In its university, founded in 1361, Alessandro Volta was teaching physics in those years (late 1770s).
11. A good personal friend of both Voltaire and Hume, Denina moved to Berlin.
12. According to Spellanzon, Alfieri, who was living in Florence when King Charles Emmanuel IV passed through that city on his way to exile in Sardinia, went to meet his former sovereign, and the King in Piedmontese dialect told him: 'Slave Alfieri, here is your tyrant'. C. Spellanzon, *Storia del Risorgimento*, vol. I (Rizzoli, 1933), p. 196.
13. The King's fateful decision followed his steady refusal to enter into any negotiation with the Jacobin emissaries offering an alliance with France to the Sardinian Kingdom in exchange for Lombardy held by the Austrians.
14. One of Charles Emmanuel's final orders was that of authorizing the Piedmontese troops to put themselves under French command. For seventeen years the King and his brother and successor, Victor Emmanuel I, were forced to live primarily in Sardinia, which Napoleon was not able to occupy as it was protected by the British Navy.
15. The existence of a personal relationship between the Sardinian King and the Emperor of Russia, Paul I, was known to Bonaparte and may explain his restraint toward Charles Emmanuel.
16. Vercelli, a provincial seat in Piedmont, is located about half way between Turin and Milan, at the center of a rich agricultural region and, at the present, of a highly industrialized area known since the early eighteenth century for its textile production and woolen mills.

17. The main branches of the family were the Avogadro di Valdengo, di Motta, di Collobiano, di Cerreto, di Quinto, di Quaregna, and di Casanova.
18. See L. Tettoni, *Notizie Genealogiche-Storiche intorno alla nobile, antica ed illustre famiglia Avogadro* (Wilmant, 1845). Also *Biella ai tempi di Amedeo Avogadro* (Biella, 1956).
19. See C. Dionisotti, *Memorie Storiche,* vol. 2 (Biella, 1860), p. 256.
20. See F. Cognasso, 'Novara nella sua storia in *Novara e il suo territorio* (Novara, 1952).
21. Biella is now the main center of the wool industry in Italy.
22. See *Dizionario biografico degli Italiani*, vol. 4 (Rome, 1962), p. 689. Also C. Dionisotti, *Storia della Magistratura Piemontese*, vol. I (Turin, 1881), pp. 387 ff. The jurisdiction of the *Magistrato della Riforma* included the Universities, appointment of teachers, tutors, spiritual directors, setting schedules for the students, and establishing Courts to pass sentences on their behavior.
23. According to the French informers, the Piedmontese aristocracy was 'extrement mécontente des Rois ... '. Only 'un très petit nombre de nobles le (the King) suivit . . . '. From G. Vaccarino, *Boll. Storico Bibl. Subalp.* 1 (1953), 37.
24. Among them were, for instance, the Marquises di Cavour, the parents of the future Prime Minister of Sardinia, and central figure in the political unification of Italy. The motives behind the collaboration of a portion of the Piedmontese aristocracy and the encouragement that they received from the French to this purpose have been examined in detail by D. Outram, 'Education and Politics in Piedmont, 1796–1814', *Hist. J.* 19 (1976), 611–633. However, the Avogadro are not mentioned in this study. According to G. Bragagnolo [*Torino nella storia del Piemonte*, (UTET 1919) p. 64] about twenty-five aristocratic families were deported to France, including the Avogadro di Casanova.
25. Antonio Manno in *Il Patriziato Italiano*, vol. II, p. 130, indicates that Amedeo had four brothers (Giuseppe, professor of law at Turin, who died in 1842; Felice, who died in 1880, Luigi and Lorenzo, who died in 1882) and three sisters (Chiara, Orsola, and Maria). The title of *Conte di Quaregna con Ceretto* passed to Amedeo upon the death of the eldest brother Giuseppe.
26. 'Essai analytique sur l'électricité', Turin, Arch. Accad. delle Scienze, memoria MS, 1803. 'Considérations sur la nature des substances connues sous le nom de sels metalliques et sur l'ordre de combinaisons auquel il parait le plus convenable de les rapporter.' Turin, Arch. Accad. delle Scienze, MS, 1804.
27. For Beccaria and Priestley, see J. Priestley, *History and Present State of Electricity*, 4th ed. (London, 1775), p. 314. Beccaria's work on the decomposition of water had attracted the attention of Cavendish. Alessandro Volta, who was only twenty-two, in 1767, took a critical position of Beccaria's *De electricitate vindice*, dealing with electrostatic induction. As for Francesco Cigna, born in 1734, professor of anatomy at Turin, he conducted a series of experiments on electrostatics which were noticed by Beccaria and were by him communicated to Franklin in 1767.
28. Antonio Maria Vassalli Eandi, born in Turin in 1761, studied under Beccaria at Turin, then became professor of natural philosophy at the secondary school of Tortona, moved back to Turin as substitute professor of physics at the university. When the French occupied Piedmont he kept his position and was called to Paris as member of the *Commission des poids et mesures*. After the battle of Marengo, he returned to Turin as full professor of physics in 1800. With the Savoyard restoration, in 1814, he lost his position and was forced to retire, although he later received some minor assignments. He died in 1825. Volta, who kept a long correspondence with Vassalli, challenged his views on galvanism.

29. Jean Claude de La Méthérie was the sole editor of the *Journal de Physique* after 1793 through to 1813. The *Journal de Physique* continued to appear even when the *Annales de Chimie* suspended its publication between 1793 and 1797, and although de La Méthérie was arrested for his political ideas of dissent from the revolutionary government. See S. Court, *Ambix*, **19** (1972), 113.
30. 'Considérations sur l'état dans lequel doit se trouver une couche d'un corps non-conducteur de l'électricité . . . ', *Journal de Physique, de chimie, d'histoire naturelle et des arts*, **63** (1806), 450. 'Second mémoire sur l'électricité . . . ' ibid., **65** (1807), 130. 'Note sur la nature de la charge électrique', *Archives de l'électricité* (suppl. *Bibliothèque Universelle de Genève*) **2** (1842), 102.
31. As M. Crosland describes in *The Society of Arcueil* (Harvard, 1967).
32. By the middle 1820s the intricate structure of Laplacian physical science had collapsed. See R. Fox, *Historical Studies Physical Sciences*, **4** (1974), 109. Yet Avogadro may be counted among those isolated 'diehards' who kept to this tradition.
33. Despite its shortcomings and the perplexities which characterized the reception of his fundamental theory, Dalton succeeded, if compared with Avogadro. He had powerful supporters in Britain, and his previous work had been recognized even across the Channel. See D.S.L. Cardwell, *Introduction to John Dalton and the Progress of Science* (Manchester, 1968), p. XIII, and F. Greenaway, *John Dalton and the Atom* (Ithaca, 1966), chap. 9.
34. The year before, Gay-Lussac, then only thirty, had been appointed to the chair of physics at the Sorbonne; his friendly relation with both Berthollet and Fourcroy undoubtedly favored his rapid rise.
35. From the *Alembic Club Reprint*, no. 4, p. 23: 'We must first of all admit that chemical action is exercised indefinitely in a continuous manner between the molecules of substances, whatever their number and ratio may be and that, in general, we can obtain compounds with very variable proportions. But then we must admit at the same time – apart from insolubility, cohesion and elasticity which tend to produce compounds in fixed proportions – chemical action is exerted more powerfully when the elements are in simple ratios or in multiple proportions among themselves . . . *in this way we reconcile the two opinions*.' (Italics, mine.)
36. See J. R. Partington, *History of Chemistry* vol. IV (London, 1964), p. 55.
37. See J. R. Partington, *Nature*, **178** (1956), 8.
38. In July 1840, Faraday, after receiving a copy of *Fisica dei corpi ponderabili* – Avogadro's most comprehensive work – thanked him in these terms: 'That work is a deep study for a philosopher and must have given you much labor, as it gives you now the higher honour.'
39. See in this connection the opinion expressed by A. P. Candolle, as reported by M. Crosland, *The Society of Arcueil* (Harvard University Press, 1967), p. 426.
40. Although of French extraction, Lagrange is considered Italian by the Italians who often spell his name *Lagrangia*.
41. See F. Cognasso, 'Le Scuole', in *Storia del Piemonte*, vol. 2 (Turin, 1960), p. 683.
42. See G. C. Faccio, 'Avogadro e le Scuole di Vercelli', in *La Sesia*, 18 March 1955.
43. According to J. R. Partington, *History of Chemistry*, vol. III (London, 1963), p. 495, de La Méthérie had expressed the same view concerning oxygen in his 'Essai analytique sur l'air pur et les différentes espèces d'air' (1785). Davy also said this in his second Bakerian Lecture.
44. The fact that Berzelius had some of his work published in the *Journal de Physique* suggests that it was not a second-rate publication, as it has been repeated by several historians of the period, in connection with the attempts to explain the neglect of Avogadro's 1811 hypothesis.

45. See, for instance, R. L. Causey, *J. Chem. Educ.* 6 (1971), 365 and M. Frické, *Method and Appraisal in the Physical Sciences* (Cambridge, 1976), p. 280.
46. See C. Dionisotti, *Memorie storiche della Città di Vercelli* (Amosso, 1864), p. 344.
47. At the foot of the Alps, with a population now of over 100,000, Biella, often called the Manchester of Italy for its many woolen mills, has become one of the most prosperous industrial centers of Northern Italy. In Biella, near Porta della Torrazza, Avogadro owned two houses, in one of them he occasionally lived with his mother, Anna Vercellone, brothers, wife, and children. [From *Biella ai tempi di AA*, (Biella 1956), Tip. un. Biellese.] Quaregna, the town which gave its name to the Avogadro family, is about ten miles east of Biella. The family chapel and the grave of Amedeo Avogadro in the village cemetery of Quaregna were spared any damage by a huge flood in November 1969. In Quaregna one can also visit the villa (no longer property of the family) where Avogadro spent many summers with his family after 1815. Here he enjoyed the simple life of the country and the companionship of his numerous relatives. Ottavio Lubatti in *Rivista Biellese*, 1956, pp. 6 and 17, describes the kind of life Avogadro was leading in Quaregna, reciting poetry in Latin, Italian, and French; he even organized a family newspaper reporting the family events.
48. Only one of them, Luigi, who became an army general, had a son, Filippo, also a general, but with no children. This branch of the family thus became extinct at the death of Filippo. He was a captain in 1911 when the grandiose celebration of the centenary of the 'Essai d' une manière' was held in Turin.
49. For a very detailed study of this topic, to which I will return again in detail in other sections of the present work, see R. Fox, *The Caloric Theory of Gases* (Oxford: Clarendon Press, 1971), p. 196 ff. and *passim*.
50. A. Cossa, *Il Conte Amedeo Avogadro di Quaregna* (Milan, 1898), reports a letter written to Avogadro by Plana in October 1819 in which he says: 'Je suis charmé d'avoir cette occasion pour vous témoigner qu'il y a bien de temps que j'ai appris à vous estimer . . . . '
51. Erasmus of Rotterdam was, up to the eighteenth century, the most famous alumnus of this university, from which he received a doctorate in theology in 1506 while traveling in Italy, where he spent time also in Venice and Rome.
52. F. Cognasso, 'Vita e Cultura in Piemonte', *Storia del Piemonte*, vol. II (Torino: UTET, 1960).
53. With the exception of the Church States, in no other European country in the nineteenth century was Latin so strongly taught as in the Sardinian Kingdom.
54. Professors in Paris were getting 6,000 Francs per year. M. Crosland, *The Society of Arcueil* (Harvard University Press), p. 203.
55. According to E. Passamonti, *Prospero Balbo e la Rivoluzione del 1821 in Piemonte* (Turin, 1923), p. 23.
56. Quoted from F. Lemmi, *Carlo Felice* (Turin, 1931), p. 182.
57. Prospero Balbo, born in 1762, had been ambassador to Paris in 1796, President of the Royal Academy of Science of Turin, member of the Council Regency in 1814, had to resign from all public position in 1815. In 1816 he was sent as ambassador to Madrid.
58. In 1817, when the chairs of economics and public law were proposed by Minister Brignole to the King, he approved them and then informed the Magistrato della Riforma of his decision. See E. Passamonti, *Prospero Balbo* (Turin: Bocca, 1923), p. 48. This was probably the same procedure adopted three years later when Avogadro received his appointment; by then Balbo was Minister of the Interior.
59. It appeared in *Mem. Reale Acc. Scienze Torino,* 26 (1821), 1–162.

60. 'Mémoire sur la manière de ramener les composés organiques aux lois ordinaires des proportions déterminées', *Mem. Reale Acc. Scienze Torino,* **26** (1821), 440.
61. The implications of these two essays and their appreciation by historians of science are discussed in subsequent portions of this study.
62. The Carignano line of the House of Savoy descended from Charles Emmanuel I's younger son. Through Charles Albert, they eventually became Kings of Italy.
63. While he inherited from the French and unreserved hate and mistrust of Austria, although he had married an Austrian archduchess, his main ambitions were, on one side, to become king and, on the other, to expand the dynasty's possessions in Italy.
64. Quite similar to the constitution granted by the Bourbons in Spain and, therefore, less liberal than the one earlier promised by Victor Emmanuel I. Significantly, eleven of the twenty-three leading citizens who advised in favor of a constitution were aristocrats. See C. Torta, *La Rivoluzione Piemontese nel 1821*, vol. 7, 1908, Bibl. Stor. Risorg., p. 232.
65. See A. Manno, *Informazioni sul Ventuno in Piemonte* (Florence, 1879), p. 205.
66. This was a good conduct certificate which, with the approval of the local police authorities, was submitted every three months to the University administration by each student.
67. From the ledgers of the University of Turin it is shown that in 1822 Avogadro's salary was 1,700 Lire per year plus 300 Lire as former teacher in Vercelli.
68. The original is in Avogadro's personal file. Archivio di Stato, Turin, 1821, Alta Polizia, Deliberazione Commissione Superiore di Scrutinio.
69. See D. Outram, 'Education and Politics in Piedmont, 1796–1814', *Hist J.,* **19** (1976), 611–633.
70. Vittorio Michelotti (1774–1842) was professor of pharmaceutical chemistry at Turin.
71. These were extensively abstracted in the *Annales de Chimie* of 1834 (respectively volume 55 and 57).
72. See *Ann. Chimie,* **73** (1840), and **3** (1841), 129. In 1866, H. Kopp reassessed these researches in terms more favorable to Avogadro. *J. Chem. Soc.* **19** (1866), 158. Avogadro's speculations on the distinction between 'true atoms', 'partial atoms', and 'atomes immédiat' did not help in clarifying the complex nature of such elements as phosphorus, arsenic, iodine, and bromine.
73. He still had this title in 1840 at the time of the second Italian Scientific Congress which took place in Turin.
74. Avogadro in this work used 'atom' and 'atomic' to refer to both elements and compounds.
75. See *Fisica dei corpi ponderabili*, vol. 2 (Turin, 1837), p. 840.
76. See below and Robert Fox, *The Caloric Theory of Gases*, chap. 6, for a discussion of Avogadro's caloric theory.
77. See for instance the first chapter of part one of the first volume of the *Fisica*.
78. For the wave theory of heat (Ampère, Melloni, etc.), see S. G. Brush, 'The wave theory of heat', *Brit. J. Hist. Sci.* **5** (1970), 153–155.
79. *Bibl. Universelle,* **29** (1840), 142.
80. Between 1839 and 1847 nine national scientific congresses were held in Italy. Avogadro, however, attended only Turin. It is difficult to explain why he did not go, for instance, to Genoa in 1846 or Milan in 1844, unless there were health reasons.
81. Both were published in *Memorie di Matematica e Fisica della Societa' Italiana di Scienze,* **23** (1844), 156 and 260. For other works on this subject see for instance A. Laurent in his *Chemical Method* (1854) and Gmelin in his *Handbook* (1852).

82. A reprint of the 'Saggio di teoria matematica della distribuzione dell'elettricitá sulle superficie dei corpi conduttori' written in 1842, and published in *Memorie di Matematica e Fisica della Societa' Italiana di Modena,* 23 (1844), 156.
83. See L. P. Williams, *Michael Faraday* (Basic Books, New York, 1965), p. 383.
84. He failed to have this work published by de La Rive in *Bibliothèque Universelle de Genève*, as appears from his correspondence with the editor, kept in the library of the University of Geneva.
85. See *Memorie Matem. Fisica Soc. Italiana,* **23** (1844), 273. One has the feeling in reading these pages that the author of the molecular hypothesis was not immune to the confusion prevailing at the time among those practicing chemistry.
86. In his 1833 and 1834 papers on specific heats, he had introduced fractional atoms of the halogens, sulfur, and phosphorus.
87. Ironically, L. Meyer in *Modern Theories of Chemistry*, Bedson and Williams trans. (London, 1888), p. 30, suggests that the application of Avogadro's hypothesis is based upon the idea that no molecule can contain fractions of atoms.
88. This subject is extensively discussed in chapter 5 of this work.
89. *Nuovo Cimento,* 5 (1856), 473. Matteucci neglects to mention also that Avogadro had been a member of *Società Italiana delle Scienze* or *Societa' dei XL* founded in Verona by A. Lorgna in 1785 and the most exclusive Italian scientific association of the time.
90. In my view, the article by A. Naumann in *Ann. Chem Pharm.* **123** (1867), 341 gives the first current interpretation of Avogadro's gas hypothesis.
91. He ascribed this notion to Gerhardt in *Nuovo Cimento,* **6** (1857), 25.
92. See *Memorie della Reale Accademia delle Scienze di Torino*, ser. 2, **16** (1857), LXI.

CHAPTER TWO

ELECTRICAL AND ELECTROCHEMICAL STUDIES

The first decade of the nineteenth century saw the physical sciences permeated by the entirely new field of research stimulated by the discovery of the Voltaic battery. In those same years Avogadro, through the guidance of Vassalli Eandi[1] and largely as a result of his own reading, became more and more deeply involved with his new intellectual pursuit. Naturally he did not fail to recognize the importance of the remarkable phenomena revealed through the application of Volta's pile.

The results of Avogadro's first speculations on the nature of electricity and the function of the metals in the Voltaic cell were discussed in two unpublished manuscripts: one written in 1804 with the title 'Considérations sur la nature des substances connues sous le nom de sels metalliques et sur l'ordre de combinaisson auquel il parait le plus convenable de les rapporter', the other, co-authored by his brother Felice, appeared one year earlier and was entitled 'Essai analytique sur l'électricité'.[2]

The *Journal de Physique* of de La Méthérie published a short essay in 1809, a few months before Avogadro accepted the teaching position in Vercelli; it discloses his first views on the relative positions that all chemical entities take in a continuous series in relation to the positive or negative electricity they assume in their mutual contact. This essay, 'Idées sur l'acidité et l'alcalinité,[3] is significant for the critical position taken by its author on one of the fundamental tenets of Lavoisier's new chemistry, that the acidic nature of a compound could be conferred merely by the presence of oxygen.

Although Avogadro had at that time only a limited knowledge of the experimental evidence, he already seemed to favor the views emerging from Humphry Davy's early conclusions in disagreement with the oxygen theory of acids.

On this occasion Avogadro introduced the notion of oxygenicity which will be discussed further in his later essays of 1811 and 1814 (Chapter 3). Some substances, for instance oxymuriatic acid, he says, seem more oxygenic or more acidic than others, but the presence of oxygen alone does not always give an acid character to a compound, as some metal oxides clearly indicate with their alkaline nature. As for the proposed existence of a neutrality point in the oxygenic series, above and below which the acids and the alkalis

Plate 6. Palace of the Academy of Sciences of Turin, facing via Santa Teresa, metal engraving by C. Bossoli, 1853.

are respectively placed, and as for the suggestion that neither absolute acid nor absolute alkaline substance exists, these two assumptions certainly represent a new and generally neglected contribution.

Very likely this short note of Avogadro's, although published in a widely known journal, escaped Berzelius' attention at the time,[4] for Berzelius' main field of interest then was the study of the combining proportions. It was only a few years later that the dualistic theory began to take shape.[5]

In 1806 and 1807 two other memoirs were submitted to de La Méthérie by Avogadro and published in the *Journal de Physique*.[6] Closely related in their subject matter – the conditions of an insulator placed between surfaces charged with electricity of opposite sign – they may be considered the first and, for many years, the only attempt by Avogadro to offer a rational explanation of the electrostatic phenomena. More specifically, the study of the behavior of non-conductors – or 'isolateurs' as he defined them[7] – was regarded by Avogadro as very important, in view of the consideration that 'il n'y a point d'électricité qui n'ait en face une électricité d'espèce contraire, avec interposition d'une couche isolatrice'.[8] Furthermore, it did not escape his attention that this topic had not yet been thoroughly examined by the physicists of the time.

It should be recalled that, in the second half of the eighteenth century, reports had been published with increasing frequency of phenomena which could be described in general terms as related to electrostatic influence between large surface conductors. In 1756 Aepinus had described his air condenser, thus disproving that the condensing effect observed in a Leyden jar should be necessarily ascribed to glass.[9] Three years later he reported the peculiar behavior of a brass ruler touched by a charged glass pipe; the far side of the ruler showed the same electric sign as the glass pipe, a fact which had not been detected earlier. Almost simultaneously (1758) Johann C. Wilcke,[10] after examining the behavior of insulators placed between conducting plates, advanced the hypothesis that such insulators were actually formed by alternate positive and negative layers.[11]

Then G. F. Cigna, a pupil of Beccaria in Turin,[12] aware of some previous observations of Symmer, conducted a series of experiments in 1765 with silk ribbons charged with an ivory ruler and noticed that they were mutually attracted. By rubbing them with a stick of sulfur, they were again charged but with opposite signs. When he approached an insulated lead plate with a charged silk ribbon, by touching the same plate with his finger, he obtained a spark, while the ribbon was strongly attracted by the lead. Upon removal of the ribbon from the plate, the former reacquired its original electric sign and the lead plate appeared charged with the opposite electricity. Using the sparks thus obtained every time the ribbon was either placed on or removed from the plate, Cigna succeeded in charging a Leyden jar.

These experiments and others repeated later by Beccaria in 1768–69 led him to argue that when two plates of opposite charges are put in contact their electricities are neutralized, but are resumed after the plates are removed from each other. He thus introduced the term 'vindex' (vindicatory) to define such electricities, a concept challenged a few years later by the young Alessandro Volta who explained the same facts in terms of electrostatic influence.[13] As a result of further investigations in 1775 of the 'vindex' electricities, Volta was able to design and construct his 'perpetual' electrophorus that he first described to Priestley, making him suddenly famous all over Europe.

In those years Cavendish was researching the inductive capacity of different condensers.[14] Although his work may have been inspired by the earlier investigations of Beccaria,[15] his interpretation of the movement of electricity in glass plates was quite original. Within the thickness of glass there are layers in which the electric fluid is alternatively movable and immovable, he explained, thus giving his theory of stratified condensers.[16]

While Avogadro in the 1806 'Considérations sur l'état' mentions Priestley's *History*, Aepinus, Symmer, Beccaria, Cigna, and Volta, all of whom very likely formed the foundation for his studies of electrostatic electricity, he was certainly unacquainted with the work of both Wilcke and Cavendish; however, as evident from his two memoirs on electricity, he arrived at a similar explanation of the behavior of an insulated body placed between two conductors of opposite sign; his 'couches' are, after all, within the frame-work of the layer structure earlier theorized by Cavendish and Wilcke.

In the first memoir[17] Avogadro disclosed that his purpose was the discussion of earlier observations (made by Haüy, Volta, and Aepinus) concerning the fact that electricity seemed to disappear from the surfaces of two bodies charged with opposite signs, when they were joined together along the same surfaces. If the separation of such bodies is attempted, however, an adhesion is observed between them, indicating that some electric charge is still present. Moreover, after complete separation, it is noticed that each body reacquires the charge it used to have before their junction. None of these facts had been observed when two conductors were involved in the experiment, instead of an insulator and a conductor.[18]

Two approaches had been followed in an attempt to interpret this behavior: in one case, two ribbons made of a very fine fabric, such as silk, inert to electric charges, were connected either together or by an insulated lead foil.[19] In the other case, experiments were conducted with much denser insulators, for instance glass plates. The substantial difference between the two systems was based on the fact that in the latter a direct dependence is achieved between the electrical state of one of the faces of the glass plates and the electrical state of the opposite face.

Avogadro recalled that among the interpretations of the reappearance of electrical charges, when insulators were drawn apart from each other, was that given by Beccaria,[20] who had proposed the existence of 'vindex' electricity, a form which – as the name indicates – would reappear as 'vindicated' by the nonconductors after their separation. In a less suggestive but simpler way, Avogadro suggested calling it latent or quiescent electricity.

Insulators made of a dense material, such as two glass plates surrounded by armatures and charged separately with electricity of opposite sign, if joined together face to face, after grounding, resist attempts to move them apart. The resistance to such action was much greater than that observed when silk insulators were used. Upon complete separation of the two glass plates, both the inside surfaces of the plates reacquired their original charge and the outside face of the armature assumed the same sign of electricity of the inside face of the adjacent plate. The same effects had been observed when the two glass plates were charged after being joined to each other.

These facts, according to Avogadro, could be interpreted on the basis of the above mentioned theory of the 'revivifaction' of quiescent electricities. More specifically, they prove that when two insulated thin glass plates are joined together to form a single body, 'each plate acquires its own charge; that is to say, on the lower surface of the upper plate, an electricity is formed of sign opposite to that present in its upper face'. The same, he added, was true for the upper and the lower face of the other plate. As a result, the charge present on such a plate is not related directly to the one of opposite sign on the upper face of the insulator, but only through the intermediate charges of the two inside surfaces directly in contact.[21] And if the charges are removed from the two plates which are then separated, it appears that their surfaces have the same sign, regardless of the fact that they had been charged separately or together; this meant, he concluded, that they were originally in the same electrical state and that this was equally modified in both cases.

The observation made by Avogadro was that there was no difference in the way charges are distributed on the surfaces of a single insulator or two insulators joined face to face. Avogadro assumed that this applied even in cases of more than two non-conductors packed together: 'le nombre ne fait rien ici',[22] because they will probably behave as though they had been charged separately. One can detect here one of the first instances of the generalizations which delineated the approach adopted by Avogadro in his scientific speculations.

At this point, two essential conclusions are reached and clearly stated: (a) it is possible to imagine any solid insulator as formed by as many layers as there are 'molécules élémentaires'[23] in its thickness, and (b) when such an insulator is charged, then each layer assumes its own charge, with the result that layers of opposite sign will face each other. Although there could be in any solid insulator an 'infinite' number of these layers, regardless of how thin, its outside faces will always have charges of opposite sign. Avogadro pointed out the originality of this concept that, he stressed, should not be confused with the conclusions reached earlier by Coulomb and Haüy after their experiments on magnetism and on the pyroelectric properties of Tourmaline.[24]

Later on, an explanation was proposed for what takes place in the same insulator, as mentioned above, when it is discharged. The infinite number of charged layers composing it is replaced by an infinite number of 'paires d'électricité', neutralized by reciprocal contact; this occurs as a result of the transformation of the electricities of different sign sustaining each other in every layer into surface electricities.

He recalled, in the concluding remarks of the *Mémoire*, the observation that the electrical state is formed again between the surface of an insulated plate and its armature when the latter is removed, after being charged and subsequently grounded. This was experimentally noticed in Volta's electrophorus and seems to indicate that an exchange of sign takes place when the armature is separated from the insulator: one assumes the charge of the other and vice versa. A possible interpretation, according to Avogadro, may be that the discharge does not nullify the electrical state of that surface of the glass plate, but instead causes the transformation of the armature electricity into surface electricity of the insulator; in agreement with the assumption made earlier, such surface electricity should then lean (*s'appuyer*) on an electricity of opposite sign which, after the discharge, appears on the inside surface of the armature.

In August 1807 the 'Second Mémoire sur l'Électricité', a continuation of 'Considérations sur l'état', was published in the *Journal de Physique.*[25] The debate between Symmerians and Franklinians, after 60 years, was not yet settled. Although his interpretation of electrical forces, as occurring through action at a sensible distance, led Aepinus[26] to regard as untenable the single fluid hypothesis, Avogadro seemed to favor Franklin's theory in this essay, with some reservations. This was despite the support given to Symmer's position by Coulomb's fundamental research twenty years earlier.[27]

The major obstacle to the Franklinian theory was represented by the observed repulsion between negatively charged bodies, a fact that Aepinus attempted to explain on the basis that the elementary particles of all bodies, when divested of electric fluid, repel each other. Avogadro, in recalling this interpretation, added that if an explanation could be found for such repulsion, then it would be superfluous to maintain the two fluids theory, for it represented only an unnecessary complication in the study of electrical phenomena.

By assuming the existence of a fluid in the air layer interposed between charged bodies, and by interpreting the electrical repulsion and attraction in terms of simple modifications of such fluid caused by actions at insensible distances, he then advanced the supposition that in the case of two identical charges (or, according to Franklin, when there is either an excess or a deficiency of fluid on both charged surfaces) they alter the intermediate fluid in the air layer in such a way as to cause a repulsion among themselves, regardless of their sign. Through the action of this fluid it was thus possible to interpret the observed repulsion of bodies having identical signs, even without recurring to a second fluid as the Symmerians do.

For Avogadro, the electrical charge must act as a 'modification chimique', because it influences all the molecules[28] of the interposed insulator. This is an interesting view not as much in relation, as had been suggested,[29] to its author's possible early appreciation of the electrochemical theory, but because it reveals that instinctive leaning toward the establishment of a close correlation among chemical and physical forces which characterized the thought of Avogadro in the following years.

For Avogadro the alteration or electrical charge which occurs in all elementary molecules consisted essentially in a 'simple shifting' (*simple déplacement*) of a substance or fluid existing among them in any body which can be charged. In the case of two conductors without an intermediate insulator and whose surfaces are superimposed and thus subject to quiescent electricities, the movement of this fluid takes place from the molecules on the surface of one conductor to those on the surface of the other, without overcoming any resistance and by just replacing a condition of equilibrium with another one. The fact that it is limited to the outside layer of molecules, and does not affect the rest of the body, means that the chemical 'combination' (earlier he had talked of chemical 'modification'), here involved, is of a 'nature particulière, inconnue jusqu'ici'.[30]

Unlike the caloric which surrounds and involves every particle, the fluid here considered affects only the hemispheres of the superficial molecules.[31] On the hemispheres of each elementary molecule (of whatever body) may be present an excess of a sparse substance which tends to distribute itself among those same molecules in relation to its affinity for them; when a surface is in contact with another, each molecule displays (on the hemispheres facing each other) an equal deficiency of such substance in comparison with the amount required by its natural affinity; as a result, the total amount of substance involved between the two hemispheres remains the same as that requested by their combined affinities.

It is not an ordinary chemical combination, he continued, since the amount of matter involved and their mutual distance remain unchanged among the molecules involved; the only shifting occurs in the boundary of fluid surrounding them. Avogadro defined this special combination as an 'electrical' one and, in the case earlier considered of a system of quiescent charges, he defined the electricity as the state wherein an increase of such sparse fluid on one of the charged surfaces corresponds to a contraction on the opposite surface.

When two charged surfaces are separated by a layer of insulating material, for instance air, a new interpretation of the electrical state, thus modified, is required. Since there is no direct contact between the surfaces, the excess of fluid on the hemispheres of one surface will not immediately find its balance in the deficiency present in the hemispheres of the opposite surface. The result then would be the mutual destruction of charges.

However, in view of the fact that the experimental evidence does not agree with the existence of such quiescent systems of charge, one may suspect that a condition has arisen preventing the movement of the insulator fluid from the hemispheres of each particle of the insulator to the adjacent particles.[32] Also an explanation is required for the fact that there is no dissipation of electrical charges. Avogadro at this point suggested that an effect of compression and expansion – caused by an electrical excess or deficiency on the hemispheres of each elementary molecule of the charged armatures respectively[33] – may displace the whole fluid from these spheres, and, as a result, cause a shift of the fluid itself from one side to the other of their surfaces.

A condition of 'excentricity' is thus created, which will successively involve, through adjacent layers of elementary molecules, the whole insulator.

For Avogadro the electrical charge consists in this excentricity of all the fluid atmospheres of the charged body. Since such condition is unnatural and antithetic to the 'law of attraction', the electrical fluid will accordingly show a tendency to be driven away. This may be the cause of the 'electrical intensity' appearing when two charged surfaces are separated by an insulator which is directly proportional to the thickness of the latter, while it is nil when the two surfaces are in contact and the whole system is quiescent.

What happens to such a system, represented by a grounded insulator placed between oppositely charged armatures, is the subject of the concluding section of the *Mémoire*. There are two ways to explain how the discharge occurs: one, through the mutual destruction of the opposite charges present in the armatures and the consequent return of the electrical fluid from its condition of displacement to the 'état naturel'. The other interpretation, coherent with Avogadro's earlier description of an electrical charge and also with the experimental evidence, consists essentially in this: upon grounding, the positive charge of the top armature[34] is transferred to the upper hemispheres of the elementary molecules of the insulator. In doing so it 'drives away' (*chasse*) the fluid from the surface of the same armature, thus determining a deficiency (or a negative charge) with which the charge enters into a quiescent state. At the same time, the fluid expelled from the surface of the top armature will cause an excess (or positive charge) on the opposite armature. Such positive charge will then be in a quiescent state with the negative one formed on the lower surface of the insulator. The positive charge present in the lower hemispheres of the first layer of elementary molecules, adjacent to the top armature, is transferred to the hemispheres of the next layer, thus creating a deficiency of fluid and, therefore, the formation of a negative charge; the result again is a quiescent state.

On the basis of such mechanism the difference between conductors and insulators could be explained, Avogadro concluded, by assuming that in the former the passage of fluid from one layer of elementary particles to the next one may occur freely through their mutually-facing hemispheres, while on the other hand no such movement is possible in the case of non-conductors.

For their attempt to explain the phenomenon of electrostatic influence[35] and for their analysis of the intimate nature of the insulators – formed by an infinite number of extremely thin layers all electrically charged through induction – these two early essays represented an extremely original contribution, anticipating by almost thirty years Faraday's classic work on the subject. This was recognized quite early by Ottaviano Mossotti during Avogadro's time.[36] For Guareschi[37] the two most important conclusions reached in these memoirs were that no electrostatic force may act at a distance, and that electrical phenomena are substantially chemical phenomena. In fact, this particular observation is not quite correct, because Avogadro draws a line between regular chemical modifications which involve the entire elementary particle and the electrical state's alterations concerning only their surfaces. Unlike chemistry, then, electricity does not modify the essence of the bodies involved.

More recently, it has been commented that in the 1807 *Mémoire* 'there is the germ of what Faraday later christened the electrotonic state, out of which Maxwell or Faraday s field theory was to develop'. In addition the originality of Avogadro's views is recognized as 'the first attempt to eliminate the electrical fluids outside the Kantian tradition'.[38]

For the following thirty-five years, with the exception of a paper describing the construction and applications of a voltmeter multiplier, one of his few experimental works,[39] published in Turin in 1823, Avogadro kept away from this field of physics. His short note of 1842 on the nature of the electrical charge has more historical than scientific value: its purpose is to reassert the priority of his ideas in the elucidation of the behavior of the dielectrics in view of quite similar conclusions reached by Faraday in subsequent researches.[40] The note appeared in Geneva's *Archives de l'Électricité*[41] as a comment on the XIth Series of Faraday's *Experimental Researches on Electricity*. These were also reported in the 1838 *Philosophical Transactions* and Avogadro probably read them in 1839. The reason for his three year's delay in coming forward to review Faraday's work is not clear; one may surmise that he had no desire to demean the merits of a man with whom he was on friendly terms.

Avogadro opened his 'Note' with a summary of the main conclusions reached in the XIth Series, namely that: (a) the induction consists of a certain polarized state of the particles of the insulator in which they are thrown by the electrified body, (b) this is a forced state of the particles and represents the intensity of the charge, (c) in the case of a Leyden jar, its discharge represents the return of such particles to their natural state, and (d) there is an inverse relation between the thickness of an insulator and the amount of inductive action, because fewer are the intervening particles and less is their resistance to the charge. Avogadro had arrived at the same conclusions thirty-six years earlier, a fact which, he continues, should be 'interesting for the history of science'.

Can this be considered a fair statement? A comparative examination of the conclusions reached by Faraday in his XIth Series and by Avogadro in his memoirs on induction reveals that there are indeed substantial and sometimes striking similarities. Although the terminology may appear different – there is a significant contrast between the lucid, concise writing of the *Experimental Researches* and the involuted, verbose, paragraphs of the 'Considérations'[42] – the main concept of the contiguous molecules strained ('l'espèce de compression'[43]) under the effect of the electric fluid and the return, after the discharge, to a system of electrical equilibrium[44] is present in the second memoir. Neglected by Avogadro, as he himself recognizes,[45] are two very important elements of Faraday's thought: that induction takes place also in curved lines and that it is specific for different insulators. Both concepts are extremely significant in strengthening Faraday's assumption, supported by his systematic and extensive experimental work, that there is no action at distance and that induction must be closely related to the nature of the particles or ultimate components of the dielectric.

As for the specific induction, it should be recalled that, in 1754, G. B. Beccaria in Turin had observed that a number of insulating materials, for instance, sulfur, pitch, colophony, and sealing wax have a different condensing power,[46] but the significance of this observation apparently failed to attract Avogadro's attention. Another point raised in the XIth Series (No. 1297) seems overlooked in the memoir of 1807: beside the attractive force existing among the particles of the dielectric and which takes place perpendicularly to the charged surfaces, there is also a repulsive or diverging force exerted in the transverse direction. What happened across his 'moléculair couches' did not seem to bother Avogadro, who, in his model of spherical elementary molecules, considered only the mutually facing hemispheres whose surfaces alone are involved in the transport of the electric fluid.

If a meaning should be given to the 'Note sur l'Électricité' – which almost certainly Faraday did not read – it is that it may help to explain an aspect of its author's personality: although of unassuming nature and definitely not a boastful man, Avogadro never hesitated to set the record straight, whenever the originality of his ideas seemed neglected by his contemporaries.

Electrochemistry

In 1809, while still an instructor at the Royal College of the Provinces in Turin, Avogadro submitted his first major essay on electrochemistry to the

Journal de Physique.[47] He would return to this subject four years later in a short note published in the *Annales de Chimie*.[48]

In terms of historical perspective, the 1809 essay, 'Idées sur l'acidité et l'alcalinité', appears between Davy's first Bakerian lecture and Berzelius' outline of his electrochemical theory in the 1812 *Journal für Chemie und Physik*.[49] While the Swedish chemist is not mentioned at all in 'Idées', Avogadro shared the view with Davy that a close link exists between chemical activity and the electromotive force observed in the contact of two bodies.

The main purpose of his investigation was to find a satisfactory answer to these questions: how can acidity and alkalinity be defined, what distinguishes acids from alkalis, and how essential is the presence of oxygen in determining the acid character of a chemical combination? In 'Idées' his position on the last one of these points is clearly indicated: oxygen is not the 'principle' of acidity, as proven by its presence in basic and neutral (water) substances and, on the other hand, by its absence in compounds decisively acid, as hydrogen sulfide.[50]

Moving from here and by admittedly relying on Berthollet's terminology, he assumes the existence of an 'acid' antagonism as distinguished and opposite to a 'basic' antagonism. In order to manifest these antagonisms, a substance must be in a particular state of aggregation, because if its cohesion as a solid or its elasticity as a gas is too great, it may effect the combination with other substances and prevent the manifestation of either its acid or alkaline nature. Such, for instance, he recalls, is the case of oxygen which, for its 'grande élasticité', does not behave as an acid when in the elemental state.

More significant, however, remains Avogadro's assumption that both antagonisms, acid and basic, should be considered as relative characteristics of substances and that the degree of acidity or alkalinity displayed by a compound is related only to the degree of acidity or alkalinity of its components. This concept, reiterated in 'Idées' and asserted again in the 1813 note, certainly represents the keystone of his electrochemical theory.

It may be interesting to examine briefly some of the ideas set forth by Berthollet in his *Statique Chimique*, because of the influence they exerted on Avogadro's electrochemical system.[51] One should recall that Berthollet in his short discussion of the nature of the electric fluid, had reached the conclusion that, in view of its behavior, it should be recognized as a chemical agent. After his attempt to establish a parallel between the action on chemical combinations of caloric and that of light, on one side, and of electricity on the other, he eventually came to the view that 'chemical effects which take place seem to be the consequence of the electric action'.[52] This effect, he rationalized as being based upon the electricity's power to overcome the cohesion forces holding the molecules together, a position quite aligned with that then held by Berzelius, but not discussed by Avogadro on this occasion.

He devoted much more attention, however, to other considerations which had been made by Berthollet in the very extensive section of his book concerned with the alkaline and acid nature of chemical substances. Berthollet had suggested that, since there are substances which behave like acids with the bases and like alkalis with the acids, then acidity and alkalinity are only relative terms. Earlier he had defined acids and alkalis respectively as those substances which can saturate the alkalis and those 'by which their union can saturate acidity'.[53] Each acid or alkali has then a different 'capacity for saturation', a characteristic which may be used to set a 'scale' of its relative power.

The essence of these concepts was later adopted by Avogadro, as a foundation, in the 1809 essay, to shape his own theory of the nature of acidity and alkalinity. By accepting Berthollet's notion that the definition of acidity and alkalinity should not be considered in an absolute way, he visualized a scale of substances according to their mutual chemical conduct and placed at its extremes those elements which act as an acid or a base in relation to all the others.

From the *Statique Chimique* Avogadro also acquired a quantitative approach to the interpretation of the electrochemical behavior of acids and alkalis which will be particularly stressed in his criticism of Berzelius' theory: their polarity is determined by that of their radicals only when an acid or a base has the same amount of oxygen. Unfortunately, other ideas that Avogadro borrowed from Berthollet only introduced unnecessary and confusing elements in his system, such as that sulfur is intrinsically acid and that it is prevented only by its great cohesion from displaying its real chemical nature, and that to hydrogen is due the alkaline character of ammonia.[54] Some of these concepts – such as the continuous attempt to correlate the physical properties of an element with its chemical behavior and to explain the latter as a function of the former – was retained by Avogadro for a long time and will reappear as a regular thread in his scientific philosophy.[55]

The French translation of Berzelius' 1812 essay ('De l'influence de l'électricité sur les affinités')[56] came to Avogadro's attention the following year. Shortly afterwards, in the 'Réflexions sur la Théorie électrochimique de M. Berzelius',[57] he took strong issue with some of the conclusions reached by the Swedish chemist. The view of the absolute electronegativity of oxygen is challenged first because, he says, other substances eventually could be found more electronegative than oxygen. It is further incorrect to consider those compounds as unable to combine with other substances, such as neutral salts, which are apparently 'saturated'; actually, Avogadro observed, they can still enter in combination with either an excess of acid or of alkali.[58]

When, under the action of a Voltaic battery, a salt decomposes and its acid and basic portions are separated, it is incorrect to define the polarity of their radicals – as Berzelius did – only on the basis of the two poles to which those portions migrate. For Avogadro the polarity of the substances containing oxygen should be related in part to this element and in part to their own radicals. He adds that not only is the chemical nature of the radical important, but also that its amount in a compound may determine the polarity of the compound. Along these lines he thus explained the alkaline or the acidic character of substances on the basis of the excessive weight in these compounds of a weakly acid element (for instance nitrogen in ammonia) or of a strongly acid element (for instance carbon in hydrocyanic acid, and sulfur in hydrogen sulfide).

The characteristic quality of a substance to impart an acidic or oxygen-like nature is defined as 'oxygenicity' by Avogadro for the first time in his 1809 essay. This property, he added, is the opposite of oxidability or affinity for oxygen, which explains why hydrogen, a very little oxygenic element, should display an alkaline character. Its presence, then, justifies the neutrality of water and definitely contributes to the basicity of ammonia.

From then on, Avogadro again and again recurred to oxygenicity as an unequivocal element of qualification of the chemical nature of a substance. It is apparent that he introduced it as an absolute term of reference and as a substitute for less valid definitions; with this yardstick at his disposal he proceeded to classify both elements and compounds according to their antagonism or chemical affinity.

The scale of acidity and alkalinity proposed by Avogadro in 'Idées' cannot be considered an electrochemical series in a strict sense. He soon realized, however, that chemical considerations alone do not offer a sufficient criterion for such an arrangement, and accordingly conceded that the relation, suggested by Davy, between chemical activity and electromotive force may be conveniently used to set the relative position of a substance in his scale. If the chemical antagonism can be assimilated to 'electrical heterogeneity', he argues, then it is possible to compare his series with that of Volta. By assuming that acids will take a negative charge and alkalis a positive one, their combination should be either favored or prevented by the effects of electricity.

Substances, such as water and neutral soluble salts which do not change the color of vegetal indicators, are placed by Avogadro in the middle of the scale. Above them he sets the acids and the oxygenic elements and below them the alkalis and the progressively less oxygenic elements. The position that Avogadro adopted more explicitly in the 'Réflexions' is evident from his discussion of oxygen. He used the conditional expression 'L'oxygène devroit être placé très haut dans ce tableau et même, selon toute apparence, à la tête',[59] indicating his reluctance to take a more definite stand on the subject.

The metals come next in the usual order established by their electrical behavior, followed by sulfur and, quite far apart, by nitrogen. He then moved to discuss the oxides of alkaline metals placed near the bottom of his scale; their basic nature is explained by recalling the small amount of oxygenicity of their radicals.

For Avogadro, the fact that not all basic substances contain oxygen (ammonia is an example, and its chemical nature can be explained by the low oxygenicity of its components) and that some radicals combine with more or less oxygen than others do, proves that Berzelius' interpretation of bases and acids in terms only of their radicals is erroneous. In fact, Berzelius had observed in 1813[60] that the combustible portion of an acid is electronegative towards a base even when the latter is not oxidized and, therefore, that substances of opposite electrochemical nature combine with each other always in the same proportion, regardless of the presence of oxygen. As an example, he recalled that sulfur and potassium enter in the same proportions in compounds such as the sulfide, the hydrosulfide, and the sulfate, in which sulfur always behaves as an acid.

The recurring theme of 'Idées' which reappeared later in the 'Réflexions' – the concept that alkalinity and acidity cannot be interpreted in absolute terms – is certainly underlined by Avogadro in the concluding observations of the 1809 memoir, when he suggested that the term acid should not be used any longer in a generic sense, because it represents only an 'accidental' characteristic; other substances, he remarked, would behave as acids, were it not for their physical structure which does not allow them to display their real chemical nature.

It should be recalled that in discussing the polarity of the elements, Berzelius had recognized the existence of a relative electrochemical value for each of them.[61] In this regard, his stand then differed essentially from that of Avogadro only as far as the absolute electronegativity attributed to oxygen.[62] He sets this element at the top of his series and he presented on the same occasion a much clearer distinction than Avogadro did at the time between metalloids[63] and true metals – such as the alkaline and the earth-alkaline elements, iron, manganese, silver, etc.

This early classification of the elements, according to their electrochemical behavior, unfortunately obscured by further futile distinctions,[64] met the sharp criticism of Avogadro in the 'Réflexions'; he considered it superfluous and based essentially on the properties of the oxides and acids rather than on those of their radicals. One cannot neglect to mention that, on one of the more important issues examined by Berzelius in the 1813 essay, Avogadro did not express any comment in the 'Réflexions'.

The question, whether electricity generated in a Voltaic battery was the cause or the result of the chemical phenomenon observed in such a system, had been solved by the Swedish chemist in favor of the view that it is the electric current which determines the conduct of chemical affinities.[65] As a proof of his reasoning, he described an experiment conducted with a copper-dilute nitric acid-calcium chloride-zinc cell clearly suggesting, in his opinion, that the passage of an electric current reversed in the metals the effect of the chemical activity.[66] Quite remarkably, a few years later, Berzelius used a similar experimental approach to reach the opposite conclusion, i.e., that the oxidation of a metal (copper in this case) is not the cause of electricity, and the oxidation of zinc (anode) is due to the movement of electrical fluid, because it does not take place when the circuit is open.[67]

An explanation of Avogadro's non-committal position on the question of the origin of electricity may possibly be found in the opening remarks of the 'Réflexions', when he states that he would refrain from discussing the topics of Berzelius' essay which were not relevant to his own theory which was mainly concerned with the relations between acidity and alkalinity. One may also reasonably assume that, at that time, he was not conversant enough with this matter to be able to discuss it and formulate an opinion. This is consistent with his personality, for he very seldom expressed his views under similar circumstances.

There was to be a gap of nine years before he again directed his inquiry to the study of electrochemical phenomena. His next paper in this field appeared only in 1823, a few months after the suppression of his Chair at Turin.

Experimental Researches in Electrochemistry: The Multiplier, 1822

Investigations into the effect of voltaic current on a magnetic needle flourished in the years that immediately followed Oersted's publication of 'Experimenta circa effectum conflictus electrici' in 1820. Ampère realized that Oersted's 'Method' could be used to measure small currents. In the fall of 1820 he first used the word *galvanometer* to define an instrument which could detect the presence and direction of an electric current in either a voltaic pile or a conductor.[68] Shortly afterwards, he designed an arrangement in which the plane of the magnetic needle would be perpendicular to the direction of the inclination needle in order to overcome the disturbing effects of earth magnetism. In 1820 Ampère greatly improved the astatic system by using two tightly connected magnetic needles of opposite polarity to counterbalance the rotational movement of the earth's magnetism upon one needle by the reaction of the second needle.

In September 1820, a few days before Ampère's first report to the Académie, Johann Schweigger showed his multiplier to the Scientific Society of Halle in Westphalia. This instrument, like that later proposed by Poggendorff, had a single magnetic needle placed in the middle of a vertical core of silk-covered wire. The current moved through the upper portion of the coil and acted upon the needle in a direction opposite to that moving through the lower portion. As a result, the effect of the current was increased.[69] Physicists now had at their disposal a very sensitive tool to detect the direction and intensity of a current.

Poggendorff's multiplier, similar to Schweigger's, was reported in Germany in the early part of 1821. It did not escape the attention of Oersted who wrote about it to Forschammer, a Danish chemist then in Edinburgh. Forschammer, in turn, informed the editor of the *Edinburgh Philosophical Journal* which published a note in July 1821. Four months later a translation of this note appeared in the Swiss *Bibliothèque Universelle*, Avogadro's regular channel of information about British scientific literature.[70] By this tortuous path the description of Poggendorff's multiplier arrived in Turin probably at the end of 1821, more than one year after Ampère had described his astatic arrangement.

At the Royal Academy of Turin on 22 April 1822, Avogadro read his 'Mémoire' on the new design of a multiplier and its use in determining the electrochemical series of the elements, which was published shortly afterwards.[71] The instrument itself had been shown at a meeting of the Academy a few weeks earlier.

During the spring of 1822 the atmosphere at the University of Turin was very turbulent. Both faculty and students experienced harsh reprisal measures as a result of the 1821 constitutional revolution.[72] Before the end of the year Avogadro would be suspended from his teaching position. Possibly he anticipated this decision because in a small community rumors spread easily. It is not difficult to imagine his personal and professional reactions while he conducted the research later reported in the 'Mémoire sur la construction'.

As has already been mentioned, Avogadro rarely based his speculations on personal empirical findings. Therefore, this short essay has particular significance in his activity. Yet, it attracted limited attention, probably because of the journal in which it appeared, but possibly also because of the modest level of its content. In itself the measurement of the 'galvanic fluid' was not new in the 1820s.[73] Only after Oersted's discovery led to the application of the electromagnetic effect did it become possible to design a reliable device capable of supplying consistent results. In view of the sensitivity of the instrument, the use of the coil, as proposed by Schweigger and Poggendorff, represented remarkable progress for the physicists of the time.

Avogadro modified Schweigger's design in at least three aspects, two of which deserve attention in the history of electrical instruments. He put the coil *only below* the needle and had it hanging from a thin silk thread, thereby reducing the mechanical resistance to the needle. Avogadro felt that he had thus greatly enhanced the sensitivity of the instrument and considered his multiplier an improvement over those constructed by Schweigger and Poggendorff. He thought that modification of the coil's form would have little effect on the instrument's characteristics. Yet, the shape adopted had the advantage of requiring a shorter wire and therefore conserved 'l'énergie d'action' of the current over the needle. This brief mention of a correlation between the conductor's length and current is interesting, although probably not original at this time.

A more detailed description of Avogadro's multiplier is necessary at this point. He used eighty turns of silk covered copper (or brass) wire[74] wound around three small glass cylinders placed 4 cm. apart on a wooden frame. The winding had a triangular shape and followed a clockwise pattern. Fastened upon it was a very thin plate of mother-of-pearl graduated in a semicircle of 180°. A 25 mm.-long magnetized needle fastened through a paper triangle to a silk thread hung 1 mm. above the plate, and the entire apparatus was covered by a glass bell. The small paper triangle through which the needle passed kept it perfectly horizontal by raising its suspension above its center of gravity. With the needle at rest, its magnetic South matched the zero position on the plate's marking.

The experimental procedure followed was very simple: two wires came from the multiplier and were connected to small cylindrical bars of the metals under examination. These were then immersed in an acid solution. Avogadro hoped to establish more accurately the series of potentials of the metals by observing the direction of the current determined by the deviation of the magnetic needle when the circuit was closed through a moist conductor. The metals' affinity for oxygen or their oxygenicity (in his view, a more precise definition of the chemical character of the substance) remained very closely linked to their 'electrical order'. Therefore, this series represented an important link between physics and chemistry.

In 1809[75] Avogadro had already set forth his main ideas on the electrochemical series of the elements, stressing the relative meaning of the concept of acidity, introducing the term oxygenicity and definitely relating chemical behavior to electricity.[76] The experiments conducted in 1822 with his colleague Michelotti[77] did little to change his interpretation of the facts.

In the 'Mémoire sur la construction' Avogadro claimed the first use of a highly sensitive galvanometer to establish the electrochemical behavior of metals. He pointed to the earlier reports of Volta and Pfaff and their use of the condenser in these investigations.

Actually, the first electrochemical series included a number of non-metallic substances such as charcoal and pyrites. Volta mentioned this series in a letter written in October 1795 to A. Maria Vassalli Eandi, Avogadro's future teacher in Turin. However, both the experimental procedure and the impurity of the substances used by Volta led to uncertain and erroneous results. Both Pfaff and Volta, in fact, limited themselves to putting the substances into direct contact. Avogadro, on the other hand, introduced a moist conductor between the metals, thus generating a continuous current whose direction is established by the electrochemical behavior of the metals in relation to each other.

Other experimental attempts to determine the direction of the current failed, because the results thus obtained were unreliable;[78] as Avogadro himself mentions, it was difficult to obtain a constant state of excitability in the neuromuscular system of the frogs' legs.

In 1822 Avogadro's experimental work began with the use of fourteen metals that, by couples, he plunged into different concentrations of nitric, hydrochloric and sulfuric acids. Avogadro observed, at first, that even when the same metallic couple and the same moist conductors were used, the needle of the galvanometer changed deviation, which suggested an inversion in the direction of the current. At the time it was assumed, in agreement with the findings of both Oersted and Ampère, that a positive current, like that flowing from copper to zinc immersed in diluted nitric acid, moved above the needle from the magnetic North, counter-clockwise, to the South. Naturally, by inverting the connection of the two metals, the needle deviated in the opposite direction (from South to West and from North to East). Thus far, the observations made with the multiplier followed this pattern. However, Avogadro soon found a rather peculiar pattern in several combinations of metals and moist conductors. First, with gold and platinum there was no current produced until the moist conductor was changed from a nitric acid solution to a sulfonitric solution (aqua regia). This seemed to confirm that without chemical attack on the metals, no voltaic electricity could be produced.

With a lead–bismuth couple (lead connected to South on the multiplier, bismuth to the North) the needle, upon immersion of the metals in concentrated nitric acid, first sharply deviated to East by almost 90°, then slowly returned to zero,[79] and finally moved in the opposite direction. This suggested to Avogadro that lead behaved at first as a more electropositive element than bismuth, then through the chemical attack of the acid, bismuth became more electropositive than lead, causing an inversion in the direction of the current. When a more diluted nitric acid was used, the deviation of the needle was of very short duration and no reversal in the direction of the current occurred. With the same couple and by immersing them in either concentrated or diluted hydrochloric acid, Avogadro observed that the needle moved to the East and remained longer in that position.

Five other couples of metals were examined in analogous conditions. In the case of tin and lead, the direction of the current remained unchanged when they were placed in diluted nitric acid and in both concentrated and diluted hydrochloric acids. The deviation of the needle observed with these acids and with the current moving from tin to lead implied that lead is electro-positive in relation to tin. Yet, by replacing diluted with concentrated nitric acid and with both concentrated and diluted sulfuric acids, a reversal of the current occurred, and tin behaved now as more electropositive than lead. With the couples bismuth and iron, and cobalt and antimony, similar reversals were observed.

When silver and gold were immersed in diluted nitric acid no electrical action occurred, but the needle deviated when the same metallic couple was put in concentrated nitric acid. When arsenic and lead were placed in diluted nitric acid, the current flowed from arsenic to lead, but when nitric acid was used in concentrated form the opposite current direction was observed. According to Avogadro, no other metal among the fourteen displayed the change in direction of current when immersed in acids of different concentrations.

As a result of these observations, Avogadro proposed the following electrochemical series. It ran from platinum, the most electronegative, to zinc, the most electropositive: platinum, gold, silver, mercury, arsenic, antimony, cobalt, nickel, copper, bismuth, iron, tin, lead, and zinc.[80]

In October 1795, Volta had described to Vassalli Eandi a table of 'first class conductors' established through the use of his condenser. As the most electronegative he placed charcoal followed by plumbago (graphite), copper, pyrite and so on, with zinc the most electropositive.[81]

A very similar series was proposed by Cristoph Pfaff shortly afterwards,[82] in which manganese oxide was placed as the most electronegative substance in place of charcoal.[83]

A comparison of Volta's and Avogadro's series of potentials showed differences in the position of some elements. Thus, mercury was set by Avogadro as much less electropositive than in Volta's revised table; gold and silver trade places. Volta put antimony next to tin and more electropositive than iron; whilst Avogadro set it as more electronegative than iron, bismuth, copper, nickel, and cobalt. In discussing antimony in its relation to tin, Avogadro again mentioned oxygenicity: antimony, then is characterized as more oxygenic than tin. The series of J. Ritter and Berzelius are included in Table 1, they are referred to in later discussion.

Table 1 Table of Potentials' Series

Volta 1795	*Ritter 1804*	*Berzelius 1814*	*Avogadro 1822*
charcoal	silver	gold	platinum
plumbago	mercury	platinum	gold
copper pyrite	gold	mercury	silver
silver	platinum	silver	mercury
gold	antimony	copper	arsenic
mercury	copper	nickel	antimony
platinum	arsenic	cobalt	cobalt
copper	cobalt	bismuth	nickel
brass	bismuth	tin	copper
bismuth	iron	lead	bismuth
iron	tin	iron	iron
tin	lead	zinc	tin
antimony	zinc		lead
lead			zinc
zinc			

Avogadro explained that the difference between his series and that of Volta was caused by experimental error. His galvanometer was more accurate than Volta's condenser. He also considered that his series was more consistent with the likely order of affinity of metals for oxygen.[84] A consideration confirmed, for instance, by the fact that platinum appears less oxidable than both gold and silver, and gold less oxidable than silver. Accordingly, the more electronegative a metal, the less affinity for oxygen it displays.

The general agreement observed between the electrochemical behavior of the elements and their affinity for oxygen encouraged Avogadro to examine Berzelius' series. Curiously, when writing the 'Mémoire sur la construction,' he neither mentioned an earlier series proposed by Ritter nor the later investigations by Berzelius, which had been published in 1814 in the celebrated 'Théorie des proportions chimiques'. Avogadro viewed Berzelius' series as based merely on 'chemical considerations'. Berzelius' and Avogadro's findings have little in common. The only relative positions upon which they have agreed are those of tin, lead, and zinc, aside from Berzelius' inclusion of some non-metallic elements. Avogadro this time did not invoke experimental error as the source of the disagreement. While he felt it possible that the degree of oxygenicity of metals might not concur with their 'electrical order', he thought the difference between the two series was too extensive. He went further and said that Berzelius' series could not easily be reconciled with that suggested by chemical considerations alone. Furthermore, he thought that the Swede had moved from his series to develop ideas that were 'scarcely admissible', casting a long shadow on Berzelius' conclusions.[85]

In the concluding paragraphs of the 'Mémoire sur la construction', Avogadro conceded that his series of experiments might also have given indecisive results because of the variations in the experimental procedure. Earlier he had recognized the merely qualitative nature of his observations by admitting that no attempt had been made to measure the degree of deviation of the magnetic needle when a current flowed through his multiplier.[86] Still he continued to stress the importance of the use of the galvanometer, a more sensitive instrument, for the investigation of both the causes of voltaic currents and the order of the affinities of the elements.

Avogadro's *Mémoire* of 1822 helped to define his position within the debate between the supporters of the contact theory and those who upheld the chemical theory of the voltaic cell. Both groups strenuously debated the issue through several decades of the nineteenth century, supporting their views with seemingly irrefutable experimental evidence. Davy, Pfaff, and Berzelius aligned themselves with Volta's original backers of the contact theory,[87] while Ritter, A. C. Becquerel, Parrot, Faraday, and Auguste de La Rive,[88] among others, became partisans of the opposite party.

Avogadro's stand on the theory of the production of current in the voltaic cell can be found within the 1822 paper. First, in the discussion of the advantages offered by his multiplier compared to the less sensitive frog's legs systems, he acknowledged that the magnetic needle of the multiplier moved only when there was 'sensible chemical action' in the conductor,[89] while the frog's legs reacted even when there was the slightest metallic contact with no chemical action to generate a current.[90] He argued that the sudden change in the needle's deviation confirmed that simple metallic contact could not be responsible for the production of the current, as Volta had advocated.

Avogadro felt that his position was validated by the crucial experiment with gold and platinum. When this metallic couple was immersed in nitric acid, regardless of the concentration, there was no current generated; no current flowed as long as there was no chemical attack on the metal. Avogadro's adoption of the chemical theory of the voltaic cell seems a rather suitable and coherent outcome of his earlier speculations linking electrochemical properties and the affinity of metals for oxygen.

The observations made by Avogadro and Michelotti with their galvanometer, generally neglected by later historians,[91] attracted the attention of Oersted. He published a short note in 1823 which described Schweigger's multiplier and one that he had constructed in a slightly different manner.[92] He used a silverplated copper wire, $\frac{1}{4}$ mm in diameter, although he suggested the use of thicker wire with higher currents. Oersted also stressed this instrument's sensitivity and its suitability for extremely small currents.

After mentioning Poggendorff's instrument, Oersted devoted most of his note to the 1822 experiments of Avogadro and Michelotti. Although he did not make any reference to the date or place of publication, it is obvious from the context that he had read the original work.[93] The arrangement of Avogadro's multiplier is not described, but Oersted did report that Avogadro's observations deserved attention and were interesting because of 'the theory of solution and that of production of the electric current'. The inversion of current had been detected by Avogadro in four metallic couples by using both concentrated and diluted nitric acids. In the case of lead and bismuth, Oersted personally repeated the experiments and, with minor variations, arrived at the same conclusions as Avogadro. He observed, for example, that the metallic plates displayed the same behavior for a short time in diluted acid as in concentrated acid. Furthermore, concentrated and diluted nitric acids acted in an opposite way upon lead and bismuth, thus suggesting lead's electronegative behavior towards bismuth in concentrated acids and the opposite behavior in the diluted acid.[94]

Oersted offered no explanation for the inversion of current. He did report an observation not mentioned by Avogadro. When two pieces of the same metal were successively immersed in an acid able to attack it chemically (such as zinc in sulfuric acid), the first plate behaved as a more electropositive metal compared to the behavior of the second. Two of Avogadro's findings seem to have especially impressed Oersted: the effect of the change of concentration of the moist conductor (the acid) upon the direction of the current, and the observation of the necessity of chemical action for the flow of current. This second observation confirmed, as Oersted himself acknowledged, the validity of the chemical theory of the voltaic cell. It is fair to conjecture that the concordance of their views on this controversial issue, in part, explains Oersted's favorable reception of Avogadro's work with the multiplier.

While Oersted's 1823 note in the *Annales de Chimie* correctly reported the main features of the 'Mémoire sur la construction', the same cannot be said of Berzelius' review which appeared two years later in *Jahresberichte.*[95] Berzelius, who read Avogadro's paper indirectly through the *Annales de Chimie*, made his comments in two short paragraphs. After describing the inversion of current found with four different metallic couples, he quite significantly pointed out that these experiments had been 'confirmed' (bestätigte) by Oersted. Thus, the rubberstamp of scientific legitimacy was applied to the surprising findings of a man with very little experimental reputation. These findings did not appear to favor Berzelius' electrical theory, which probably accounts for his indifferent reaction to the Avogadro paper.

Berzelius' 1825 review hardly rates mentioning except that this was the first time that the name (albeit misspelled) of the author of the molecular hypothesis appears in the *Jahresberichte.*[96] In the next twenty-two years Avogadro published thirty articles, but under the editorship of Berzelius only four reviews of his work appeared in that prestigious journal.[97]

Electrochemical Investigations (1836 – 1838)

In reply to a specific question on the subject of electricity, Faraday wrote Auguste de La Rive in March 1854. 'I have never contested the point because . . . I have not since found any reason to add or alter . . . '.[98] This was in reference to experiments he had conducted twenty years earlier, indicating that electric conduction without decomposition could occur even through solutions of electrolytes, provided that the intensity of the current was reduced below the level necessary to cause the chemical change.[99] Thus, by merely lowering the intensity of the current, water could conduct electricity (produced by a series of less than eight voltaic elements) without undergoing chemical decomposition.[100]

Intensity, then, appears to be the key characteristic in the distinction between conduction in ordinary metals, where no chemical action occurs, and electrolytes where, provided there is sufficient intensity, chemical decomposition will always accompany the passage of current. Yet, Faraday hypothesized that even metals might be decomposed by a current of very high intensity.[101] He admitted that the nature and characteristics of this property of the current were still imprecise, and should be examined along with the effects of poor conductors on the lowering of intensity.[102]

Faraday's VIIIth Series of *Experimental Researches in Electricity*, which contained these observations, appeared between a series of investigations conducted by G. D. Botto in 1834[103] and those conducted by Avogadro together with Botto two years later. The results of their research were reported at the Royal Academy of Turin in February 1838 and published later on in the Academy's *Memorie*.[104]

In their joint work, Avogadro and Botto indicated that the distinction made by Faraday in 1834 had no reason to exist because evidence indicated that a chemical decomposition of the electrolytes could take place even with currents of very low intensity. As a possible explanation of their argument, they suggested that a 'peculiar' change caused by the passage of the current would take place in the molecules of the liquids.

Another purpose of Avogadro and Botto's memoir was to attempt a comparison of the different kinds of currents in terms of both their intensity and their 'electrodynamic force', the former defined by them as a quality and the latter as a quantity of current. To better understand what Botto and Avogadro meant by this terminology, it should be recalled that in his essay in the fall of 1820, Ampère clearly stated the notion of a dynamic electric current as distinct from that of static electricity; the first had a definite effect on the chemical decomposition of electrolytes and on the magnetic needle's deviation, the second was detectable mainly by the common electrometer and had no effect on chemical solutions. Ampère also indicated that, by adding a salt to water in which two connecting wires were dipped, the chemical decomposition immediately took place; thus suggesting that while the electrical tension decreases at the extremities of the two wires (as the saline solution begins to conduct), the electric current increases and, therefore, this factor only is responsible for the chemical decomposition occurring in the solution itself.

It was Faraday's turn to draw the line between electricity of tension and motion electricity which he did in the IIIrd Series of his *Experimental Researches* (1833). In the course of this series Faraday intended to identify electricities from different sources: he defined ordinary (static) electricity as characterized by great *tension or intensity*; this could be diminished – as he proved experimentally – by the insertion of poor conductors, while its quantity remained unchanged.[105] Faraday, furthermore, observed from the behavior of various electricities that intensity appeared to be very low in thermo-electricity and very high in magneto-electricity.[106] From these investigations, he succeeded in firmly establishing two major propositions: (a) that voltaic electricity always involves far greater quantities of electricity than those detected in static electricity; and (b) that both magnetic force and chemical action are in direct proportion to the absolute quantity of electricity flowing through.[107]

In 1834, in his study of electricity generated by the voltaic pile (VIIIth Series *Experimental Researches*), Faraday qualified the intensity of the pile as proportional to the intensity of the chemical affinities involved in the production of such electricity;[108] he also equated intensity with the power of the current,[109] a peculiar property of current which may be increased without increasing the quantity of electricity. Later on (XIIth Series *Experimental Researches*) in discussing the induction theory, he used the term intensity (again as a synonym for tension) to describe the peculiar condition of the particles of the dielectric which he assumed to be in a special polarized state.[110]

As previously mentioned, a second purpose of Avogadro and Botto's 1839 memoir was the comparison of different kinds of current. More specifically, they wanted to establish a relationship between the intensity of the electromagnetic currents – used by Botto even in his earlier research, as well as by Barlow[111] – and the voltaic currents, and to discover the factors which influenced such intensity. Their design consisted of lowering the intensity of an electromagnetic current to the level of very weak voltaic currents incapable, according to Faraday, of chemical action.

Intensity, as defined by Avogadro and Botto in their paper, is the property which allows a current to overcome transmission through such poor conductors as, for instance, liquids.[112] To test the validity of this definition, they first passed the currents under comparison through a metallic circuit to bring them to the same 'dynamic force' (detectable by the galvanometer). The loss of 'dynamic force' could then be determined when either the voltaic or the electromagnetic current is allowed to flow through a liquid conductor placed in the circuit.[113]

Should the dynamic forces of the currents under examination and flowing through metallic conductors be different, a wire of sufficient length would be inserted in the circuit to equalize them. The next step would be to pass the two equal currents (voltaic and electromagnetic) through a liquid conductor, sulfuric acid in water, for instance. As a source of induced currents, Botto and Avogadro used a soft iron core with several windings of silk-covered copper wire,[114] arranged so that it would be possible to contact them end to end, either through their equal or opposite signs (i.e., in parallel or series). Thus, by modifying the connections of the windings around the soft iron core, they had a way to vary the amount of electromagnetic current produced, as indicated by the reading of the galvanometer.

Another experimental variable they adopted was that of placing the windings, connected end to end, at first closer and then at greater distance from the core; this was also done to vary the amount of current formed. They observed that by changing the length of the windings, i.e., by varying the amount of induced current produced, the conductivity through the liquid conductor was not affected. They then prepared a coil by joining three wires by their ends of the same sign; this coil, having the same length but a section triple that of the single wires, did not display a greater dynamic force in the current obtained than that produced by each separate winding. As Avogadro and Botto pointed out, this indicates the independence of the inductive action from the cross-section of the wire used.

To demonstrate the validity of this observation, they decided to extend throughout the whole circuit and under the galvanometer the wires joined together in the way described above. If their view was correct, a decrease in the total inductive action should be observed with the new arrangement. On the other hand, if one winds three separate wires on the same soft iron core, it is possible to detect whether the action of one of them (on which the galvanometer is inserted) would decrease when the ends of opposite sign of the other two windings are joined together to form a separate circuit. The experiment indicated, however, that the deviation of the galvanometer's needle caused by each of the three separate windings was about the same, regardless of whether the circuit formed by the other was open or closed. Therefore, the conclusion reached earlier was incorrect. For Avogadro and Botto the decrease of the current's force observed when several coils were joined together could be explained only by assuming that the separate wires carrying the current join into a single wire of about the same size as each of them.

This may be understood by considering the case of two separate coils subject to the same electromagnetic action and whose ends, after being joined together, go to the galvanometer. Here, two interpretations are offered. In the first one, the current formed in each coil would meet a resistance approximately equivalent to that which the current would have encountered only in one coil whose wire size is the same as each of the two wires; in this instance, then, the force of the current formed would depend on the cross-section of the wire. In the second interpretation, the current formed in each coil, upon its arrival at the junction with the galvanometer's wire, would find two paths, one going to the galvanometer, the other returning through the second coil. Should this be the case, while the current in each wire of a double coil would appear weakened by the other current moving in the opposite direction, the current passing through the galvanometer would be composed of the sum of the two partial currents.

The resultant effect was that the current in each wire of the coils retains only one-half the force it would have had if the two wires were in separate circuits. Avogadro and Botto argued that the division of each coil's current between the wire of the galvanometer and the wire of the other coil will take place in different proportions, according to the resistance offered to the passage of the current by the two paths.[115] This consideration certainly deserves notice, because it reveals the understanding that the two physicists had at the time of the relation between current and resistance.

No significant difference was observed by Avogadro and Botto in the intensity of the inductive currents – measured by their conductivity in liquids – as a result of changing either the size of the coil or its position in relation to the core. To decrease the magnetic force of the inductive current, tests were conducted whereby the contact between the soft iron core and the poles of the permanent magnet was diminished. However, no conclusive results were obtained by comparing the inductive currents thus produced with those formed by a perfect contact of the magnet's poles with the core.

In the next series of experiments an electromagnetic machine was used provided with a commutator similar to that adopted by Botto in his previous researches.[116] This arrangement had the advantage of producing a continuous induced current with the least possible fluctuations of the galvanometer. Again Avogadro and Botto introduced in the circuit, at first wholly metallic, a cell containing a solution of sulfuric acid in water. The deviations of the galvanometer's needle were recorded. To compare the intensity of the induced current with that of a voltaic pile – both in a metallic circuit – they varied the immersion of the platinum–zinc couple in the pile's electrolyte, until both currents had the same intensity. The effects of the passage of two equal currents of different origin through a liquid conductor were measured,[117] and it was observed that the intensity of the voltaic current always remained higher than that of the induced current. To Avogadro and Botto this indicated that the intensity of the latter appeared to be smaller than that of the current which, according to Faraday, may pass through an electrolyte without decomposing it. They concluded, therefore, that even for such a small current, a chemical decomposition, although not apparent, always accompanied electrical conduction.

The same variables introduced earlier were then tested with the electromagnetic apparatus provided with the commutator. No change in the intensity of the induced current occurred when the contact between the core and poles of the magnet was varied.

From their observations the authors drew the following conclusions. The intensity of an induced current is: (a) directly proportional to the number of windings around the core of an electromagnetic machine, (b) independent from the force of the magnetic action, and (c) generally smaller than the intensity present in the voltaic current produced by only one Pt–Cu couple.

Avogadro and Botto replied in the affirmative to the question of whether intensity should be qualified as a real property of the currents which differentiates them regardless of their 'absolute force'. Not only did their experimental evidence seem to support this view, but also that presented by Peltier in 1836,[118] which indicated that the decrease in the force of a current as a result of the insertion of a poorer conductor, is due to the 'retrogression' of a portion of such current through the same conductors it traveled before the poorer conductor was inserted. Peltier had also observed that by increasing the number of coils in the electromagnet, the intensity of the current produced grows as well as its ability to overcome the resistance of the liquid conductors.[119] By thinking in terms of 'division' and 'retrogression' of the currents, Avogadro and Botto also explained why, with two coils connected by the ends of the same sign (in parallel), the deviation of the galvanometer does not increase; as earlier mentioned, they assumed that each current divided into two paths, the first one following the galvanometer's wire and the second one returning through the other coil.

The primary purpose of Avogadro's and Botto's 1839 memoir remained, however, that of proving that there is always a chemical decomposition when a current, even of low intensity, flows through an electrolyte; to explain this behavior, they suggest that low intensity currents apparently cause a modification in the molecules of the liquid conductor.

For their investigations, Avogadro and Botto recurred again to the electromagnetic apparatus as the source of current.[120] In his earlier research of 1833 and 1834[121] Botto had connected the terminals of an electromagnetic coil, with two small cups full of mercury joined by two platinum electrodes of equal surface, to a third cup containing the electrolytes under examination; a double-needle galvanometer was also inserted in the circuit.

Throughout his experiments conducted with several liquids, Botto had observed that voltaic polarization took place in the platinum electrodes, as the galvanometer clearly indicated.[122] There was, however, an unusual fact which attracted his attention: a sudden, rather conspicuous impulse of current took place whenever the core with the electromagnetic coil touched a pole of the magnet, and one of the terminals of the electromagnetic coil was simultaneously withdrawn from the mercury cup, thus breaking the circuit. This phenomenon could be interpreted as the effect of the formation of a secondary current produced by a 'fortuitous' ('accidentelle') polarization of the negative electrode resulting from an electrochemical decomposition. It was apparent that this secondary current, flowing in the direction opposite to the principal, partially suppressed it, as long as the circuit was closed. Conversely, no secondary current would form by opening the circuit and, therefore, the main current would encounter no obstacle in its flow; the abrupt, large deviation of the galvanometer's needle confirmed this to be the case.

When Avogadro and Botto resumed their research in 1836, they took care to rearrange the apparatus in such a way as to assure that the formation of the inductive currents and the breaking of the circuit took place at the same time. To this purpose, they placed a small insulating whalebone spring between the electromagnetic coil and the pole of the magnet so that when the former was suddenly attracted by the latter, as a result of the action of the inductive current, the whalebone automatically inserted itself into the circuit, causing its interruption.

Thanks to this simple device, Avogadro and Botto were now able to confirm and extend the earlier observations made by the latter. While suspecting that the secondary current was of electrochemical nature, they sought to discover whether the sudden surge observed by opening the circuit would also occur when the liquid conductor was replaced by an all-metallic one. Consequently, they inserted a very long copper wire (neither diameter nor length are given) and, as a result, upon the breaking of the circuit, the needle of the galvanometer deviated much less than with the closed circuit. It appeared then that the peculiar upsurge of the galvanometer, observed when the circuit was opened, might have its source in the liquid conductor; and more specifically, that an electrochemical decomposition or, at least, *something similar*, occurred in the liquid conductor, although with no visible manifestation.

The experimental procedure they adopted consisted in testing several electrolytes[123] through which a current would flow of so low an intensity that no apparent electrochemical decomposition seemed to occur. Faraday, it was known, had ascribed to the electrolytes in these conditions a metal-like conduction.

To establish the nature and the source of the secondary current observed was then the focal point of their investigation. Should this source be of electrochemical nature, Avogadro and Botto argued, then Faraday's position, denying the electrochemical decomposition of solutions by currents of very low intensity, would become untenable.

As mentioned earlier, one could assume an alternative interpretation to electrochemical action. Schönbein,[124] too, had independently thought that a simple 'displacement' of the components of the solutions might occur both within the electrolyte and near the surfaces of the electrodes, followed by a return to their original position after the principal current stopped flowing. For Avogadro and Botto this sequence would then be the reason for the secondary current detected in their experiments. It is important, however, to stress that they refrained from fully endorsing this interpretation and, as a matter of fact, left the reader to reach his own conclusions.

The electrolytes used in the tests ranged from distilled water to solutions of different strengths of sulfuric, nitric, hydrochloric, acetic acids to solutions of bases, to a rather long series of salts (bicarbonates of Na and K, potassium nitrate and chlorate, ammonium chloride, lead acetate, etc.). By using the whalebone spring system they could, in the course of the experiment, have either a continuous or an interrupted flow of current and, accordingly, examine the behavior of the galvanometer.

It may be a matter of surprise that for most of the electrolytes examined, Avogadro and Botto recorded only two observations: one with, the other without current flow interruption; so that obviously the results obtained appear to have a very limited meaning, a fact recognized by the authors themselves. In the case, for instance, of concentrated hydrochloric acid and of solution of potassium hydroxide, the variation in current flow between the closed and open circuit seems quite large (almost 30 degrees of deviation of the galvanometer needle), but these results, it was conceded, could also have been affected by changes in the experimental conditions.

The tests conducted with solutions of sulfuric acid indicate that the effect of the interruption of the circuit is greater when the concentration increases; conversely, there is less action in the secondary current with weaker solutions of sulfuric acid. When a solution with the same concentration of sulfuric acid in water (25%) is tested again and again, the effects of the interruption of the circuit on the galvanometer continuously decrease; yet, the relative difference in the deviations of the galvanometer, with an open and a closed circuit, remains constant. This appears to be true also for a solution of higher concentration of sulfuric acid (37%). The data reported in the memoir indicate that the largest secondary current can be detected in the solutions of 75% hydrochloric acid: however, no rationalization is offered for this occurrence.[125] Tests conducted with solutions of acetic and oxalic acid showed no effect for the former and a rather large one for the latter, compared to that observed with the 25% sulfuric acid solution.

Among the bases, potassium hydroxide[126] had the greatest effect on the galvanometer, with a deviation of 29 degrees between closed and open circuit. Comparable results were given for saturated solutions of potassium bicarbonate and nitrate and ammonium chloride. For all these substances, however, only single tests were conducted, and again their very limited meaning was conceded.

In this section of their memoir dealing with the anomalous behavior of an induced current flowing through a liquid conductor, Avogadro and Botto referred only briefly to polarization. Antoine Becquerel's attempt to link it to metals involved in electrochemical phenomena is recalled, together with his suggestion that polarization seems to take place even in the absence of the same phenomena. In discussing Botto's previous work in which, by using induced currents, he had observed the decrease of current flow when its direction in the liquid changed, Avogadro and Botto theorized that this event could be interpreted as the result of the 'peculiar condition' of the platinum plates carrying the current. They alluded to the existence of a deposit on those plates of very thin layers produced by the electrochemical decomposition of the liquid surrounding the electrodes.[127] Botto's observation of a sudden impulse in the galvanometer's needle any time he broke the circuit could then be explained by an 'accidental' polarization of the negative electrode.

Such attempts by Avogadro and Botto to arrive at an acceptable explanation of the observed facts should be placed within the framework of the time. Voltaic polarization, although detected as early as 1803 by Ritter, had not yet, thirty years later, found an adequate explanation, despite the investigations conducted in the late 1820s by A. de La Rive, Marianini, and Fechner. Furthermore, the experimental conditions followed by Botto and Avogadro made it very difficult to obtain current of constant intensity, despite the use of the commutator; moreover, it had been observed by then that polarization phenomena were less evident when one used electromagnetic rather than voltaic currents.

That Avogadro and Botto were reluctant to take a definite position is evident from their suggestion that the secondary current detected with their apparatus could derive also from something similar to the induction Faraday had noticed by interrupting the flow of the current in a circuit; obviously, in this case, the presence of a liquid component in the circuit would become superfluous.[128] However, the experiments conducted in an all-metallic circuit apparently indicated to them that the secondary current was not due to induction. While they seemed familiar with the researches of Marianini, Becquerel, and de La Rive on this subject, the observation made by Ohm in 1830 concerning the origin of polarization had failed to attract Avogadro's or Botto's attention. Using electrodes of gold or platinum in a solution of sulfuric acid, Ohm had noticed the quick decrease in the electromotive force of the battery; he explained such behavior in terms of a 'countertension' (gegenspannung) caused by the formation of a negative layer on the positive platinum electrode and of a positive layer on the negative electrode.[129] This resulted in an electromotive force moving in a direction opposite to the primary current and of such strength that the overall current became weaker and weaker. In the same article, Ohm suggested a way to overcome the problem and to build a constant voltaic cell.[130]

The total lack of interest that, outside Germany, had characterized this work and also Ohm's earlier and more significant findings found no exception in Avogadro and Botto. The French, (with the exception of *Ferussac Bulletin des Sciences*), the Swiss, and the British journals ignored the 1825–26 essays as well as *Die Galvanische Kette* the following years, where Ohm's theories appear in their most comprehensive form.

In those years, it should be recalled that Avogadro's attitude towards the main issues of electricity aligned him closely with his acquaintance Auguste de La Rive, who happened to be the most vigorous supporter of the chemical theory of the voltaic pile.[131] On the other hand, one identifies, among the proponents of the opposite view, Ohm himself as well as those few German physicists – Fechner, Schweigger, Pfaff, and Poggendorf – who soon recognized the merit of his work. In 1829 Fechner reaffirmed the validity of Ohm's Law and he used it extensively afterwards[132] as did Lenz, Weber, and Gauss. Outside Germany, Berzelius even earlier had found *Die Galvanische Kette* a 'meritorious enterprise',[133] although with some reservations. Just at the time Avogadro and Botto's memoir appeared in Turin, Poggendorf quite pointedly lamented that as a result of their ignorance of Ohm's findings, a large number of meaningless studies were conducted by physicists both in Britain and in France.[134] Had he known of its existence, he probably would have included the work of the two Piedmontese among them.

The conceptual chasm then dividing the contactists from the supporters of the chemical theory of the pile contributed significantly to making the ideas proposed by Ohm less acceptable to the latter group and thus delayed their acceptance for several years.[135] Only in 1841 did de La Rive's journal, the *Archives de l'électricité*, publish an article on Ohm by E. Wartmann, followed by comments from the editor. By then, very likely, Avogadro took note of them.[136]

Closely related to the issue of Avogadro's awareness of Ohm's investigations, is the fact that he had read the report submitted by Pouillet to the Paris Academy in October 1831. In this Pouillet discussed his observations on the correlation between the force of a current and the resistance of the circuit. Although he admitted having seen a summary in Ferussac's *Bulletin* of Ohm's work, he reasserted the originality of his own research and held the view that, after all, only a mathematical demonstration of Ohm's Law had been given. This, as expected, created a useless controversy at the time.[137] Avogadro, who mentioned Pouillet in a footnote in the 1839 memoir with Botto,[138] seemed unaware of this episode and limited himself to remarking that for Pouillet the electrodynamic force of a current would remain constant even if the length of a metallic circuit was varied, provided that the distribution of the force at different points of the circuit was taken into consideration. A view, Avogadro concluded, which was not too far from his own interpretation of the same facts.

In the second section of the 1839 memoir Avogadro and Botto came as close as possible, in their train of thought, to an appreciation of Ohm's famous relationship linking electric tension to the length of the conductor (Figure 1). Let's assume, they said that two currents *ABCD* and $A'B'C'D'$ of equal force produced by induction in two coils wound up around the same

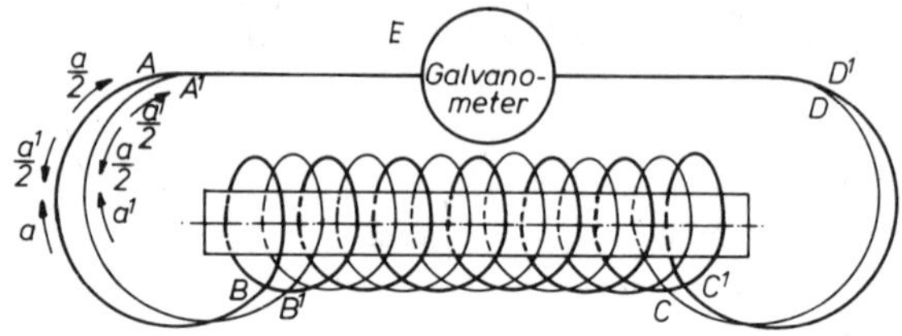

Fig. 1.

core and whose terminals of the same electrical sign are joined through a galvanometer.[139] At the moment the current *a* following through coil *ABCD* arrives at point *A* to enter the galvanometer wire in *E*, it faces two paths; now, assuming that the resistance to its flow through such paths is the same, the current divides into two partial and equal currents: $a/2$ which flows through the galvanometer and the rest of the circuit, and the other current, also $a/2$ which flows in the opposite direction through coil $A'B'C'D'$. Conversely, the current a', flowing through coil $A'B'C'D'$, once arrived at A' will also divide into two equal currents of force $a'/2$, one going through the galvanometer and the other flowing through coil *ABCD*. Consequently, only one current $a = a' = a/2 + a'/2$ enters the galvanometer, while the current in the wires wound up around the core is only one half of what it would be if the two wires were forming separate circuits.

Avogadro and Botto conceded, however, that the current at points *A* and A' did not divide into equal parts, but rather according to the size, length, and nature of the two wires.[140] If the total current appears to be about the same as that flowing through each coil, this meant, they argued, that the resistance offered by each on one side and by the galvanometer and the rest of the circuit on the other must also be the same.

By stating these concepts, it is evident that Botto and Avogadro had perceived to some extent the correlation between current and resistance.[141] Their comprehension was still confused as to the behavior of the intensity of an electromagnetic current compared with that of a voltaic current, when either was flowing through a liquid conductor. As previously mentioned, Avogadro and Botto had observed in their experiments that less than half of the electromagnetic current came out from the liquid conductor as compared with the voltaic current, and that this difference further increased when two voltaic piles were used in series.

Like de La Rive and most other adherents to the chemical theory of the pile, Avogadro focused his attention on the concept of intensity of current; for him an inherent property which enabled the current itself to overcome the resistance of the conductors. He also employed the expression 'dynamic force' ('force dynamique') to define the action of the voltaic current on the galvanometer. By analogy he used 'electrodynamic force' when the current had been produced by induction.[142] Through the introduction of conductors of appropriate length, both metallic and non-metallic, he had appreciated the importance of equalizing the currents (voltaic and electromagnetic) to the same dynamic force. Both Avogadro and Botto, like de La Rive and Faraday, used the galvanometer primarily as an instrument they thought the most pertinent to detect and measure the action of the current. Here, their divergence from the experimental procedure followed by Ohm clearly emerges. First of all, very little attention had been paid by Avogadro and Botto to accurate measurements of the length and diameter of the metallic conductors; secondly, even under the same experimental conditions, the intensity of the current available was anything but stable.

To obtain meaningful and coherent results from all this seems an almost impossible task, and one may wonder why the authors of the 1839 memoir failed to realize it. Apparently, they were more interested in detecting an anomaly in Faraday's system than in a strict quantitative experimental approach. The meaning of Peter Barlow's 1825 article[143] abstracted in Ferussac's *Bulletin* – usually read by Avogadro – and in which he established a relationship between the dimensions of the conductor and the intensity of the current, had apparently escaped the attention of both physicists when they conducted their own research.[144] Barlow's work was quite remarkable in observing that a constant factor persisted throughout the circuit or that the electromagnetic effect remained unchanged at every point of the metallic conductor; an observation certainly missing in Avogadro and Botto's 1839 memoir.

By using the thermoelectric couple, suggested by Poggendorf in 1826, Ohm had available a source of steady current strength and low internal resistance, thus removing the serious experimental hurdle which had hampered his earlier work as well as that of his contemporaries.[145] As for the magnetic force caused by the current in the metallic conductor, Ohm measured it with a modified torsion balance electrometer rather than with the multiplier or a regular galvanometer.[146] In this way he had at his disposal a very sensitive and reliable instrument which enabled him to appraise the electroscopic force of the open pile. He identified this with the electromotive force of the closed pile, the varying factor that Ohm related to each point of the metallic conductor. Furthermore, thanks to his electrometer, he established that the *Elektrischen Spannung* of the open pile increased by increasing the number of elements in the voltaic pile: a consideration that escaped de La Rive, Avogadro, and all the followers of the chemical theory of the pile. Quite likely they could not detect the electromotive force, because they failed to use the proper instrument to do so. But even more likely, perhaps, their conceptual leaning in favor of the chemical theory prevented them from correctly defining both the notions of quantity and intensity of current.[147]

Avogadro and Botto's 1839 memoir, although abstracted in the *Annales de Chimie* of 1839, apparently failed to attract the attention of those who at the time were working in this field of electricity, the only exception being a review written by a non-physicist. Berzelius, in fact, published his comments in the physical section of the 1841 *Jahresbericht.*[148] This, by chance, was also the year which saw the full recognition of Ohm's work when he received the Copley Medal. As mentioned earlier, Berzelius, although impressed had also criticized some of the conclusions reached by Ohm, that in 1832 he defined as 'quite inadequate'.[149] However, in the following decade, sufficient evidence had been gathered to justify a change of mind, and Berzelius became an enthusiastic advocate of Ohm's theory. Now, in reviewing in 1841 the memoir written by Botto and Avogadro, he stressed that the authors with their investigations had proved once more the long neglect of Ohm's Law; he thus supported Poggendorff's opinion that physicists outside Germany had no idea of such a law nor of its implications. Berzelius, who based his review on the content forcibly reduced to an abstract in the *Annales de Chimie* (only fiften pages out of thirty-eight in the original), limited his comments to the section of the memoir in which Avogadro takes exception to Faraday's idea that no electrochemical action occurs when a very low intensity current flows through an electrolyte. He concluded by saying that the final question raised by the authors in their essay proves that 'they had anticipated the possibility of explaining these results in a way coincident with Ohm's theory, though completely unknown to them. . . '.[150]

Avogadro and Botto, like other supporters of the chemical theory of the pile, could not fit Ohm's Law within their conceptual framework, even if they had known of it, as few did. Quite likely, by apparently blurring the distinction between current electricity and tension electricity, Ohm himself was responsible for creating around his research an unfavorable atmosphere which slowly disappeared only after its full significance was grasped.[151]

And Berzelius, as Schweigger pointedly wrote to Ohm in 1832, had once more proved with his comments on Avogadro and Botto's article that 'he had no feeling for physics . . . .[152]' He could record and often criticize publications dealing with physical subjects, but rarely interpret them.

NOTES

1. See chapter 1, note 28.
2. This is in manuscript form at the Biblioteca Civica, Torino.
3. *J. Physique, Chimie, Hist. Naturelle,* 69 (1809), 142. From here on referred to as *J. Phys.*
4. Berzelius, the same age as Avogadro, had by 1807 been appointed professor at the University of Stockholm.
5. The original Swedish edition of the 'Versuch über die Theorie der Chemische Proportionen' appeared in 1814. However, a first outline of the dualistic theory had been given by Berzelius in 1811. The French translation of this appeared in *J. Phys.* of the same year, by coincidence in the same volume (73) in which Avogadro's 'Essai d'une manière' had been published.
6. 'Considérations sur l'état dans lequel doit se trouver une couche d'un corps non-conducteur de l'électricité . . . ', *J. Phys.* 63 (1806), 450. 'Second mémoire sur l'électricité . . . ', *J. Phys.* 65 (1807), 130.
7. *J. Phys.* 63 (1806), 451.
8. Ibid.
9. *Hist. Acad. Roy. Sciences: Berlin* (1756), p. 105.
10. J. C. Wilcke, from Uppsala, a former pupil of Aepinus at Rostock. His two memoirs were published by the Swedish Academy of Sciences. He worked with Aepinus in 1756 in Berlin.
11. See M. von Laue, *History of Physics* (Academic Press, 1950), p. 43.
12. To Cigna is attributed the discovery of the electrophorus, thus anticipating Volta. See also chapter 1, note 27.
13. See 'De vi attractiva ignis electrici', 1769, written in the form of a letter addressed to Beccaria, in which Volta denies that there is a passage of electric fluids from one face of the insulator to the other.
14. Which remained unpublished at the time.
15. In his fifth letter to G. B. Beccari in Bologna. See note 46 below.
16. Wilcke's memoirs were in Swedish and escaped the attention of Volta, among others.
17. Title: 'Considérations sur l'état dans lequel doit se trouver une couche d'un corps non conducteur de l'électricité, lorsq'elle est interposée entre deux surfaces douées d'électricité de différent espèce.' The manuscript was submitted and published by the Academy of Sciences of Turin in 1805 with Avogadro's brother, Felice, as co-author.
18. Since it is practically impossible to join the surfaces of the two conductors simultaneously at all points, then as soon as the first contact is established, the whole electrical state of both surfaces is annulled.
19. The two ribbons acting as armatures in relation to the layer of air between them.
20. These researches were published by Beccaria between 1760 and 1766. He died at the age of fifty-five in 1771.

21. This concept can be more clearly explained with the following notations: if S_1 and S_2 are the surfaces of the upper plate, and S_3 and S_4 the surfaces of the lower plate and the two plates are considered as joined together face to face, then if S_1 has a positive charge, S_2 will be negative, S_3 in turn will be positive and finally S_4 negative again. In this way, the conclusion is reached that the charge on S_4 is not directly related to that on S_1, but only through the intermediate charges on S_2 and S_3.

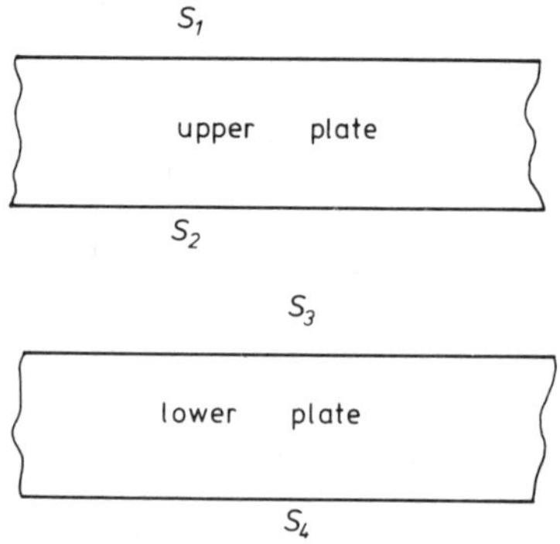

22. *J. Phys.* 63 (1806), 458.
23. In the terminology adopted five years later in his fundamental 'Essai d'une manière', 'molécule élémentaire' has the current meaning of atom.
24. The possible influence of Coulomb on Avogadro will be discussed later.
25. Full title: 'Seconde Mémoire sur l'électricité ou suite des considérations sur l'état ou se trouve une couche de corps isolateurs interposée entre deux surfaces douées d'électricité d'espèce contraire', *J. Phys.* 65 (1807), 130.
26. As reported in the *Tentamen theoriae electricitatis et magnetismi* (St. Petersburg, 1759).
27. For instance, in the footnote on p. 132, he raised a question concerning which one of two electricities (vitreous and resinous) contains an excess or a deficiency of electric fluid. An answer to this, he added, was outside the purpose of his essay. He criticized, however, those who, in supporting the single fluid theory, adopted as positive the vitreous electricity. This choice had been done perhaps by chance and "sans faire réflexion'.
28. Although Avogadro avoids in general any further qualification of the term molecule in this essay, it is legitimate to assume that he still refers to 'molécule élémentaire' which he considers as having a spherical shape.
29. See V. Cappelletti and M. Alippi, *Dizion. Enciclop. Biogr. Italiani*, vol. 4 (Treccani, 1962), p. 690.
30. *J. Phys.* 65 (1807), 136.
31. He does not rule out, however, the possibility that such fluid may be the caloric itself acting through this new kind of combination.
32. 'Particle' here is used as synoymous with elementary molecule.
33. When two charged surfaces previously in contact are separated, thus allowing a layer of air to intervene, they then become the armatures of such layer.
34. Avogadro generally referred to upper and lower layers of molecules, in relation

to the horizontal disposition assumed for the dielectric system, and he assumed that the positive armature is the one on top.

35. The terms 'dielectric' and 'polarization' were never used by Avogadro at this time. W. Whewell in 1836 suggested the word 'dielectric' to Faraday.
36. Mossotti's theory of dielectrics (1850) was later accepted by Clausius.
37. I. Guareschi, *A. Avogadro: Opere Scelte* (Torino: Accad. Sci.), 1911.
38. Both quotations in this paragraph are from L. P. Williams, *Hist. Sci.,* **1** (1962), 9. Helmholtz, in his introduction to the *Prinzipien der Mechanik* (Leipzig, 1894) of H. Hertz, attributed to this physicist the concept of dielectric polarization of the insulators, but does not mention Avogadro's prior speculations, quite likely unknown to him.
39. This also should be recalled as the first article of Avogadro to be reviewed by Berzelius in the *Jahresbericht*, 1825.
40. *Experimental Researches on Electricity*, XIth series (November, 1837), paragraphs 1161 ff.
41. *Archives de l'Électricité, Supplem. à la Bibliothèque Brit.* **2** (1842), 102.
42. One may argue that the style of scientific reporting had decidedly changed in the course of thirty years; this is true for Avogadro himself, whose writings in later years are much clearer.
43. *J. Phys.* **65** (1807), 140.
44. Ibid., p. 143.
45. See 'Note sur la nature', *Arch. Électricité, Supplem. à la Bibl. Brit.* **2** (1842), 102.
46. As he reported in a letter to G. B. Beccari in Bologna. See M. Gliozzi, *Storia della Fisica*, Vol. II (UTET, 1965), p. 185.
47. *J. Phys.* **69** (1809), 142. As mentioned earlier, this had been preceded in 1804 by an unpublished manuscript, 'Considérations sur la nature des substances connues sous le nom des sels metalliques et sur l'ordre des combinaisons auquel il parait le plus convenable de les rapporter'.
48. Together with two memoirs on the specific heats of solids and liquids (1833 and 1834), this was the only work of Avogadro originally published in the prestigious French journal *Ann. Chimie*, **87** (1813), 286.
49. See vol. 6, p. 119.
50. Berzelius arrived at similar conclusions in 1818 in his 'Forsok till en theorisk asigt af Laran om de chemiska proportionerna . . . ", French translation (1819) entitled 'Essai sur la théorie des proportions chimiques' which was extensively commented on by Avogadro in the 'Nouvelles considérations sur la théorie des proportions determinées . . . ' (1821) which is reviewed in another section of this work.
51. *Statique Chimique* appeared in 1803 and reached Avogadro in Turin while he was still holding his position with the French Administration. Quite likely it represented in those years one of the main instruments of his chemical education, if one considers the reputation of its author at the time.
52. Berthollet, *Chemical Statics*, trans. B. Lambert, vol. I (London: Mawman, 1804), 163.
53. Ibid., p. 47.
54. The assumed compound nature of both nitrogen and hydrogen, which was widely supported in those years, did not contribute to a clarification of the chemical behavior of ammonia. In both his 1811 and 1814 chemical essays, Avogadro, on the other hand, adopted the correct formula for this compound.
55. Despite this trend, he later criticized, as mentioned in other parts of this study, both the conclusions of Mitscherlich on isomorphism and of Dulong-Petit on the relation between atomic weights and atomic heats.
56. *Ann. Chimie*, **86** (1813), 146.

57. *Ann. Chimie,* **87** (1813), 286. For Avogadro, an unusually short (six pages) essay. It was translated into German and appeared in *Ann. Phys.* **46** (1814), 248.
58. A few years later, Berzelius revised his position in admitting the existence of weakly positive and negative salts which may further combine to form the so-called double salts such as Alum.
59. *J. Phys.* **69** (1809), 145.
60. See 'De l'influence de l'électricitè'. *Ann. Chimie,* **86** (1813), 167. Article translat. by Vogel, origin. appeared in *Journal für Chemie und Physik,* **6** (1812), 119. Berzelius uses the term 'électropositif', but, as Avogadro pointed out in a footnote on page 289 (*Ann. Chimie,* **87**), this should be understood as electronegative.
61. 'De l'influence de l'électricité', p. 154. See previous note.
62. Berzelius used at the time the opposite term in referring to oxygen and the other elements. Avogadro who discusses the question in a long footnote in the 'Réflexions', had adopted the correct terminology following the prevailing trend. See note 60 above.
63. A term not used by Avogadro at the time. Metalloids and 'acidifiable' metals were classified by Berzelius as 'electropositive' (electronegative in our meaning).
64. He used the term 'indifferent' and 'alternative' to define the peculiar behavior of silicon and chrome, respectively; the lower oxides of the 'alternative metals' are basic, while the higher oxides behave as acids. (See 'De l'influence . . . ', p. 156, note 60 above.)
65. When Davy in the *Elements of Chemical Philosophy* (1812) said 'nothing is further from the hypothesis which I have ventured to advance . . . than the idea that chemical changes were occasioned by electrical charges', very likely he was not referring to Berzelius whose article in the *Journal für Chemie und Physik* had certainly not reached him.
66. Many chemists in those years favored the chemical theory, while the opposite position was in general supported by physicists. J. W. Ritter's work suggesting a relation between the electrochemical activity of metals and their affinity for oxygen, while recalled by Davy, was neither mentioned by Berzelius on this occasion nor by Avogadro.
67. The cell described in the *Traité de Chimie,* trans. Jourdan, vol. I (Brussels: Esslinger, 1845), p. 86, is zinc-zinc sulfate-copper-nitric acid, with copper as a cathode. Berzelius in the *Traité* (p. 85) mentioned Faraday and de La Rive among the supporters of the chemical theory of the pile. To their assumption that electricity was produced by the oxidation of one of the two metals of the couple, he asked how they could explain the electricity produced by metals such as gold and silver.
68. Ampère gave six lectures at the Académie of Paris between 18 September and 6 November, 1820, on his electrodynamic investigations. They were published in *Ann. Chim. Phys.* **15** (1820), 59.
69. Poggendorff's multiplier is reported in *Gilbert Ann.* **67** (1821), 422 and Schweigger's in *Allgem. Litter. Zeitung* (1820), 296.
70. Avogadro himself relates this *iter*, see *Mem. Reale Accad. Sci. Torino,* **27** (1823), 43. (From here on indicated as *MRAST.*)
71. Full title: 'Mémoire sur la construction d'un voltimétre multiplicateur et sur son application à la détermination de l'ordre des métaux relativement a leur électricité par contact', *MRAST,* **27** (1823), 43.
72. They are described in another section of this work.
73. Stephen Robertson in 1800 had attempted to determine the intensity of current from the amount of its chemical effect. Maréchaux in 1802 had described a galvanometer based upon the decomposition of water. See *Gilbert Ann.* **11** (1802), 123.
74. Whose diameter is not given.

75. 'Idées sur l'acidité et l'alcalinité', *J. Phys.* **69** (1809), 142.
76. This subject is discussed in detail in another section of this work.
77. See Chapter 1, note 70. In 1823 he published an essay on 'Influence of chemical action in producing electric current', *MRAST*, **27** (1823), 1. The first part of the essay was published in *MRAST*, **26** (1822), 365.
78. Among them, Le Hot and Bellingeri are mentioned by Avogadro. *MRAST*, **27** (1823), 60–61. Carlo F. Bellingeri (1789–1848) was a physician in Turin, Medical Councilor of the Royal Household, Medical Councilor of the University of Turin, member of the class of physical sciences of the Royal Academy of Turin after 1819.
79. This relates to the zero position of the semi-circle traced on the plate of mother-of-pearl.
80. Avogadro mentions that he had also considered studying the behavior of other, newer metals, but had difficulty in obtaining them in sufficient amounts.
81. Volta's table appears in his third letter to V. Eandi, published in *Brugnatelli Annali*, **11** (1796), 84.
82. *Neus Journ. d. Physik* (*Gren*), **8** (1794), 196.
83. According to the Italian historians, the priority of Volta over Pfaff in establishing the potential series is beyond question. See *Opere scelte di A. Volta*, ed. M. Gliozzi (Torino: UTET, 1967), p. 581. In his earlier letter to G. Aldini in November 1792 published in *Giornale di Brugnatelli*, **1** (1793), 63. Volta reported that very dry charcoal was not only an excellent conductor, but that by connnecting tin and charcoal he was able to detect a stronger electrical action on his tongue than with any other combination of two metals. Thus, he considered charcoal much superior to silver and gold. As is known, Volta often used the tongue as a sensitive detector of currents.
84. See *MRAST*, **27** (1823), 79.
85. See 'Réflexions sur la théorie électrochimique de Berzelius', *Ann. Chim.* **87** (1813), 286. It may be interesting to note that these negative comments on Berzelius' series do not appear in the main text, but in a long footnote. See *MRAST*, **27** (1823), 81.
86. Ibid., p. 75.
87. In the 1845 edition of his *Traité de Chimie*, Berzelius still supported the contact theory and gave experimental proof of it. See vol. 1, p. 85, ed. Firmin Didot.
88. Although Auguste de la Rive was only twenty-one years old when the 'Mémoire sur la construction' appeared, it is quite possible that he read it and was influenced by Avogadro's considerations.
89. *MRAST*, **27** (1823), 53.
90. Ibid., p. 61.
91. See F. I. E. Hoppe, *Geschichte der Elektrizitat*, (Leipzig, 1884), Rosenberg, *Geschichte der Physik* (Braunschweig, 1887), but recalled in M. Gliozzi, *Storia della Fisica* (Torino: UTET, 1965), p. 264, who gives the wrong year. J. R. Partington quotes Oersted as mentioning Avogadro's series, but gives the wrong article by Avogadro. See *Hist. of Chemistry*, vol. IV (Macmillan, 1964), p. 133. M. Crosland discusses it in his article on Avogadro in *Dict. Scient. Biogr.*, vol. 1 (Scribners, 1970), p. 347.
92. *Ann. Chimie*, **22** (1823), 358. A summary of Avogadro's essay appeared also in *Bulletin des Sciences de Ferussac*, **1** (1824), 30, 345.
93. There is, for instance, a verbatim quotation of Michelotti's idea that definitely there existed a difference between the current produced through the chemical action and the 'fluid' causing the motion of the frog's limbs.
94. An analogous behavior was found by Avogadro for arsenic and antimony, as recalled by Oersted; who pointed out that, travelling at the time, he did not have

the opportunity to study in detail the other experimental findings of Avogadro.

95. *Jahresbericht*, 4 (1825), 29.
96. Avogadro's name is misspelled (Avogrado), as Ampère had done in his letter to Berthollet in 1814.
97. For more details on the subject, see Chapter 4 of this book.
98. The letter was dated 1 March 1854. See *Selected Correspondence of Michael Faraday*, vol. 2 (Cambridge University, 1971, ed. L. P. Williams), p. 716.
99. *Experimental Researches Electricity*. VIIIth series, para. 983 (also *Phil. Trans.* June 1834, p. 425).
100. Ibid., para. 985. Faraday says here that the conductivity would be the same regardless of the fact that pure or acidulated water is used.
101. Ibid., para. 987. Assuming that Faraday referred to pure metallic conductors, he then seems to suggest that chemical elements themselves might be separated into simpler constituents through the use of electricity.
102. Ibid., para. 983.
103. 'Esperienze sull' azione chimica delle correnti', abstracted in *Bibl. Univ.* 1 205, n.s. Giuseppe Botto was then professor of physics at the University of Turin, while Avogadro still held the chair of mathematical physics. Surprisingly, despite his hostile attitude to science, Charles Felix sent Botto to Paris and London in order to acquire equipment for the physical laboratory of the University of Turin. Certainly, Botto did not share Avogadro's fate. See *Descrizione di Torino*, a cura di G. Pomba, Turin, 1840.
104. *MRAST*, 1 (1839), 179. Full title of the essay: 'Mémoire sur les rapports entre le pouvoir conducteur des liquides pour les courans éléctriques et la décomposition chimique qu'ils en éprouvent'.
105. *Exper. Res. Elec.* III series, paras. 284 and 295 (read January 1833). In 1834 W. Snow Harris, quoting Haüy's *Philosophie Naturelle* and moving from the position that electricity is an elastic fluid, defined *tension* as the elastic force of a given quantity of fluid accumulated in a given space. Intensity which, he warned, should not be confused with tension, can be detected by an electrometer and changes in relation to the square of the density of the stratum (*Phil. Trans.*, 16 (1834), 222–23).
106. *Exper. Res. Elec.*, III series, paras. 349 and 343.
107. Ibid., para. 377.
108. *Exper. Res. Elec.*, VII series, para. 916.
109. Ibid., para. 908.
110. *Exper. Res. Elec.*, XII series, para. 1370. W. Snow Harris, to whom Faraday referred on this occasion, wrote an article in 1863 (*Phil. Mag.* 26, 504) with the purpose of clarifying terms such as intensity and tension; he defined the former as the quantity of electricity at a given point of a charged surface acting on the electrometer (p. 506). As for tension, he referred again to a state of electrical strain such as is present in a dielectric medium, as Faraday had mentioned in his XIIth Series.
111. *Edin. Phil. J.* 12 (1825), 105.
112. *MRAST*, 1 (1839), 198.
113. Faraday had adopted a very similar experimental design in 1833 in his IIIrd Series and quite likely Avogadro and Botto followed it.
114. Quite surprisingly, no data are given in the paper concerning the size of the wires used. The 1839 memoir is characterized by a considerable lack of experimental details, which is typical of Avogadro's style.
115. See *MRAST*, 1 (1839), 204 ff. This section of the 1839 memoir will be examined in greater detail further on.
116. See *MRAST*, 37 (1834). This report read by Avogadro before the Academy on 10 March 1833.

117. For instance, a liquid conductor is represented by a glass cup containing a solution of sulfuric acid in water, in which two platinum plates of 4 sq. cm. each are immersed. This is one of the few dimensions given in the memoir.
118. See *Ann. Chim. Phys.*, ser. 2, **62** (1836), 422. Avogadro and Botto became aware of this article only after they had finished their experiments.
119. This is probably due to the difficulty that the current meets in its retrogression as a result of the greater length of wire required for the windings; so that it is easier for the current to flow through the liquid conductor.
120. Barlow in 1825 had also used the electromagnet in his work. See note 111.
121. See *Bibl. Univ.* **1** (1834), 205.
122. Avogadro and Botto recalled in their memoir that polarization had been known since Ritter had observed it in 1803, and that both Marianini and A. de La Rive had investigated this phenomenon giving their conclusions in 1828, *Ann. Chim. Phys.* **37** (1828), 256.
123. I wish to emphasize that the term electrolyte is – with one exception – not used in this paper by Avogadro and Botto. It had been introduced earlier by Faraday.
124. *Bibl. Univ.* March and April 1838. This reference to Schönbein appears in a footnote (*MRAST,* **2, 1** (1839), 185) and, therefore, is subsequent to the date of the reading of the memoir at the Turin Academy. The authors apparently introduced it before printing the memoir in order to prove the validity of their argument.
125. Only two tests were conducted and reported for this solution.
126. No concentration is given.
127. See *MRAST,* **1** (1839), 184.
128. Apparently Avogadro and Botto were not aware of John Daniell's description of his cell which had been published in *Phil. Trans.* **126** (1836), 107 and 125.
129. See *Schweigger J.* **60** (1830), 32.
130. See also *Repertorium der Physik,* **1** (1832), 392.
131. In the letters by Avogadro to A. de la Rive after 1829 he never touched on the subject of electricity.
132. See his *Lehrbuch des Galvanismus und der Elektrochemie* (Leipzig, 1829).
133. *Jahresbericht,* **7** (1828), 15.
134. *Poggendorf Ann.* **47** (1839), 123.
135. See article by K. Caneva on Ohm in Scribners *Diction. Scient. Biographies*, 1974, vol. 10, p. 192. After examining the two volumes of Faraday's *Selected Correspondence*, Cambridge, 1971, ed. L. P. Williams, I found only one reference made by Faraday about Ohm in a letter he wrote to W. Thomson on 28 October 1859.
136. *Arch. Elec.* **1** (1841), 31, 67. In 1841 Ohm was also elected member of the Turin Academy of Sciences.
137. Hoppe mentions in his *Geschichte*, Leipzig, 1884, pp. 261 ff., that it is beyond doubt that through Poggendorf, Pouillet knew of Ohm's findings, so that the whole story of Pouillet's priority seems 'comic', to say the least.
138. *MRAST,* **1** (1839), 217.
139. Ibid., pp. 204 ff.
140. Ibid., p. 205.
141. Ibid., p. 219. Avogadro and Botto were familiar with Becquerel and Barlow's work on this topic, which appeared in 1826 and 1825, respectively. On the other hand, they seem unacquainted with William Ritchie's report to the Royal Society (1833) on his experimental research on electromagnetism. In discussing electric conduction, *Phil. Trans. Royal Soc. London,* **15** (1833), 313, he first criticized Barlow and Pouillet for assuming that in general the tangents of the deflection were proportional to the deflecting forces; this would be true only in extreme cases. After giving experimental proof of his point, Ritchie examined Becquerel's conclusions

of 1826 (conductivity is directly proportional to the section of the conductor and inversely to its length) and dismisses them as based upon erroneous reasoning. Finally, he gives his own thoughts on conduction. In carrying out his experiments, Ritchie employed a torsion galvanometer rather than a common galvanometer and he made sure that the voltaic cells used were of the same size and contained acid of the same strength. He concluded that the conductivity of a wire seemed inversely proportional to the square root of its length, and that the ratio of the deflecting forces between short and long wires of the same diameter (or between) wires of same length but different diameter would change when the concentration of the acid in the voltaic pile was either increased or diminished. It should be mentioned at this point that, while Ritchie had apparently grasped the significance of the electromotive force in its relation to conductivity, he never introduced the terms intensity or current in his discussion, thus indicating that he was still quite removed from the train of thought followed by Ohm whose work Ritchie, too, certainly appeared to ignore.

142. See *MRAST,* **1** (1893), 197.
143. See note 111 above.
144. Barlow's article was also translated in *Schweigger J.* **44** (1825), 367, where Ohm saw it.
145. M. I. Schagrin in his article, *Amer. J. Phys.* **31** (1963), 36 on the reception accorded Ohm's Law, argues that his use of the thermopile only made Ohm's calculations easier, but that he would have arrived at the same conclusions even through the use of currents of unstable strength.
146. As Schagrin mentions (ibid., p. 544), Ohm considered the multiplier more a hindrance than a useful tool. He thus dismissed the relevance of this instrument in helping him.
147. See K. Caneva, ibid., p. 190. Faraday himself had expressed doubts on the 'delicacy' of the results obtained with the electrometer and did not encourage its use (*Exper. Res. Elec.*, para. 1808).
148. *Jahresbericht,* **20** (1841), 87.
149. Ibid., **11** (1832), 21.
150. Ibid., **20** (1841), 88.
151. See Schagrin, ibid., p. 545.
152. The letter was related to Berzelius' criticism of an article that Ohm had published on *Schweigger J.* **59** (1830), 385, and is quoted by H. J. Winter in *Phil. Mag.* **35** (1944), 376.

CHAPTER THREE

THE MOLECULAR HYPOTHESIS: THE 1811 AND 1814 ESSAYS

The decade which preceded the elaboration by Amedeo Avogadro of the manuscript of his 1811 'Essai d'une manière' was one of the most fertile periods in the history of the physical sciences. It had begun with Volta's letter to Sir Joseph Banks, in March 1800, describing his 'artificial electric organ' and displayed a long array of discoveries and far-reaching observations ranging from optics to chemistry, from mineralogy to electricity.

For a young man, newly interested in natural philosophy, such as Avogadro was in 1800, there was so much scientific information available that it would not be surprising if he had lost his bearings. In fact, this state of bewilderment may be detected in Avogadro if one considers the varied nature of his first five scientific contributions, from 1803 to 1809, having as the only underlying feature a tendency to link electrical to chemical phenomena.[1]

A new direction to his work emerged after his receipt in 1809 of the newly published French translation of Thomas Thomson's *System of Chemistry* and the second volume of the *Mémoires de physique et de chimie de la Societé d'Arcueil*. Mainly through these works, Avogadro had now at his disposal all he needed for the formulation of his own chemical and physical theory.

Since the manuscript of the 'Essai d'une manière' was submitted to Jean Claude de La Mètherie[2] in the spring of 1811, one may conclude that for almost two years Avogadro labored to prove that the apparently irreconcilable chasm between Gay-Lussac's volumetric approach and Dalton's atomistic path could indeed be bridged. While Gay-Lussac had concluded in his 1809 memoir that gaseous substances 'présentent des lois simples et regulières . . .' without going further, Avogadro had taken the next step by assuming in 1811 the existence of '. . . rapports très simples entre les volumes des substances gazeuses et le nombre des molécules . . . .' And by arguing for the exclusion of other solutions to the problem, he was able to draw the conclusion now known as Avogadro's Law.

A reading of the 'Essai d'une manière'[3] indicates that, beside Gay-Lussac, Avogadro was, by 1809, familiar with the work of other French scientists of the time: Haüy, Laplace, Biot, and Thenard, among others, were quoted again and again. Moreover, through summaries and translations mostly appearing in *Bibliothèque Britannique*, he was aware as well of scientific progress in Britain. As a result, Humphry Davy's and Dalton's findings are, next to Gay-Lussac's, those most referred to by Avogadro. This familiarity with British work must be taken into account before reaching conclusions on the effect that French physics – and more specifically the so-called mechanico-molecular tradition within it – might have had on the author of the gas hypothesis. While those associated with the Laplace–Berthollet circle in Arcueil seemed reluctant to express views which could be considered favorable to Dalton's atomism, this is not the case with Avogadro. Through Riffault's 1809 translation of Thomas Thomson's *System of Chemistry*, he acquired a general idea of Dalton's system. Probably he noticed the observation attributed by Thomson to Dalton that it would be possible to determine the relative densities of atoms entering into gaseous combinations.[4]

In striking contrast with Gay-Lussac, who in his 'Mémoire sur la combinaison' had emphasized Berthollet's strong criticism of the Daltonian doctrine – calling it 'l'idée ingénieuse de M. Dalton . . .' – Avogadro, in the concluding section of the 'Essai d'une manière', emphasized that there was a difference only of degree between his own system and Dalton's. Yet, in the course of the same article, one finds several instances of more than marginal disagreement. In an earlier section of the 'Essai d'une manière', for instance, Avogadro asserts the accuracy of his approach in determining chemical composition:

> Dalton, on arbitrary suppositions which seemed to him the more natural as the most likely relative number of molecules in compounds, has endeavored to fix ratios between the masses of the molecules of simple substances. Our hypothesis, supposing it well-founded, put us in a position to confirm or rectify his results from precise data, and, above all, to assign the magnitude of compound molecules according to the volumes of the gaseous compounds, which depend partly on the division of molecules of which this physicist [Dalton] has no idea at all.[5]

Obviously, Avogadro could not accept the rule of simplicity adopted by Dalton.[6] And thus he challenged the composition attributed by Dalton to ammonia, water, and the oxides of sulfur and phosphorus. As for the weights of ammonia and water, Avogadro pointed out that the apparent consistency of his figures and Dalton's were only the result of compensating errors.[7] However, as the first section of the 'Essai d'une manière' indicates, the difference in the interpretation that the two men had of the structure of matter in the gaseous state was more substantial. Both thought of gases as formed by particles of roughly globular form, whose size was represented by a hard center surrounded by an atmosphere of caloric.[8] A repulsive force, inversely proportional to the particles' affinity for caloric, balanced their mutual attraction; a lessening in the mutual repulsion of the particles would then correspond to an increase in affinity for caloric. For Dalton such a decreased repulsion meant a condensation of volume of the atmospheres of caloric and a contraction in the size of the particles, while no variation occurred in the actual amount of heat surrounding them. Since different gases had different affinities for caloric, Dalton argued, their particles had to have different sizes and, therefore, they must be in different numbers in a given volume.

Avogadro, aware of Dalton's views on the subject only through Thomson, discussed them in the introductory part of his 'Essai d'une manière'. He contended that by assuming them as correct, it would be impossible to explain the very simple ratios found in the combinations of different gases reported by Gay-Lussac. For him – while he conceded that the 'Laws' concerning the amount of caloric surrounding the molecules are still 'unknown'[9] – the more likely and logical answer lay in supposing that in gaseous bodies the intermolecular distances are so large that no mutual action between such molecules could take place; under these conditions, a change in the attraction for the caloric displayed by each molecule might affect the *amount* of caloric condensing around it, but not the *volume*. The distances between molecules of different gases were not influenced by their uneven attraction for the caloric. Thus, it was reasonable to assume that, under equal volumes (or under equal temperature and pressure) there was always the same number of molecules.[10]

When Avogadro was preparing the manuscript of the 'Essai d'une manière,' he did not know[11] that in the second chapter of the *New System* in 1808 Dalton had admitted that:

> At the time [in 1801] I formed the theory of mixed gases I had a confused idea, as many have I suppose at this time, that the particles of elastic fluids are all of the same size, that a given volume of oxygenous gas contains just as many particles as the same volume of hydrogenous . . .[12]

In a rather revealing admission, Dalton added that, if this were not the case, then we would have 'no data from which the question could be solved'.

What caused Dalton to change his position so radically? In 1803, after a careful review of the data on the physical properties and on the composition of water, Dalton concluded that equal volumes of different gases do not contain equal numbers of particles. These considerations, which had appeared in his notebooks five years before they were published in the *New System*, became one of the tenets of his doctrine:

> If equal measures of azotic and oxygenous gases were mixed and could be instantly united chemically, they could form nearly two measures of nitrous gas having the same weight as the two original measures; but the number of ultimate particles could at most be one-half of that before union.[13]

In addition to water and nitrous gas, Dalton quoted many other examples which intrigued him: not only was steam lighter than oxygen, but ammonia was less dense than nitrogen, and carbon monoxide less dense than oxygen. In some cases, no contraction occurred: in others, a dilation of volume was observed. Inasmuch as volume ratios were not reliable he used the densities to prove that the equal volumes–equal numbers generalization could not be supported. And, furthermore, one may understand his skepticism towards any correlation between specific gravities of gases and the weights of their 'ultimate particles', and his rejection of Gay-Lussac's conclusions on the simple ratios of combining gases. Dalton's inability to link atoms to volumes, also in the case of gaseous elements and compounds, went beyond his criticisms of the French chemist's experimental results, because it is known that he never changed his position on this issue, even after the techniques in handling gaseous substances greatly improved.[14]

In 1814, towards the end of an article written in reply to Berzelius' 'Essay on the cause of chemical proportions', Dalton summarized his thoughts on the subject. After indicating a 'wait and see' attitude and the intention not 'to say much at the present', he wrote:

> I own I do not see how we are to remove the difficulties attending the atomic theory by substituting the term 'volume' for that of atom, nor how [and here he quotes Berzelius] we can figure to ourselves a demivolume, while in the theory of atoms a demiatom is an absurdity.[15]

Thus expressed, it was absurd to talk of a fraction of an atom; on the other hand, the experimental evidence did support the existence of fractions of volumes. Even Dalton's quizzical remark could not disprove the argument that, unlike atoms, volumes could at least be directly measured.

Gay-Lussac's memoir on combining volumes was central to the formulation of the molecular hypothesis and, in turn, Gay-Lussac came under the intellectual influence of C. L. Berthollet, the *grand patron* of French science during the Empire, who in his major work, *Essai de Statique Chimique*, discussed extensively the gaseous state and the chemical behavior of gases.

Experiments on compression and expansion suggested to him that in gases only the action of caloric had an effect on their physical changes.[16] On this position Berthollet had the support of Laplace, who confirmed that the gaseous molecules were sensible to the repulsive force of heat, which was much greater than the affinity action, while the repulsive force of the molecules of gases depended on the temperature and not on their mutual distance. Shortly afterwards[17] Laplace modified his earlier view, by conceding that the repulsive force of two neighboring molecules of a gas was a reciprocal function of their mutual distance and proportional to the temperature. Further elaborating on this, Berthollet stressed that caloric increased the molecular distance[18] and that all gases at the same temperature acquired a *quantity of caloric* proportional to the volumes determined by the pressure.[19] The fourth section of the *Statique Chimique* contained a detailed review of gases chemically combining with other gases and with solids and liquids. When two gases combined, their properties changed, Berthollet pointed out, because their mutual action overcame the affinity for caloric; as a result, a condensation of their volumes occurred, followed by new chemical and physical characteristics. It was at this point that Berthollet found a first major obstacle; how to account for the condensation, sometimes 'considerable', observed in the gaseous combinations; he explained this in terms of a mutual affinity of the gases which was 'greater than their difference of specific gravity'.[20] This condensation, much larger than that observed in liquids, was such that water vapor, for instance, occupied less volume than hydrogen and oxygen producing it, while both ammonia and nitrous gas displayed 'specific gravities' greater than that of a simple mixture of their elements.[21] Such 'coming together' (*rapprochement*) of the molecules involved upon formation of the gaseous compounds, was invoked by Berthollet to explain the experimental evidence.[22]

However, another problem arose; how to reconcile the fact that an increase of caloric could affect, for instance, the combination of hydrogen and oxygen, while the theory required that a great quantity of caloric should instead be eliminated?[23] Berthollet thought he was able to explain this by borrowing from Monge[24] the argument that the caloric '. . . only causes, by the dilation of one part of the gas, a compression of that which is less heated, but the total effect is owing to the sudden *rapprochement* of the molecules produced by the combination. . . '.[25]

In the next chapter of the same section of his *Statique*, Berthollet attempted to explain the condensation of gaseous substances upon chemical combination by introducing other considerations as well: for instance, sulfurous acid containing less oxygen than sulfuric acid was less condensed and, accordingly, more volatile and as such less easy to change from gaseous to liquid state.[26]

In the course of his lengthy discussion of the proportion of the elements in combination, Berthollet returned to examine the issue of chemical behavior of elastic substances; the way they seemed to combine, by definite proportions, did not agree with his prevailing view of the subject. He argued here that, as a result of an increase in their mutual action, gaseous substances united in proportions in which their action had the greatest force, an argument to which Gay-Lussac referred in the concluding paragraphs of the 'Mémoire sur la combinaison'. In Berthollet's view, more uniform proportions occurred in combinations of elastic substances than in those of liquids, for instance, because in the former the much greater contraction caused a much greater obstacle to the establishment of other proportions; hence, the kind of progressive combinations observed in 'acidulous salts' could not be found when gaseous substances united chemically.[27]

There is no doubt that by the time Avogadro was preparing the manuscript of the 'Essai d'une manière', he was thoroughly familiar with Berthollet's major treatise and with his approach to the problem of chemical combinations among gases.[28] One may speculate that he was especially impressed by the predicament in which Berthollet had found himself in trying to reconcile his interpretation of molecular distances in terms of the caloric theory with the assumed molecular *rapprochement* often occurring when gases chemically combined. As appears from the 'Mémoire sur la combinaison', Gay-Lussac had realized before Avogadro that a *rapprochement* of the elementary molecules could not account for the decreases in volume observed upon chemical union of certain gases. Yet, in 1808, five years after the publication of the *Statique Chimique*, the thorny question of the *contraction apparente* had not been solved by Gay-Lussac.

The regular behavior of gases whenever submitted to expansion or compression had certainly impressed Gay-Lussac, and he thought that such behavior could be explained in terms of the absence of intermolecular attraction in gases; and thus, by moving from strict physical considerations, Gay-Lussac felt confident in extending them to a possible interpretation of some *proprietées nouvelles* of gases having to do with the very simple proportions in which they enter into chemical combination, and also with the regularity of the contraction of their volumes as a result of these combinations.[29]

By 1808, a great deal of experimental data on chemistry of gases was at Gay-Lussac's disposal beside his own findings, and he certainly made excellent use of them all. Unlike Dalton, he saw beyond the limitations of the empirical observations; he rounded the figures, when necessary – as Avogadro would do in many occasions, following the same tradition – but he did not fail to recognize the consistency and the recurrence of the results before him. However, Gay-Lussac, although coming very close to making the next step in his inductive process, fell short of doing so. In the 'Mémoire sur la combinaison', it is easy to detect his sense of puzzlement whenever he found an unexplainable variation in the expected volumes of gaseous substances after combination. He recalled, for instance, that the density (relative to air) for ammonia was 0.596. Using its composition, correctly interpreted, he then calculated a density of 1.188, close to double that experimentally observed. He reasoned, 'if we suppose the apparent contraction to be half of the whole volume, we find 0.594 for the density'.[30] This confirmed, but did not explain, the formation of two volumes of ammonia – from the combination of one volume of nitrogen with three of hydrogen. However, the subject was pursued no further,[31] although Gay-Lussac himself was struck by the closeness of the figures for the calculated densities and those experimentally found. After assuming that the volume contraction in the combination of hydrogen and oxygen is solely due to the volume of the latter, he devoted some space to a discussion of the density of its vapor. He disagreed with the figure (too high, as reported by De Saussure) and mentioned other experimental findings suggesting that the correct value for water vapor density should be 0.625 (relative to air). This was in accord with the experimental evidence, and indicated that a contraction of volume of the combining gases was equal to the whole volume of oxygen added.[32] Gay-Lussac argued that the density of carbonic gas (carbonic acid) is equal to that of carbonic oxide (carbon oxide) plus half the density of oxygen, and vice versa that the density of carbonic oxide is equal to that of carbonic gas minus one-half the density of oxygen.[33] The same approach was used for determining the amount of oxygen present in sulfur oxide. Yet, Gay-Lussac was at a loss in explaining a situation whereby a gas, reacting with a solid, yielded an even lighter gas. Such was the case for carbonic oxide formed by one volume of carbon and half a volume of oxygen.[34]

The more typical example, however, was nitrous gas (nitric oxide) which could be obtained either from the direct combination of equal volumes of the elements with no contraction of volume, or from one volume of nitrous oxide with half a volume of oxygen with an expansion of volume. This seemed to contradict, as Gay-Lussac pointed out,[35] the assumption made by Berthollet that elementary molecules got closer to each other, as a result of the contraction in volume following chemical combinations. The example of nitrous gas is very important, because it was mentioned by Avogadro, when he discussed the division of integral molecules. Berthollet had offered the explanation that the presence of hydrogen in carbon dioxide contributed to its low density. After discarding this interpretation, Gay-Lussac concluded, 'we must admit that the condensation of the molecules of two combining substances, in particular, of two gases, has no immediate relation to the condensation of volume. . .'.[36] A very remarkable statement indeed, and one which suggests that he could not find any way to reconcile the Daltonian view of multiple proportions with the variations observed in the volumes of combining gases. Accordingly, condensation of atoms followed one course; condensation of volumes followed another.[37]

This impasse attracted the attention of Avogadro, as clearly indicated in 'Essai d'une manière'. A few short paragraphs near the end of Gay-Lussac's 1809 memoir had a very profound impact on the author of the molecular hypothesis, and in effect they led him to overstretch its applications. In the 'Mémoire sur la combinaison', Gay-Lussac had suggested the possibility of calculating the vapor density of an element capable of entering into combination with oxygen. The procedure was simple: the composition by weight of any oxide, as found through the analysis, could be transformed into proportions by volumes; by further assuming the metal or 'radical' to be in the gaseous state and combining with oxygen in simple ratios, one would then be able to calculate the relative density of the radical. As he put it: '. . . it is sufficient to convert the oxygen into volumes and also the combustible, under the condition that its vapor shall be equal to the volume of oxygen, or else double or half this value.'[38]

As an example, Gay-Lussac computed in this way – a fact generally neglected by historians – the weight of mercury relative to oxygen three years before Avogadro did it in the 'Essai'.[39] However, careful to avoid any further extension of his combination law beyond the gaseous substances, Gay-Lussac did not elaborate this line of thought; the reason he gave was that these calculations were founded only on analogies easily multiplied.[40] In reality, he had found himself in the unpleasant situation of either confining the application of his combinations law or entering into serious disagreement with the essentials of the system set forth by his mentor and protector Berthollet.[41] This was a risk that Gay-Lussac was not ready to take; one may assess the great relevance of 'patronage' for the advance of science, especially in France at the time, in this particular instance. And the situation may also help to explain in part why Gay-Lussac did not reach the same conclusions three years earlier as Avogadro did, at least as far as 'les masses relatives des molécules élémentaires' are concerned. It should be said, however, that Berthollet's ascendancy would not have prevented Gay-Lussac from proposing either the equal volumes–equal numbers generalization or the concept of structured molecules.

The 'Essay on a manner of determining the relative masses of the elementary molecules of bodies and the proportions in which they enter into combinations' is divided into eight sections.[42] In the first two Avogadro presented the reasoning behind his fundamental assumption. Hence these sections are the most important and require a detailed review. By reading these pages, one can appreciate that, as a result of his style, some of Avogadro's key concepts may not have been fully grasped. While the logic is rigorous and the conceptual path fully coherent, the construction of his sentences is occasionally quite elaborate and obscure.[43] The validity of this criticism is not altered by the fact that Avogadro's writings may be easier to understand today than at the time they were written.

In the opening paragraphs of the essay, after briefly recalling Gay-Lussac's 'Mémoire sur la combinaison', Avogadro introduced his famous generalization. He argued that a simple and regular relationship between volumes of gases in chemical combinations could not be observed by supposing that equal volumes of different gases contain different numbers of molecules. Therefore, the opposite inference was true, and the number of integral molecules of any gas was always the same in equal volumes or always proportional to the volumes.

As mentioned previously, Avogadro arrived at his central conclusion from the notion that there was no actual interaction among gaseous molecules and that, therefore, their distances and their number in a given volume remained unchanged. This was clearly revealed by the 'simple and regular laws' (as they were defined by Gay-Lussac in 1809) which the gases followed in their physical and chemical behavior.

In view of the importance for its acceptance that a fundamental scientific statement be clearly enunciated, it may be proper to recall here that three years after the 'Essai d'une manière' Avogadro restated in the most lucid and concise way his gas generalization:

In my [1811] essay I have submitted a very natural hypothesis – as it seems to me – not superseded so far, to explain the discovery by Gay-Lussac that the volumes of gaseous substances mutually combining and those of compound gases thus obtained are always in very simple ratios. This hypothesis

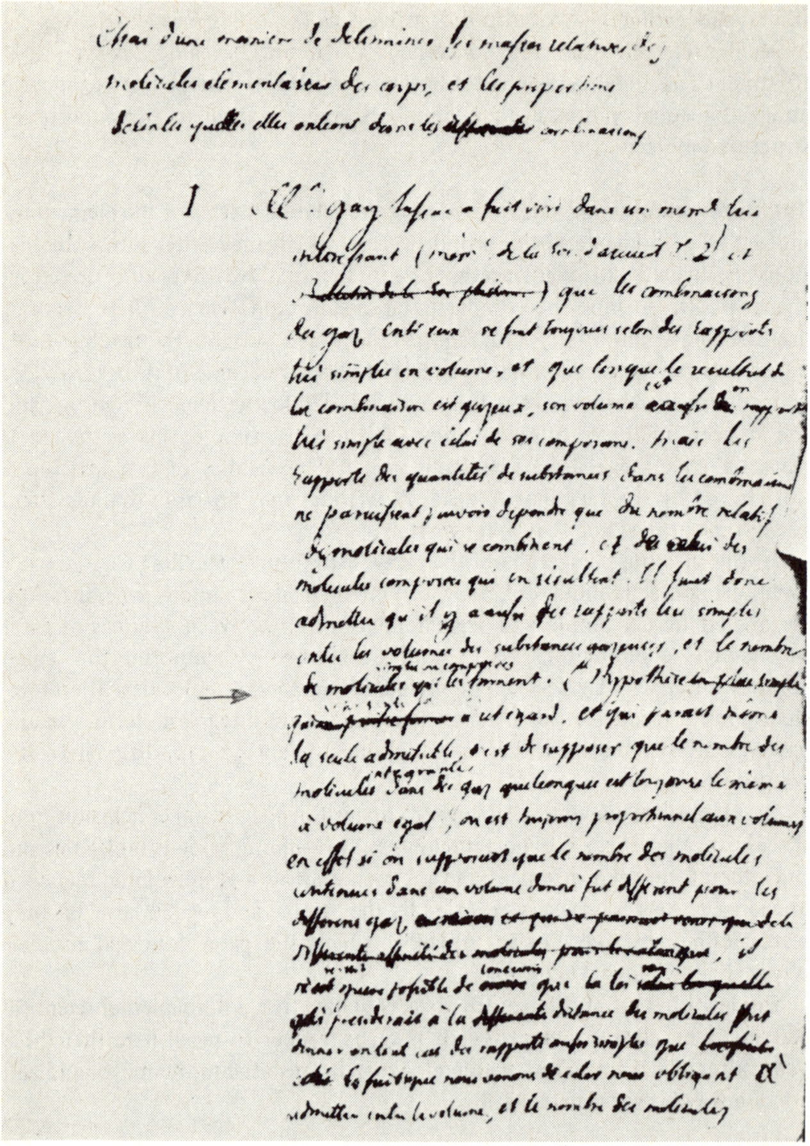

Plate 7. Reproduction of the first page of the original draft of the 'Essai d'une manière de déterminer les masses relatives des molècules élémentaires des corps et les proportions selon lesquelles elles entrent dans les combinaisons' (1811). The paragraph containing his famous generalization begins half way down the page (at the arrow). (From the collection of Avogadro's scientific manuscripts in the Biblioteca Civica di Torino).

states that equal volumes of gaseous substances, under same pressure and temperature, represent equal number of molecules; hence, the densities of different gases are the measure of the molecular masses of these gases and the ratios of volumes in the combinations are nothing else than the ratios among the numbers of molecules which combine to form the compound molecules.[44]

This is the opening paragraph of the 'Mémoire sur les masses relatives' which was submitted to de La Métherie towards the end of 1813 and appeared in February 1814 in the *Journal de Physique.* Generally overlooked by historians, it represents, together with the 'Nouvelles considérations' of 1821, the most comprehensive attempt made by Avogadro to prove the validity of his gas hypothesis. As for the above quotation, it indicates the fallacy of the thesis that its author was himself unable to comprehend the far-reaching importance of his own conclusions.[45]

The 1814 essay, even more than the 'Essai d'une manière' was devoted to a detailed description of the procedure adopted by its author for calculating the relative densities of several elements and their compounds, and for assigning formulas to the latter.

As a corollary to the gas generalization, Avogadro argued in fact, if it is assumed that in equal volumes V_A, V_B, V_C of gases A, B, C, the number of particles contained therein n_A, n_B, n_C is the same, then the ratios of their densities (relative to the same standard substance) stand as the ratios of their relative masses.[46] The combining volumes indicate also the number of molecules entering into combination. Thus, from the densities (relative to air) of a number of substances – both elements and compounds – Avogadro could calculate their relative masses,[47] and from the volume ratios reported by Gay-Lussac, he established the number of simple and compound molecules present in those same combinations.

For several gaseous substances, however, the experimental evidence seemed to indicate – as Dalton himself had pointed out – [48] the absurdity of the equal volumes–equal numbers generalization. While the example of nitrous gas given by Dalton ('the number of ultimate particles could at most be one-half of that before the union')[49] to prove his point is well-known, many other gaseous compounds led to similar conclusions.

Avogadro mentioned water, the volume of which is twice and not equal to that of the combining oxygen. He then suggested, in a little known but certainly very remarkable understatement, that 'il se présente *assez naturellement* un moyen d'expliquer les faits de ce genre' (italics mine). Immediately afterwards, in a breath-taking, twenty-line long sentence, he introduced his 'way of explaining' with the formulation of an idea so unorthodox that its real meaning would escape the comprehension of his contemporaries for over half a century:

> ... we suppose namely, that the constituent molecules of any simple gas whatever, i.e. the molecules which are at such a distance from each other that they cannot exercise their mutual action, are not formed of a solitary elementary molecule, but are made up of a certain number of these molecules united by attraction to form a single one, and that when molecules of another substance unite with the former to form a compound molecule, the integral molecule which should result splits up into two or more parts or integral molecules composed of half, a quarter of the number of elementary molecules of which was formed the constituent molecule of the first substance, combined with half, a quarter, etc. the number of constituent molecules of the second substance that ought to enter into combination with the total molecule, or what comes to the same thing, with a number equal to that of half molecules, quarter molecules of this second substance; so that the number of integral molecules of the compound becomes double, quadruple, etc. what it would have been if there had been no splitting up, and exactly what is necessary to satisfy the volume of the resulting gas.[50]

In short, he said, elementary gaseous molecules[51] are complex and divide at the moment of their combination with other elements. Gaseous compound molecules also divide to yield the required volumes of the final products of gas combinations. Thus, when two or more constituent molecules[52] – each formed by a certain number of 'elementary molecules' – combine, there is first the formation of an integral molecule. This then splits into two, four, or more integral molecules, composed of one-half, one-fourth, etc., the number of elementary particles present in the original combining molecules. Without this division of the integral molecule it would have been impossible for Avogadro to reconcile his main hypothesis with the experimental evidence. It sounded like a *phantasiespiel*, as Berzelius would have said. Nevertheless, the solution had been found by an imaginative mind.

In order to place both 1811 and 1814 essays in their historical context, the issue concerning the terminology adopted by Avogadro requires attention. First, in neither paper, even when referring to Dalton's work, was the term atom used, while Gay-Lussac had done so in his 1809 memoir. At the same time, Avogadro made abundant use, at the risk of being confusing, of molecules in all its qualifications: simple, integral, constituent, elementary, compound. This was especially true in the early part of the 'Essai d'une manière'; later on in the same article, he dropped any further qualification of molecule. Three years later, in the 'Mémoire sur les masses', he adopted molecule alone, throughout the entire paper, occasionally interchanging it with 'volume' in the same context.

Why this sudden change in terminology? One may speculate that, after setting forth his theory in the first part of the 'Essai d'une manière', Avogadro found it superfluous to repeat elaborate qualifications. Another explanation is that, in the lapse of time between the 1811 and 1814 essays, he had become aware of the prevailing chemical nomenclature of the time – especially that of Berzelius, which was appearing more and more in the *Annales de Chimie* – and he wanted to be more consistent with it. Later on, in his major chemical essay of 1821, the 'Nouvelles considérations' Avogadro would return to using such qualifications of molecule as 'simple', 'partial', 'total', and 'integral'.

Whether the adoption of rarely used terms may have contributed to the sparse attention paid by his contemporaries to Avogadro's 1811 and 1814 essays has been discussed by a number of historians. Quite likely at the time he set forth his fundamental assumption, he could have expressed it with simpler and less equivocal phrases. Yet, despite this, to find fault with him for the protracted confusion between atom and molecule which well outlasted his lifetime would be historically incorrect.[53]

A few important aspects of the first section of the 'Essai d'une manière' deserve careful examination now. Foremost is the concept of molecular division and its etiology. How Avogadro had arrived at his most original contribution? Should it be considered an extempore perception or was it an *ad hoc* solution to the deadlock created by his own equal volumes-equal numbers generalization? Unfortunately, there is very little evidence in support of either thesis. We are left with a paragraph in which Avogadro indicated that the existence of compounds with a very large molecular mass would seem unlikely and in violation of Nature's laws.[54] Therefore, a splitting into smaller parts appeared appropriate and, he added, could be expected *a priori*. Thus, at first Avogadro invoked an harmonious or aesthetical view of Nature to account for the molecular division; but shortly afterwards, on less metaphysical grounds, he indicated that only through such division could one explain the contraction of volume observed in many gas combinations. An occurrence which, he pointed out, could not be the result of a simple 'rapprochement moléculaire', as Gay-Lussac had conceded.

While on the issue of the number of *integral molecules* involved in combinations Avogadro was very clear and assertive, he remained vague in both the 1811 and 1814 essays on the structure of the *constituent molecules*. They definitely divided; but how? Whether they were formed by two, four, eight, or more elementary molecules, he left quite undetermined until at least the 1821 'Nouvelles considérations', where he assumed as 'likely' the existence of four 'partial molecules' in each molecule of a gaseous element. However, a few paragraphs earlier, he seemed to indicate that in some gases integral molecules could be formed by two or even by one partial molecule.[55]

There is no doubt that in 1811 and 1814 the mass, shape, and number of integral molecules concerned Avogadro more than their 'subdivisibility', a term he used in the 1821 essay. Characteristically, in the 1814 'Mémoire sur les masses' the effect of the division of a molecule of elongated shape, for instance carbon, on its physical properties is discussed, but not what kind of division occurs.

On another point, which some historians consider of critical importance for the rejection of his assumption, Avogadro remained very vague in both the 1811 and 1814 essays. How did 'elementary molecules' stay together in the 'constituent molecules' of a gaseous element? In the 'Essai d'une manière' he just hinted at an 'attraction' not otherwise qualified. In the 1814 essay he did not elaborate further, and the question of how unwieldy it was to explain his molecular structure in terms of Berzelius' dualism and electrical attraction and repulsion never arose even in the extensive 1821 article.[56] Thus, one is left in the dark with regard to such an important issue; the observer is left with the feeling that, in Avogadro's system, electrical forces had no relevance at the molecular level.

After setting forth his generalizations – the divisibility of molecules and the equal volumes–equal numbers hypothesis – Avogadro's next object was to confirm their validity. First of all, he checked that the molecular masses which could now be calculated with his assumptions would agree with those found by means of their densities.

He seemed confident of having found the way to 'confirm or modify Dalton's results from precise data . . . ', and accordingly he disputed Dalton's conclusions for a number of gaseous elements and compounds. Quite skillfully, he used the experimental data of Gay-Lussac, Biot, Arago, among others, to prove the superiority of his own approach over that adopted by Dalton. He emphasized that there was no evidence, for instance, that the rule of greatest simplicity would apply to gaseous substances: ammonia represented the most erroneous among the compositions proposed by the Manchester naturalist and the one most irreconcilable with the combining volumes found by Gay-Lussac. No more easily could Dalton explain, in Avogadro's view, the combinations of sulfur with oxygen; specifically, sulfuric acid (sulfur trioxide) which required one and half volumes of oxygen and which could scarcely dovetail with the 'fausse' ratio of two atoms of oxygen per atom of sulfur that Dalton had adopted. On the other hand, by using the data on gas densities available, in 1811 Avogadro arrived at results which were far more correct than anything then available. He thus found that the molecular mass of sulfurous acid calculated from the masses of its components in a correct ratio, agreed with the experimental density. And he concluded that sulfur, as a vapor, should have a density 2.32 times that of air or 31.7 times the mass of hydrogen.[57]

Avogadro next directed his attention to three more nonmetallic elements: phosphorus, carbon, and oxymuriatic acid. For phosphorus, he shared Davy's explanation of its oxides as having the same oxygen ratios as those observed in the sulfur oxides. From the analytical data available for phosphoric anhydride (phosphorous anhydride was not discussed) the mass of phosphorus relative to hydrogen was calculated. Previously Davy had put it at 16.5 and Avogadro – relying on Heinrich Rose's figures quoted by Gay-Lussac – set it at 38, a figure whose accuracy could not be verified against the yet undetermined density of the element in the gaseous state. Needless to say, both Avogadro and Davy were misled by the wrong analogy with the sulfur dioxides, so that the problem of determining the elementary mass of phosphorus lay, in 1811, beyond the application of Avogadro's generalization.

In examining the combinations of carbon with oxygen in his 1809 memoir, Gay-Lussac had reached a number of conclusions: the volume of oxygen remained unchanged when carbonic acid (carbon dioxide) was formed by direct union of the elements, but it doubled in the formation of carbonic oxide (carbon monoxide) and also when oxygen combined with the same volume of carbonic oxide to give carbonic acid. From these observations, Gay-Lussac had been able to calculate the density of carbonic oxide, which he found in fair agreement with that experimentally obtained by Cruickshank.

Avogadro took the next step, by inferring that, if the volume of carbonic acid was equal to that of the oxygen forming it, then the volume of combining carbon in the gaseous state should be one-half that of oxygen, and the integral molecule of carbonic acid would be formed by one molecule of gaseous carbon and two molecules of oxygen. From the densities of carbonic acid and oxygen known to him, he then proceeded to compute the molecular mass of carbon. The low weight, 11.36, of this element which showed 'la solidité de son agrégation dans les températures les plus élevées ...'[58] definitely puzzled Avogadro. How could carbon be lighter than gaseous elements such as nitrogen and oxygen and at the same time have an extremely high physical stability? He addressed these questions and he did not exclude the possiblity of doubling and even quadrupling carbon's molecular mass.[59] In order to explain this anomaly of carbon and also that of mercury, which despite its large mass vaporized at a much lower temperature than iron, Avogadro suggested the presence of 'some unknown property of substances' which, aside from the size of the molecule, affected its physical state.

Although generally neglected, Avogadro's conclusions on the composition of oxymuriatic and muriatic acids and his evaluation of their masses should be mentioned for their remarkable intuition. By then, Gay-Lussac's and Thenard's researches on oxymuriatic acid[60] indicated their failure to decompose this substance by means of carbon. As a result, Avogadro assumed as correct the elementary nature of oxymuriatic acid, while Gay-Lussac and Thenard, out of respect for Berthollet's position, remained vague on the issue.[61] At the time, the French chemical school had no intention of abandoning what they considered one of the supporting pillars of Lavoisier's system, that all acids contain oxygen, and outside France, Berzelius, in agreement with the prevailing opinion, supported the muriaticum theory. Familiar with Davy's early findings,[62] and yet before the fifth Bakerian Lecture, Avogadro, while drafting the 'Essai d'une manière', had reached the same conclusion: there is no oxygen in oxymuriatic acid which, therefore, should be considered as a simple substance, like hydrogen. From Davy's figures for the density of oxymuriatic acid gas, he calculated that its mass should be 32.8 times that of hydrogen, a figure later changed to 33.9, after considering the density of muriatic acid reported by Gay-Lussac and Biot.[63] Thus, through the application of his hypothesis and his acceptance of Davy's views, Avogadro had proposed by 1811 the correct elementary mass for chlorine.

As earlier mentioned, Gay-Lussac in his 1809 'Mémoire sur la combinaison' had hinted at the possibility of calculating the densities of all 'combustible' (oxidizable) substances by assuming them in the gaseous state, and he had also given mercury as an example of how this could be done. A possible explanation for his reluctance to pursue this speculative path further was the public disagreement with Berthollet that this would entail. The question then remains: why did Avogadro in his 1811 and 1814 essays immediately proceed to extend his gas generalization to solid and liquid elements and their compounds? A first answer is that he, unlike Gay-Lussac, had no conceptual restraint in accepting a translation of volumetric into gravimetric expressions; by concentrating on density as the characteristic property of a given substance, he thought of overcoming the contradiction between its weight and its volume. A second answer is that he presumed that he had at his disposal the perfect instrument to resolve what already appeared to be two major difficulties in the orderly development of the young science of chemistry. What he offered was an unequivocal way of determining both the atomic-molecular weights of chemical elements and their combining proportions.

On this point, in stressing the importance of his contribution, he was especially assertive in the 1814 essay when he stated:

> This hypothesis, once admitted, by partially confirming the results to which Dalton, Davy, and others have been led by particular considerations on the masses of molecules of several known substances according to their proportions in the combinations, provided us with a general method for modifying these results and for thus improving the *theory of definite proportions, which is or will be the foundation of all modern chemistry and the origin of its future progress.*[64] (Italics mine.)

It appears that at this time and at least until 1821, Avogadro had very clear ideas about his program for putting the chemical house in order. In 1814 he was still confident that his hypothesis would be useful and acceptable to the chemists. Quite likely, he failed to realize that his largely speculative approach to the solution of empirical problems could not be received favorably. Seen from the perspective of his own time, the method adopted by Avogadro must have seemed at least suspect and unwarranted to the practicing chemists. Unfortunately, there is no record of how they reacted to the content of the 1811 and 1814 essays, but in all probability those who read them dismissed their content and above all the attempt to extend the gas hypothesis to non-volatile substances as a combination of reveries and futile computations. The experimental evidence in support of such a hypothesis was debatable, and when it grew stronger, several important facts, such as anomalous vapor densities remained unexplained until quite late in the nineteenth century.[65]

Among historians – from Wurtz and Meldrum to Lowry and more recently to Partington and Coley – the opinion prevails that Avogadro was not successful in his endeavor to evaluate the masses of solid and liquid elements and their compounds.[66] This conviction should be somewhat tempered after a detailed review of the facts. In discussing metals in the 1811 and 1814 essays, Avogadro limited his observations mostly to their combinations with gaseous elements, and he drew his conclusions on the basis of the findings of some of the best experimental chemists of his time. Yet, in the concluding paragraph of the third section of the 1814 memoir, after his discussion of the 'Métaux ordinaires', he pointed out:

I will refrain, for the time being, from extending my calculations to other metals, either because there are not yet available analyses as precise as the preceding for their oxides, sulfides, etc. or because they offer in these combinations only one proportion well determined, and therefore I have doubt about the application of the volume theory.

In the 'Essai d'une manière' only six metallic elements were discussed: mercury, iron, potassium, lead, silver, and copper, and for the last three Avogadro gave only their masses. To explain his approach, one example may be used; for Davy, potassium oxide contained one part of each component, and since the equivalent of oxygen was for him 7.5, he figured for potassium an equivalent weight of 40.5. Avogadro, on the other hand, by giving oxygen a weight of 15, had tentatively arrived at 81 as the mass for potassium. He conceded, however, that there was not sufficient evidence to support the combining ratio of 1:1 rather than 1:2 or 2:1 between the metal and oxygen. In the 1814 essay, he returned to this subject: by then he had available both Davy's and Berzelius' analytical data for potassium oxide. He argued now that the potassium/oxygen 1:2 ratio seemed unlikely, because it would result in a mass of about 160, very large for such a light substance as potassium. As for the opposite ratio 1:0.5, (he had discarded by now the 2:1 ratio earlier suggested) this would make it difficult to explain the higher potassium oxide, found by Thenard, having three times the amount of oxygen present in the lower oxide or a 1:1.5 ratio; not compatible, Avogadro added, with the usual combination by volume.[67]

Through analogous considerations based upon the compositions by weight of their oxides, Avogadro arrived in the 'Mémoire sur les masses' at the relative weights of sodium, magnesium, barium, strontium, and aluminum.

As for the heavier metals, in 1811 he had given lead a mass of 206. At that time he assumed lead oxide to be formed by equal 'volumes' of the components. Three years later, on the basis of Berzelius' analyses, it seemed established that one 'volume' of lead could react with either two, three, or four volumes of oxygen to form its oxides; since in any case the lowest oxide would contain at least twice the amount of oxygen earlier considered, Avogadro set the 'density' of lead at 391.5, abandoning the 1811 figure.

For one more example, it may be interesting to follow Avogadro's reasoning in some detail. In the 1814 memoir, after discarding as unreliable both Berzelius' and Stromeyer's findings on silica, he discussed fluosilicic acid, the weight of which John Davy had set at 48.8, and he assumed that this gas was formed by equal parts of fluoric acid and silica.[68] Earlier in the same article he had assigned to the former compound a weight of 20.08. Since fluoric acid was considered as composed of two parts of oxygen and one part radical, by subtraction he arrived at the weight of silica. As an oxide, he assumed this as formed by two volumes of oxygen and one of silicon and he computed the weight of the latter as 27.33. Consequently, Avogadro has been recognized by some historians of chemistry as the first to suggest the correct composition of silica.[69] However, it is fair to say that he himself considered his finding as conjectural, due to the conflicting analytical data which were at his disposal.[70]

In relation to this, Avogadro seldom accepted *tout court* the experimental figures available to him, but in general discussed them; when he reached his own conclusions, they were not necessarily in agreement with those found by practicing chemists. In an overall assessment of his work, out of twenty-two elements discussed in 1814, his results were faulty in fourteen cases; yet for twelve of them, the same is true for Berzelius.[71] (See Tables 2 and 3.)

As for the elements and their compounds which are gaseous under ordinary conditions and represented the most valid application of his hypothesis, Avogadro had extensively discussed them in the 'Essai d'une manière', and very little was added in the 1814 memoir. However, newly mentioned was a gas obtained by the combination of two volumes of chlorine and one of oxygen, and a liquid compound of chlorine and nitrogen which had been discovered in the meantime respectively by Davy and Dulong. As the experimental data reported by Davy indicated, the former compound did not display the whole number ratio generally observed by Avogadro in binary combinations. Nitrogen chloride, on the other hand, apparently followed the rule of the even ratio of combining volumes.[72]

The other gaseous substances examined were olefiant gas (ethylene) and carburetted hydrogen (methane). Here Avogadro recalled that the combustion of one volume of the former with three of oxygen gives two volumes of carbon dioxide and one of water. Accordingly, one volume of olefiant gas has to be formed by one volume of carbon and two of hydrogen, as had been confirmed by the experiments of Theodore de Saussure and also by the calculation of the specific gravity of olefiant gas.[73] Among binary gaseous compounds this represented an exception to the prevailing rule observed also in the formation of carburetted hydrogen and pointed out by Avogadro in the 'Essai d'une manière' that they double in volume in relation to the component entering in the smaller proportion.

Phosgene, obtained by John Davy from equal volumes of chlorine and carbon monoxide, was one of the few ternary compounds described in the 1814 memoir. Its composition, correctly given, indicated that there was no

Table 2. Atomic and molecular weights of elements and compounds examined by Avogadro in the 'Essai d'une manière', compared with those adopted by Dalton

Elements and Compounds	*Dalton*		*Avogadro*
	1810	*1827*	*1811*
Carbon	5.4	5.4	11.36
Copper	56	28	123
Chlorine (oxymuriatic acid)	–	29 or 30	33.9
Hydrogen	1	1	1
Iron	50	25	94
Lead	95	90	206
Mercury	167	84	181
Nitrogen	5	5 or 10	13.24
Oxygen	7	7	15.074
Phosphorus	9	9	38
Silver	100	90	198
Sulfur	13	13 or 14	31.73
Ammonia	6	6 or 12	8.12
Carbon dioxide	19.4	19.4	20.75
Carbon monoxide	12.4	12.4	13.2
Hydrochloric acid	–	30.3	17.45
Nitrous acid	–	19 or 38	57.54
Nitrous gas	12	12 or 14	14.16
Nitrous oxide	–	17	20.78
Sulfurous acid	20	28	30.94
Water	8	8	8.53

Dalton figures from *New System*, 1810 and 1827 edit.

need for a further doubling of volume in the formation of gaseous ternary compounds resulting from gases – carbon monoxide in this case – which have already doubled their volumes in relation to one of their components.

In the seventh section of the 'Essai d'une manière', Avogadro examined the possibility of applying his hypothesis to the combination of acids with bases. In 1808, Gay-Lussac had suggested that, whether forming neutral or basic salts, acid and alkaline substances always combined in whole number ratios.[74] He indicated that ammonia, united with its own volume of either muriatic, carbonic, or fluoboric acids, formed neutral salts, or united with one-half its own volume of the last two, gave 'subsalts'. To Avogadro this meant that acids and bases could unite, molecule by molecule or half molecule, to yield salts. Thus, one molecule of ammonia combined with one molecule of carbonic acid, and the result was a salt with a molecular mass of

Table 3. Elements and compounds examined by Avogadro in 'Mémoire sur le masses relatives'

Elements	*Compounds taken into consideration by Avogadro in 1814 and their suggested composition*[a]	*Density of the element (rel. to H) as calculated by Avogadro in 1814*	*Atomic weight reported by Berzelius in 1814*[b]	*Atomic weights reported by Berzelius in 1818*
Aluminum	Al + 2O	34.28	34.3	54.76
Antimony	Sb + 2O Sb + 3O Sb + 4O Sb + 2S	162	243	258
Arsenic	As + 2O As + 2S	89	126	150.52
Barium	Ba + 4Cl Ba + 2O	261	257.5	274.22
Boron	B + 4O	60	11.04	11.15
Calcium	Ca + 4Cl Ca + 2O	79.4 77.3	76.8	81.93
Carbon	2C + 4H C + 4H CO + 2Cl	[c]	11.3	12.05
Chlorine	2Cl + O 3N + Cl	[c]	21 (Muriatic radical)	–
Copper	Cu + O	123[c]	121.4	126.62
Fluorine	F + 2O	10	9.6	12.00
Gold	Au + O Au + 2O	374	374.3	397.76
Iron	Fe + 2O Fe + S	106.5[d]	104.5	108.55
Lead	Pb + 2O Pb + 3O Pb + 4O	391.5[d]d	391	414.24
Manganese	Mn + O Mn + 2O Mn + 3O Mn + 4O	107.6	107.2	113.85

Table 3 Cont'd.

Elements	*Compounds taken into consideration by Avogadro in 1814 and their suggested composition*[a]	*Density of the element (rel. to H) as calculated by Avogadro in 1814*	*Atomic weight reported by Berzelius in 1814*[b]	*Atomic weights reported by Berzelius in 1818*
Magnesium	Mg + 2O	47	47.5	50.66
Mercury	Hg + 2Cl[d] Hg + Cl	370 408[d]	381	405.06
Phosphorous	2P + 3O P + 3H P + 3Cl P + 2Cl	20[d]	25.2	52.77
Potassium	K + O K + 2Cl	75.78	147.3	156.77
Silicon	SiF + 2O Si + 2O	27.33	32.6	47.43
Silver	Ag + O Ag + 2Cl Ag + S	198[c] 206[d]	405	432.51
Sodium	Na + 2O Na + 4Cl	90	87.3	93.08
Strontium	Sr + O	201	213.7	175
Sulfur	S + 2O 2H + S C + 2S	30.65[d]	30.2	32.19
Tin	Sn + 2O Sn + 3O Sn + 4O	222	221.5	235

[a] In writing these combinations, I followed the graphic system adopted by Berzelius in 1814, that Avogadro never used in his chemical writings of the time.
[b] These have been recalculated from oxygen = 100 used by Berzelius in *Ann. Phil.* 3 (1814), 362.
[c] Had been given in the 1811 essay.
[d] Revised in 1814 from the 1811 value.

57.75. Since the one-to-one volume ratio was assumed, then the integral molecule obtained divided itself by two in agreement with the general behavior observed in this group of compounds.[75]

Gay-Lussac had also suggested that 'if all acids and all alkalis could be obtained in the gaseous state, neutrality would result from the combination of equal volumes of acid and alkali.'[76]

For this notion of neutrality, Avogadro now substituted one which he defined as 'more precise'. Whenever an acid and an alkaline substance combined, there was a well-defined level of 'oxygenicity' corresponding to a certain degree of neutrality, and it was related to their combining proportions by weight and the oxygenicity of their components. Oxygenicity, a concept already introduced by Avogadro in his 1809 essay 'Idées sur l'acidité et l'alcalinité',[77] was redefined here as the property by which 'subtances are ranked in an [electrochemical] series at whose top are placed those behaving as acids in relation to others'. This meant that the more a substance (an element or a compound) was electronegative or closer to the top of the electrochemical series, the more oxygenic it was.

When two substances combined, the oxygenicity of their components was more or less well-defined. Therefore, Avogadro added, the neutral state was determined mainly by the excess of weight of the component displaying acidic or basic characteristics. In other words, among the simple ratios observed in the combination of molecules, there was one which is closest to the level of oxygenicity corresponding to neutrality. Consequently, any change in such combining ratios would represent a deviation from the neutral point. It followed that in the case of two acids combining with two different bases, if their respective oxygenicities were not too far apart and the ratios of their masses were also quite similar, then the ratios of their molecules, conducing to the neutral point, would be the same in both combinations. On the other hand, this did not occur when the combining ratios were different. In conclusion Avogadro recognized that only the experimental evidence would be able to confirm or refute the validity of his observations.

It is not surprising that this final, substantive portion of the 'Essai d'une manière' has been neglected in general by historians.[78] After the formulation of his fundamental assumption on equal numbers–equal volumes and on the polyatomic nature of molecules, and after an extensive demonstration of their application, Avogadro introduced further speculative ideas on the combination of acids and bases. Although these ideas may, at first, seem to have little relevance to the main subject of Avogadro's paper, Gay-Lussac himself had dealt with simple ratios in the combinations of acids and bases in his 1809 memoir and from them he had drawn some general conclusions. Avogadro rejected these conclusions not because of his main hypothesis, but on the basis of considerations he had expressed earlier concerning the electrochemical behavior of substances in general.

This last section represents one of the more obscure parts of the 'Essai d'une manière'. Although some concepts were stated in a very involuted form, the main, recurring theme of Avogadro's speculative system was still discernible. The concept of oxygenicity was for him a means of linking, after all, the chemical affinity of an element with its relative position in the electrochemical series and with its chemical character. The higher the affinity of an element for oxygen – or its tendency to be oxidized – the lower was its oxygenicity and the farther down was its rank in the electrochemical series. In other terms, the chemical affinity between two elements increased in direct proportion to their mutual distance in such a series. Even the physical state of an element was influential in determining its chemical nature. Properties such as 'elasticity' or 'cohesion' (typical of fluids and solids respectively) might tend to conceal the intrinsic acid or alkaline character of a substance.[79] In the 1814 essay he further elaborated his ideas by arguing that the cohesion of many elements, their specific gravities, and the size of their 'molecules' were closely linked. Furthermore, oxygenicity was in general directly proportional to the elementary weights, and therefore to the specific gravities and the cohesion of the elements, and was inversely related to their affinities for both caloric and oxygen.

Thus, to quote him: 'tin, mercury, and lead will have in their solid state less density than they should according to the size of their molecules, compared with that of silver and gold, because they are much less oxygenic or more electropositive than the metals just mentioned. . . .'[80]

On the other hand, the compactness and great cohesion observed in carbon found verification in the fact that, when heated, unlike other elements, it passed directly into the gaseous state. Like silicon and aluminum, carbon had a relatively low weight, which might be related to the shape of its elementary molecules, long and narrow, which allowed a reciprocal attraction to take place at their extremities only; due to the oblong form of the molecules, however, such attraction appeared uneven. The opposite was true for a very heavy gas, chlorine for instance, whose molecules were assumed to be spherical and, accordingly, provided with a mutual attraction which was the same in every direction. In a similar way the large mass observed for the elementary particle of mercury, a liquid under ordinary conditions, could be explained.

These considerations, which occupy only a few paragraphs of the 'Mémoire sur les masses'[81] are very important because they represent the only time that, in his first two chemical essays, Avogadro linked the chemical and physical properties of his elementary particles with their geometrical form. He underlined here the importance of both mass and shape in determining the degree of intermolecular attraction. Unlike Dalton and Wollaston, however, he did not elaborate on the specific geometrical form assumed by the particles. Those of very cohesive elements had 'une forme allongée' with centers where their masses were concentrated. When they came together to form the solid state, the distances between these centers were still too large to allow their masses to influence and attract each other. So that the very oblong molecules of carbon displayed great cohesion and hardness more as a result of their mutual position and shape than as an effect of gravitational attraction.[82] When subjected to the action of heat, one may assume that their geometrical form again did not allow the free movement in every direction which characterized the liquid state.

The explanation offered by Avogadro for what occurred when the molecules of very cohesive and light elements – such as silicon and carbon – entered into chemical combination, is very revealing. They divided themselves in combining, he said, and such divisions took place lengthewise, leaving them more spherical. It would thus be easier for a substance to acquire 'the place [in the classification of the physical properties] that the size of its compound molecule will assign to it.'[83] At this point nothing more is added on the subject, which seems to indicate a certain uneasiness about further discussion.

It is quite apparent that a theory of molecular geometry, like that proposed by Ampère in 1814,[84] was not present in Avogadro's considerations at the conclusion of his memoir of the same year. Although in 1821 he took a noncommittal position in his first extended comments on the ideas expressed by Ampère, he never actually adopted a lattice-like intepretation of the molecular structure.[85] Later on, in his most comprehensive discussion of the subject, this was very clear when he defined the integral molecules – representing the smallest portion of a body still provided with all the external properties characteristic of the same body – as formed by the union of simple molecules, which were more closely bound together in a 'determined' position.[86] What this position was, Avogadro did not say; his concern lay more with the masses and the mutual distances of the integral molecules than with their form since these were the factors responsible for the density and the physical state of the integral molecules.

At the end of the 'Mémoire sur les masses' there is a very short supplementary note which should not be overlooked for its historical and human relevance. In the November 1813 issue of Geneva's *Bibliothèque Britannique* appeared an extract of Humphry Davy's *Elements of Agricultural Chemistry* which had been published earlier that year. It attracted the attention of as regular and interested a reader of the Swiss journal as Avogadro. So upon completion of the 1814 memoir, he did not neglect the opportunity to emphasize that the results and the conclusions of Davy were, in many instance, in close agreement with his own. As the first example, he quoted the composition of water – two proportions of hydrogen for one of oxygen – proposed in 1811 and now restated by Davy, in conflict with Dalton's one-to-one interpretation. And he recalled also the elementary weights of carbon, potassium, lead, etc. whose values, as determined by Davy in their combination with oxygen, agreed with those he had calculated.[87]

Although Avogadro admitted that Davy did not use 'mon hypothèse sur la constitution des gas', he clearly indicated his satisfaction by observing that this concordance represented the best demonstration of the validity of the generalization he had set forth in 1811. The position of Avogadro was quite evident. As an outsider and as a man with a modest scientific reputation, he was looking for recognition. He realized that the ideas stated in the 'Essai d'une manière' had either met with indifference or had been totally ignored. Now, three years later, he had the opportunity of expressing them again and – he felt – of supporting them with the experimental evidence offered by one of the most renowned chemists of the time. It was an occasion that could not be missed, and hence the purpose of his 'Note additionelle' to the 1814 memoir, probably forwarded to the editor after the manuscript had been submitted.

It is quite possible that Avogadro had hoped for an answer from Davy which apparently never came. But there is also another side to the problem which definitely escaped Avogadro's attention: at least from the scientific point of view, the relations between Davy and the leading men of French chemistry – Gay-Lussac, Thenard, and Berthollet – were not the best in those years. By officially praising Davy's achievements, he certainly did not endear himself to the Arcueil circle, which was already at loggerheads with de La Méthérie and his journal.[88] Needless to say, besides this unintentional *faux pas*, more substantial points in the 1814 memoir may have caused a great deal of reservation and criticism among those in France who had the opportunity of reading it. Another historical note should be mentioned here. Just by coincidence, shortly after the 'Mémoire sur les masses' appeared, the same man quoted by Avogadro for having apparently proven with his results the soundness of the gas hypothesis, passed through Piedmont: Humphry Davy, accompanied by Faraday, even stopped in Turin for a few days during his

Italian journey. He was on his way to Rome, coming from Montpellier, where Jacques Berard had put at his disposal for almost a month the chemical laboratory of the local university. There is no evidence that Avogadro, then living in Vercelli, had the opportunity of meeting Davy on that occasion;[89] he was not present at the meeting of the Turin Academy of Sciences which Davy attended. He never encountered Faraday either, although eventually they corresponded with each other.

The Neglect of the Molecular Hypothesis

Amedeo Avogadro's fundamental 'Essai d'une manière de déterminer les masses relatives des molécules élémentaires des corps et les proportions selon lesquelles elles entrent dans ces combinaisons' appeared in the July 1811 issue of the *Journal de Physique, de Chimie et d'Histoire Naturelle*. Half a century, however, went by before the importance of this essay was recognized. Most of Avogadro's contemporaries did not appreciate the significance of his equation: volumes equal molecules. Avogadro also suggested that, through the densities of gaseous molecules, the molecular weights could be determined as well as the correct chemical composition of both inorganic and organic compounds. These ideas were generally either overlooked or misunderstood by chemists and physicists. Men of scientific reputation, such as Berzelius, Dumas, and Gay-Lussac, were certainly aware of Avogadro's generalization, but failed to interpret it correctly.[90]

Even after Cannizzaro's 'Sunto' and the Karlsruhe Congress, the molecular hypothesis did not find unqualified endorsement especially from the chemists. The kinetic theory of gases provided a theoretical basis for the molecular hypothesis, but dissociated it from the conclusions reached by Avogadro. Approval and recognition came through physical chemistry, but not until the work of van't Hoff, Arrhenius, Nernst, and Perrin confirmed irrefutably its general validity.[91] The etiology of the prevailing lack of reaction to such a very important development in the history of chemical and physical doctrines has been extensively examined. To most historians and scientific observers of the last century it seemed inadmissible that Avogadro himself could have formulated the molecular hypothesis which, in fact, was credited, until the end of the century, to someone else.[92]

Some of the recurring reasons found in the history of science for this neglect are very relevant, although they should be evaluated within the scientific attitude of the time. For instance, the exceedingly speculative nature of Avogadro's generalization is often mentioned. Such an approach represented a cardinal sin for the empiricism then prevailing, and may also explain the lukewarm reaction to Dalton's atomism in those same years. Another valid criticism focused on the unacceptable – from the electrical point of view – notion of polyatomic elemental molecules; a concept reproposed by Clausius in 1857 and earlier by Gaudin and Laurent, but which lacked a reasonable theoretical explanation until the rise of quantum mechanics.

The anomalies found, in the 1820s, in the vapor densities of several elements and compounds could not be explained for fifty years, and remained one of the bulwarks of criticism to Avogadro's assumption even after the Karlsruhe Congress. Further serious hindrance originated in the misleading speculation proposed by Berzelius in 1813 which equated volumes and number of atoms (and, accordingly, identified atoms and molecules in elemental gases). This was followed in 1814 by the publication of Ampère's letter to Berthollet containing a generalization similar to Avogadro's, but much more complex and confusing. When Dumas attempted in 1826 to explain the relation between gaseous densities of elements and compounds, their weights and their combining ratios, he jumbled Avogadro's ideas with those of Ampère and reached absurd and unacceptable conclusions. This blunder provoked the sharp criticism of Berzelius, who shortly afterwards ridiculed the then young French chemist in the *Jahresbericht*. As a result, the equal volumes –equal number of molecules assumption came to a dead end in 1828.[93]

Other explanations, less critical than those just mentioned, have also been considered by historians through the years: Avogadro's elaborate style of writing which often obscures his ideas, the irrelevancies in the text,[94] and the recurrent inconsistencies in terminology.[95] The last factor seems to be even more prevalent in his later writings and cannot be easily forgiven.[96]

It has been said that none of Avogadro's major chemical memoirs was published in chemical journals. This is true; however, the *Journal de Physique*, where the 'Essai d'une manière' appeared, contained articles in every field of science, although on a different level than those published in the more prestigious *Annales de Chimie.*[97] It is also true that the attention of the scientific community in the years immediately preceding and following the publication of the molecular hypothesis seemed diverted towards events much more remarkable and *éclatant*, such as the isolation and identification of new chemical elements: sodium and potassium in 1807, chlorine in 1810, iodine in 1813.[98]

Strangely missing from the etiology of the neglect of the gas hypothesis is the observation that the eventual use of chemical symbols, such as those proposed by Berzelius in 1813, would have helped to clarify Avogadro's thought, as they did for Gaudin twenty years later. These symbols were not adopted even in the major chemical essays of 1821, the 'Nouvelles considérations' and 'Sur la manière de ramener'.

Avogadro weakened the plausibility of his calculations by excessive recourse to analogies. This resulted in faulty atomic and molecular weights for non-gaseous substances.[99] Analogy, certainly a very important tool in scientific reasoning, should not be abused and here Avogadro was at fault. By using this inferential process again and again – as he did in 1811, 1814, and 1821 – he reached conclusions often at odds with and inferior to those of the experimenters.[100] No one can blame the practical chemists of the time (Gay-Lussac, Thenard, and Berzelius, among others) who had very little patience for and appreciation of his mental images.

A review of what has been written on the subject in the past one hundred years indicates that not enough attention has been paid to the possible effects that the peculiar personality of Avogadro, and the events surrounding his life, exerted on the neglect of the 1811 hypothesis. For example, A. Meldrum does not devote a single line in his monograph, [101] probably the most thorough among the English works for over half a century, to the human and historical aspects of Avogadro's life.

When Icilio Guareschi[102] published *Il discorso storico critico*[103] in 1911, he had a specific purpose, that of vindicating, once and for all, the singular scientific merits of a man who was little known not only beyond the Alps, but even in his own country. In this respect he certainly succeeded, but not without infringing occasionally on that impartiality which should characterize every endeavor of the kind. Here again, it is quite surprising that Guareschi gives only a short description of the distinctive traits of Avogadro's character and a bare outline of the main events of his life. Yet *Il discorso* remained for sixty years the unparalleled exegesis of the molecular hypothesis and of its proponent.

Shortly after *Il discorso*, a very accurate and extensive review of the history of the development of the molecular hypothesis, from 1811 to its application by van't Hoff to the diluted solutions, was presented by the German organic chemist Carl Graebe.[104] Although the predominant tone of this essay was one of praise, Graebe appeared more objective than Guareschi and consequently his work was more valuable in reporting the position of all those involved in the debate over chemical systematics before and after the Karlsruhe Congress. He justified the complete inadequacy of the biographical information in the 'Entwicklungsgang' on the basis that Guareschi had already covered the subject sufficiently in his *Discorso*, which is not true. Among those few who attempted to identify the reasons behind the long-delayed acceptance of the molecular hypothesis, by suggesting the importance of human elements, was Charles A. Wurtz. For twenty years at the Faculté de Medicine at the Sorbonne, he stood out as the leader of the French chemical school in the 1870s following Dumas.

Wurtz, whose historical work is generally less known than his accomplishments as a chemist, went through a few interesting *renversements* in his study of the molecular hypothesis. Four years after attending the Karlsruhe Congress, he published *Leçons sur quelque points* in which he constantly referred to both Ampère and Berzelius as responsible for proposing the 'equal volumes–equal numbers of atoms' generalization. Returning to the subject again in 1868, in the *Histoire des doctrines chimiques,*[105] he mentioned Avogadro and his concept of polyatomic molecules. It is surprising that the author of the often quoted chauvinistic statement 'la chimie est une science française' referred to Avogadro as a 'chimiste italien'. In asking how 'un conception si juste et si simple semble avoir échappé à l'attention des contemporains', Wurtz raised the point that its author not only had 'discredited' it through the extension to nongaseous bodies, but also lacked '*l'authorité nécessaire pour la faire adopter*'. It is not clear how much he knew about Avogadro at the time; in pointing out, however, the paltry weight of the Piedmontese's scientific work before 1811 and his modest position as a teacher in a small town secondary school, Wurtz definitely raised a very important point. In 1877 he devoted six articles to the atomic-molecular theory.[106] Shortly afterwards Wurtz, in discussing the atomic theory, dropped further references to Ampère's contribution and devoted all praise to Avogadro for his original interpretation of the molecular constitution of gases.[107]

Hermann Kopp, a better known historian who was also present at Karlsruhe, never mentioned the molecular hypothesis in his *Geschichte der Chemie.*[108] A quarter of a century later[109] he returned to this topic. After recalling that Avogadro had submitted his theory without adding any experimental proof, and this certainly being the main reason for the scant attention it attracted, he adds that 'had this theory been tied with the explanation and the exposition of findings as important as those reported in Gay-Lussac's Law on volumes and had it contributed in some way to experimentally answer questions deemed basic, then it certainly would not have been entirely forgotten as it was in fact'. And he draws a conclusion which is significantly close to that earlier expressed by Wurtz: 'Avogadro's Hypothesis was considered as a *mainly conjectural* endeavor on a topic concerning chemistry and made by a man who, in this field of human knowledge, had contributed *no other work* so well known that it might encourage others to consider his opinion.' (Italics mine.)

There is no doubt that the 1811 generalization on gaseous molecules was the result of a mainly speculative approach and it is also beyond question that Avogadro's contribution to the physical sciences was modest until that time, but one may wonder how valid these considerations are from the historiographic point of view. That scientific progress has been bred by the often fortuitous intertwining of experimental data with pure conjecture[110] and that some men have brought forth their best intellectual achievements early in their lives – when still unknown – are such well-recognized truths that it comes as a matter of surprise to see them disregarded by an authoritative historian.

Kopp charged that Avogadro had not supported his hypothesis with 'new facts'. He seems, however, to forget that Avogadro constructed his assumption upon the conclusions drawn from wide experimental evidence and from those reported by Gay-Lussac in 1808. A similar criticism, after all, could also have been made by Kopp about John Dalton, who added very little empirical proof, but had, nevertheless, the imaginaton and the ability to formulate his atomic doctrine. Ernest von Meyer, whose *Geschichte*[111] was so violently attacked by Meldrum for its excessive partiality towards Berzelius and its disparaging criticism of Dumas,[112] shared with Kopp the view that one of the reasons for the neglect of Avogadro's hypothesis was the lack of experimental data. Recently, the explanation offered by Kopp has again been mentioned by Partington.[113]

A more original interpretation of the failure of the molecular hypothesis to be recognized was offered a few years ago by Linus Pauling[114] when he suggested that the real significance of the 1811 generalization probably escaped even its author. This seems proved, he added, by the considerable time and effort Avogadro devoted to investigating much less important subjects in unrelated fields of the physical sciences.

It is true that after the 'Essai d'une maniére' Avogadro conducted extensive theoretical work in thermophysics, especially on calorimetry. He turned his attention also to electric phenomena and to their relation with chemistry. It is only fair to say, however, that ten out of his fourteen essays concerning chemical problems revolve around the equal volumes – equal numbers assumptions. This is clear from the second volume of his most far-reaching work, the *Fisica dei corpi ponderabili.*[115] In addition, Avogadro's tone, when discussing this subject, is too resolute and unswerving for a man not fully convinced of the correctness of his position and the validity of the gas hypothesis.

Quite recently, a very original explanation for the neglect of Avogadro's hypothesis has been offered by Nicholas Fisher.[116] He advances the view that chemists in the first half of the nineteenth century, having no use for the molecular hypothesis, therefore neglected it. For Fisher the question is: would Avogadro's assumption have helped chemistry at the time? This is an intriguing approach, and one's immediate temptation is to agree with Fisher and reply in the negative. Many chemists did become or remained equivalentists almost to the end of the nineteenth century, a fact which is historically unchallengeable and which had profound implications.

However, on further consideration of Fisher's premise, one realizes that Avogadro's hypothesis offered more than a method for determining atomic and molecular weights. Without its application, physical behavior and chemical combinations involving gaseous elements, or more simply Gay-Lussac's Volume Law, could not be explained. Gay-Lussac himself was at a loss in his attempt to understand why, for instance, the density of carbon dioxide should be smaller than oxygen's, and why two volumes of ammonia were formed by the combination of one volume of nitrogen and three of hydrogen. To argue that at the time chemistry was dealing with only a handful of compounds and elements, gaseous at room temperature, would be accurate but superficial. In his 1811 hypothesis Avogadro stated simply that equal volumes of any gases contain the same number of integral molecules. Three years later he used the expression 'under equal temperature and pressure' which does not necessarily mean room temperature; thus, molecular weights could also be obtained from the vapor densities of solids and liquids, as Dumas understood so well.

Furthermore, by applying his gas hypothesis, Avogadro was the first to convey the correct formulas for water, ammonia, hydrochloric acid, hydrogen sulfide, sulfurous and sulfuric anhydrides, among others. In 1810 he calculated from the density of oxymuriatic acid the atomic weight of chlorine after assuming its elementary nature. The chemists of the time, therefore, had uses for the molecular hypothesis which they failed to explore; yet, for this neglect, they should not be blamed.

One may rather offer the interpretation that the cultural climate in the first decades of the nineteenth century had not reached the maturity necessary for the acceptance of ideas as hypothetical as those proposed by Avogadro. The thesis of science's neglect of an idea because it was not deemed immediately necessary makes it difficult to explain the favorable consideration given other theories proposed almost simultaneously. Did chemists, for instance, need the atomic theory when Dalton offered it in 1807? The quantitative laws of chemistry preceded the atomic doctrine although they implied its validity. The doctrine was widely recognized as was its proponent; its reception, however, was cool, especially in France, and also in Scotland and England. If chemists could do without atomic weights – as did many of Dalton's contemporaries – maybe they had no need of an atomic theory.

It is then possible to argue that chemistry in the early part of the nineteenth century had little sympathy – as the failure, too, of Prout's hypothesis indicates – for mere speculation. If viewed in these broader terms, the attitude of chemists towards the molecular hypothesis becomes more understandable.

Even in Italy the recognition of Avogadro's hypothesis came very late. As earlier mentioned, in their 1856 obituary, the editors of *Nuovo Cimento* recalled the studies of Avogadro on the specific heats, the law of compression of gases, the atomic volumes, electrochemistry, and the chemical theory of electricity.[117] The four-volume treatise of physics was mentioned, but nothing was said of the 1811 essay formulating both the equal volumes–equal numbers generalization and the existence of polyatomic molecules.

At the same time in the courtyard of the University of Turin a bust was dedicated to Avogadro's memory. According to Guareschi,[118] no Italian scientist present pronounced a single word of praise for the achievements of the person being honored. One is tempted to conclude that even Stanislao Cannizzaro who, just the year before had been made professor of chemistry at the University of Genoa, was, in 1856, still unaware of the relevance of Avogadro's work. His silence at the latter's death cannot be explained otherwise. From 1851 to 1855, Cannizzaro was teaching in Alessandria, a town only 50 miles from Turin, where the proponent of the molecular hypothesis was living after his retirement.

Also of interest was an excerpt published in *Nuovo Cimento* in 1857 for a course in organic chemistry.[119] The course was offered by Raffaele Piria,[120] a pupil of Dumas, and given at the University of Turin, the same institution where Avogadro had taught until 1850. It dealt with the biatomic nature of the chemical elements and with the behavior of such biatomic molecules when they combine. Rather surprisingly, these concepts are described as part of a 'General Law which had been vaguely expressed earlier by many chemists and especially by Gerhardt'. Under the influence of the French chemical school, Piria, while quite familiar with the views expressed by Gerhardt and his friend Laurent, seemed to have completely overlooked that the polyatomicity of elementary molecules had been proposed almost half a century earlier by Avogadro.

In his excerpt Piria rationalized the chemical reasons behind his assumption of the existence of diatomic molecules in hydrogen, chlorine, and nitrogen. When discussing, for instance, the action of hydrochloric acid on hypochlorous acid (at the time considered ClO), he said that it was necessary to recognize that a compound may act more vigorously upon another compound than if it were in a free state. This explained why a very stable compound such as HCl was decomposed by the oxygen present in hypochlorous acid to give water and chlorine, a reaction showing, for him, the biatomic nature of the latter.

To illustrate the biatomic nature of hydrogen, Piria mentioned the reaction between copper hydride and hydrochloric acid, giving copper chloride and free hydrogen in molecular state. In 1811 Avogadro had, more simply, used the combination of hydrogen and oxygen to form water, as a proof that *both gases* are polyatomic. However, in Piria's time, water was still considered HO and, moreover, the biatomic nature of oxygen had not yet been fully established; in fact, Piria did not even include oxygen among the biatomic elements he reviewed.

He accounted for the presence of two like atoms in the molecules of hydrogen and chlorine by considering one atom as electropositive and the other as electronegative, without further elaborating this idea which, of course, was still highly debatable.

Giua,[121] in examining the sources of Cannizzaro's 'Sunto' which appeared in March 1858 shortly after Piria's article, intimated that the two may have discussed the subject, particularly since they were on friendly terms.[122] If this was the case, it shows that not until 1857 had Cannizzaro, the first among Italian scientists, rediscovered the originality of Avogadro's hypothesis and understood its meaning. This was proven the following year when, in his famed essay,[123] Cannizzaro successfully solved some of the major chemical problems of his generation.

If would be unfair to justify Avogadro's very limited reputation south of the Alps on the basis that his major papers were written in French. Not only did Avogadro publish extensively in Italian, including his voluminous treatise of physics,[124] but also one should remember that in the first half of the nineteenth century, French was well-known and used throughout the peninsula by most of the educated people. The widespread knowledge of French may in part account for the surprising delay in the Italian translation of the 'Essai d'une manière' which appeared only in 1901.

The limited circulation of the *Journal de Physique* in Italy has also been suggested as a reason for the long neglect of the molecular hypothesis. Though this is true, one can still assume that the 1811 article reached a fair number of concerned chemists and physicists of the country.[125]

A further explanation may be found in the unstable political situation which afflicted the entire peninsula for almost half a century after 1812. It was a period of continuous upheaval. The restoration of the 'anciens régimes' was followed by the first liberal revolutions in the twenties and their repression, and finally by the unity and liberation of Rome in 1870. Supported by only a small minority of young, middle-class intellectuals,[126] the idea of making a country out of seven component States took a long time to attract even a limited following. Even the unification reached under the French during Avogadro's early years was too short-lived to help modify a situation which had originated in the fall of the Roman Empire.

As a result, it is possible to understand why, throughout most of the nineteenth century, the Italians, unlike the British and the French, had no national consciousness. Men, such as Berthollet, Gay-Lussac, Ampère, Laplace, Davy, Dalton, and Faraday, were widely recognized and appreciated in their respective countries, while as early as the seventeenth century Marcello Malpighi, and later on Lazzaro Spallanzani, Luigi Galvani, Giovanni B. Beccaria, and Alessandro Volta, among others, were better known beyond the Alps than in Italy.

Only in 1839 in Pisa did the first meeting of scientists from different Italian regions take place; in the following years similar gatherings were convened in Turin[127] and again in the Grand Duchy of Tuscany, then ruled by one of the less narrow-minded of the Hapsburg princes, Leopold II. Quite characteristically, however, most of the participants came from the northern regions, with only two percent from Naples and Sicily.[128]

The situation of political disarray and wide cultural fragmentation, such as that observed in Italy in the first half of the nineteenth century, may explain the lack of any cultural manifestation at the national level, from painting to theater, from architecture to science. The thesis may even be supported that had Avogadro's scientific merits been fully recognized, his death would still have been scarcely noticed by his countrymen. The provincialism of the Italians did not encourage them to look too far beyond the borders of their own regions. In the case of Avogadro, the aversion to travel, even inside Italy, limited his personal relationships and these he maintained mostly through correspondence. He was, however, on friendly terms with Ottaviano Mossotti, a Piedmontese emigré,[129] and Carlo Matteucci[130] who were both teaching at Pisa. Years later, in discussing his own theory of dielectrics, Mossotti quite fairly recognized[131] that the concept of polarization of dielectrics at the molecular level had been proposed by Avogadro as early as 1807.[132] It is quitely likely, however, that he became aware of these old essays only after Avogadro had established his priority over Faraday in 1842.[133]

Without taking into proper perspective also the human and historical components which may have affected, throughout his long lifespan, Avogadro as a man and as a scientist, it would be difficult to fully understand the etiology of the neglect of the molecular hypothesis. Pure scientific considerations are just inadequate for this task.

NOTES

1. 'Essai analytique sur l'électricité', ms, 1803, and 'Considérations sur la nature des substances connues sous le nom de sels metalliques . . . ', ms, 1804. The only two items which Avogadro wrote jointly with his brother Felice. See chapter 1, note 26.
2. The editor of the *Journal de Physique, de Chimie et d'Histoire Naturelle*. Born in 1743. His modest rank in Imperial France's scientific circles, in view of his strong criticism of Lavoisier's new chemistry, is described by M. Crosland, 'H. Davy: An alleged case of suppressed publication', *Br. J. Hist. Sci.* 6 (1973), 308. His obsolete scientific stand may explain why the leading French chemists of the time did not publish their works in de La Métherie's *Journal* and why Avogadro found easier access to this publication than to the more respected *Annales de Chimie*. The *Journal* ceased to be published in 1823, not too long after de La Métherie's death in 1817.
3. Full title: 'Essai d'une manière de determiner les masses relatives des molécules élémentaires des corps, et les proportions selon lesquelles elles entrent dans ces combinaisons', *Journal de Physique, de Chimie et d'Histoire Naturelle*, 73 (1811), 58.
4. And denied by Dalton himself in 'Remarks on the essay of Dr. Berzelius on the cause of chemical proportions', *Ann. Phil.* 3 (1814), 175.
5. This is my own translation. I found the *Alembic Club Reprint*, no. 4 translation less true to the original text.
6. In an 1815 manuscript, Berthollet, too, criticized Dalton's rule of simplicity, while taking an agnostic position on the number of combining atoms. See M. P. Crosland in D. S. L. Cardwell ed., *John Dalton and the Progress of Science* (Barnes and Noble, 1968), p. 280.
7. In the case of water, for instance, Dalton calculated the weight based upon one atom of hydrogen and one of oxygen: $6 + 1 = 7$; for Avogadro, on the other hand, two molecules of hydrogen plus one of oxygen gave a mass of 17.074 for two molecules of water and thus 8.537 for each molecule.
8. For Avogadro's views on this, see also his 'Mémoire sur les masses relatives . . .', *J. Phys.* 78 (1814), 154, and for Dalton, *Ann. Phil.* 3 (1814), 175.

9. *J. Phys.* 73 (1811), 59.
10. In a memoir written in 1817 and published three years later in *Mem. Mat. Fisica Soc. Ital. Scienze,* 18 (1820), 153, Avogadro further elaborated this point of his theory. He was attempting to correlate the amount of caloric present in a gas with the affinity of the same gas for caloric. He assumed that at a given temperature and pressure, the spheres of the molecules in gases were the same, and that the same molecules placed at the center of these spheres occupy a certain portion of their total volume. These molecules were placed at the center of the spheres, where the density of the caloric must be indefinitely large. He assumed also on this occasion that the volume of the molecules was likely proportional to their mass, which meant, he added, that the density of the matter of the molecules of all bodies was the same and that they differ only by their mass inasmuch as they differ in their volume. Therefore, the volume occupied (site) by each gaseous molecule within the caloric sphere would be proportional to the mass of said molecule. [If, in $(m/v) = d$, d is the same when m and v are proportional]
11. I think this contradicts P. Achinstein's statement about Avogadro in *Law and Explanation* (Oxford Press, 1971), p. 152.
12. *A New System of Chemical Philosophy*, part I, 1808, p. 187.
13. Ibid., pp. 70–71.
14. Dalton, quite strong-minded about his scientific views, was prone to definite statements. See, for instance, *A New System*, part I, p. 213.
15. *Ann. Phil.* 3 (1814), 180. The Berzelius paper was published in *Ann. Phil.* 2 (1813).
16. See *Essay on Chemical Statics*, transl. B. Lambert (London, 1804), p. 117.
17. Laplace's comments appear as Notes V and XVIII at the end of the *Statique Chimique*; they are not dated.
18. *Essay on Chemical Statics*, p. 129.
19. Ibid., p. 181.
20. Ibid., p. 218.
21. Ibid., p. 219.
22. However, as Gay-Lussac mentioned in his 'Mémoire sur la combinaison . . . ', p. 227, this *rapproachement* could hardly account for another experimental result, i.e., the fact that carbon oxide is lighter than oxygen, although it is formed by hundred volumes of 'gaseous carbon' combined with fifty volumes of oxygen. Besides, he added, there are other instances of gaseous combinations, whose constituent molecules would be very 'condensed' upon each other, and yet in which one observed either no decrease of volume or even an increase of volume upon their occurrence. See *Mem. Phys. Chim. Soc. Arcueil,* 2 (1809), 227.
23. Berthollet, *Chemical Statics*, p. 221.
24. He quotes a memoir of 1788 submitted by Monge to the Académie.
25. Berthollet, *Chemical Statics*, p. 22.
26. Ibid., p. 229.
27. Ibid., p. 276.
28. Among his manuscript notes I have found an undated handwritten reproduction of several entire sections of the *Statique Chimique*. It appears as if at the time he had only limited access to the printed work and, as a result, he took extensive notes for reference.
29. *Mem. Phys. Chim. Soc. Arcueil,* 2 (1809), 208.
30. Ibid., p. 226. Also *Alembic Club Reprint*, no. 4, p. 20.
31. Dalton discussed ammonia and its composition in vol. 2 of the *New System*, but omitted any reference to the value of its density in relation to those of its components. For Dalton, ammonia was formed by one atom of nitrogen and one of hydrogen. He seemed to propose three weights for this compound: 6, 12, 13

(p. 353). The last one could stand for NH_3 if one takes nitrogen = 10 instead of 5, as indicated in the table.

32. *Mem. Phys. Chim. Soc. Arcueil,* 2 (1809), 224. The composition of water had been extensively studied by Gay-Lussac and Humboldt and their conclusions submitted to the Institute in January 1805.
33. Ibid., p. 219.
34. Ibid., p. 227.
35. Ibid., p. 228.
36. Ibid., p. 229.
37. Ibid.
38. Ibid.
39. Gay-Lussac was more correct than Avogadro, because he assumed for mercurous oxide a metal/oxygen volume ratio = 2/1, while Avogadro took 1:1. Since the mercury/oxygen ratio by weight gave 100:4.16, then for Avogadro mercury weighed 24.02 times oxygen or 362 times hydrogen, twice the figure assigned to it by Gay-Lussac.
40. Somehow, this argument seems odd in line with what Gay-Lussac himself had earlier and almost triumphantly characterized as 'the era when we shall be able to submit the bulk of chemical phenomena to calculation' *Mem. Phys. Chim. Soc. Arc.* 2, (1809), 208.
41. This has been pointed out by John Bonner in his 1974 Ph.D. dissertation (p. 150), Johns Hopkins University.
42. The MS of the 1811 essay was examined in detail by M. A. Morselli in *Ambix* 27, (1980), 147.
43. Recently L. Nash and F. Greenaway have shared this opinion, while Ihde takes a less critical position. C. Graebe, Der Entwicklungsgang der Avogadroschen Theorie, *J. Prakt. Chemie,* 87 (1913), 145, has no comments on this point, which is not raised by I. Guareschi, M. Giua, and V. Cappelletti. For Cappelletti see *Dizionario Biografico degli Italiani,* IV, 689. (Ediz. Enc. Ital. Rome, 1962.)
44. *J. Phys.* 78 (1814), 131.
45. As indicated by L. Pauling in his commemorative speech in Rome, June 1956, and published in *Science,* 124 (1956), 708–13.
46. If $V_A = V_B = V_C$ and $n_A = n_B = n_C$ then $dA/dB = mA/mB$ and $dA/dC = mA/mC$ and if $m_A = n_A \cdot w_A$ and $m_B = n_B \cdot w_B$, then $mA/mB = wB/wB$, etc. (where V, n, d, m, and w represent volume, number of molecules, density, mass, and weight of each particle of gases A, B, and C).
47. Oxygen and hydrogen, for instance, had densities relative to air of 1.1035 and 0.0732, and for the same volume of the two gases, the ratio of their densities represented the ratio of their masses. Thus, oxygen had a mass 15.074 times that of hydrogen taken as 1. And in the same way, nitrogen had thirteen times the mass of hydrogen, nitrous gas 14.156, nitrous oxide 20.775, etc.
48. Dalton included Gay-Lussac among the supporters of the equal volumes–equal numbers hypothesis (see *New System,* vol. 1, part II, 1810, p. 558).
49. Ibid., pp. 70–71.
50. This is a corrected version of the *Alembic Club* translation.
51. 'Molecula', in the sense of a small mass, as a form of aggregation of atoms or as an intermediate entity between large bodies or masses and the atoms, had a long tradition in French science. In the first section of his *Sintagma philosophicum* (1658) and more specifically in the third book ('De materiali principio'), Gassendi introduced the term molecule, following the Lucretian tradition: 'Sed uno verbo diceret ille, non fieri omnia ex omnibus quod non omnes Atomi uniusmodi sint neque propterea capacitatem ad corpora eadem conformanda habeant. Heinc ex Atomis confermari primum moleculas quasdam inter se diversas quae sint semina

rerum diversarum . . . '. (*Opera omnia*, 1658, vol. 1, sec. 1, Liber III, p. 282.)

Gassendi's atoms were not only simple and indestructible, but also different in size, shape, and weight. In their continuous movement favored by the vacuum surrounding them, they collided frequently with each other, and the molecules were then the result of these collisions. In turn, molecules mutually combined to form bodies which could then be perceived by our senses.

Almost a century after Gassendi, the term molecule appeared in a work written by one of the staunchest French supporters of Flogiston, Antoine Baumè. In his *Manuel de Chymie*, published in Paris in 1763, he discussed chemical combinations in which 'integrant molecules' were placed close to each other. He also called them 'parties integrantes', since they represent the smallest molecules of a body still conserving its properties (see p. 18 of *Manuel*). This definition was eventually adopted to indicate the compound molecules in general.

Molecule, without further qualification, was found in Lavoisier's *Mèthode de nomenclature chimique*, published twenty-four years after Baumè's *Manuel* (see also Pierre Macquer: Lavoisier mentioned 'molecules of elastic fluids' and 'molecules of caloric', meaning, needless to say, only small masses. On the other hand, Guyton de Morveau, in his *Elemens de chimie thèorique*, vol. I (Dijon, 1777), p. 20, referred to 'partie', rather than to molecule. He said: 'le plus petit atome d'un sel est une partie intégrante du crystal salin', which was a restatement of the earlier definition given by Baumè. Guyton, however, also used 'parties constituants'.

After the turn of the century, the distinction between 'molécules intégrantes' and 'molécules constituantes' appeared in the *Système des connaissances chimiques*, written by Fourcroy and published in 1801. He defined as 'intégrantes' those molecules which were all of the same chemical nature and were kept together by the aggregation force (vol. I, p. 65), while the 'constituantes' were those of the 'principles' from which a compound was formed and which could be separated by chemical analysis.

A few years later, in the *Mémoire sur la combinaison*, the term 'molécule élémentaire' was used by Gay-Lussac. He said that these, by approaching each other, formed a chemical combination according to the 'law of chemical affinity'. In the same essay and in a similar context, Gay-Lussac also spoke of 'molécules constituantes'.

It is worth remembering that towards the end of his *Mémoire sur la combinaison*, Gay-Lussac, in mentioning 'Dalton's ingenious idea that combinations are formed from atom to atom', added: 'the various compounds which two substances can form would be produced by the union of one molecule of the one with one of the other' *Mem. Phys. Chim. Soc. Arcueil*, **II** (1808), 231. The implication is clear enough. He seemed to equate the term 'atome' with 'molécule'. This represents, probably, an early instance of a misuse of the two terms, leading to the confusion which would become more apparent in the following decades and would have serious repercussions on the acceptance of Avogadro's hypothesis. Even Avogadro, in some of his physical essays after 1830 confused these terms.

52. See later on for Avogadro's qualification of molecule.
53. Greenaway in *John Dalton and the Atom* (Cornell Press, 1966), p. 175, correctly examines this point.
54. *J. Phys.* 73 (1811), 61.
55. Ibid., p. 14. B. W. Mundy in *Chymia*, 12 (1967), 152, examined a few examples of 'submolecularity' of molecules of compounds among those cited by Avogadro. Mundy's conclusion that, on the basis of the molecular constitution of nitrous acid, both oxygen and nitrogen had tetratomic molecules seems to agree with the facts. In the case of gas euchlorine (chlorine monoxide and dioxide) which was discussed by Avogadro in the 1814 memoir on the basis of Davy's unreliable analytical data, the assumption of a pentatomic molecule for chlorine made by Mundy is unjustified

in my view, as the formation of hydrochloric acid suggested by Avogadro in 1811 indicates. Moreover, in 1821 ('Nouvelles considérations') he set aside the possible division of a molecule by thirds or fifths as not compatible with his system.

56. Only in 1837, in the first volume of the *Fisica dei corpi ponderabili*, in the first chapter of the first book, did Avogadro discuss molecular attraction in detail; however, even on this occasion, with the exception of a note of the speculations proposed a few years earlier by Ampère on the nature of the oscillation of atoms within integral molecules, he did not give an account of his own views.
57. The mass of the integral molecule of sulfur is determined from the densities of sulfurous acid and oxygen and by assuming a division of the molecule of sulfurous acid formed: $(d_S + 2d_O)/2 = d_{SO_2}$ thus $d_S = 2d_{SO_2} - 2d_O = 2.324$ and $2.324/0.073 = 31.73$ (mass relative to hydrogen).
58. *J. Phys.* **73** (1811), 60. *Alembic Club Reprint*, p. 42.
59. In order to do so he would have to further divide the carbonic acid molecule by four or by eight.
60. See *Mem. de Phys. et Chim. Soc. Arcueil,* **2** (1809), 339.
61. Ibid., 357.
62. *J. Phys. Chim. Hist. Nat.* **71** (1810), 326.
63. How Avogadro arrived at this result is characteristic of his procedure: ratio of density of muriatic acid to density of hydrogen = 1.278/0.0732 = 17.45, mass of one molecule of muriatic acid. But he knew that one volume of hydrogen and one of oxymuriatic acid gave two volumes of muriatic acid; therefore $2 \cdot 17.45 = 34.90$ and $34.90 - 1$ (hydrogen) = 33.90 mass of oxymuriatic acid.
64. *J. Phys.* **78** (1814), 132.
65. In his 'Entwicklungsgang der Avogadroschen Theorie', Graebe reviewed in detail the objections to Avogadro's hypothesis still being raised seventy years after its formulation. The main criticisms were based on the anomalous vapor densities observed in a number of solid and liquid compounds. See *J. Prakt. Chemie,* **87** (1913), 145.
66. A. Meldrum, *Avogadro and Dalton* (1904), p. 39; T. Lowry, *Historic Introduction to Chemistry* (1915), p. 343; A. Wurtz, *Histoire des Doctrines Chimiques* (1868); J. R. Partington, *History of Chemistry*, vol. IV (1964), p. 216; N. G. Coley, 'The Molecular Hypothesis', Dissertation (Leicester, 1964), Appendix V.
67. *J. Phys.* **78** (1814), 147.
68. John Davy's researches on 'fluorium', as Avogadro had called the radical of fluoric acid, had been published in *Annales de Chimie*, May 1813. Only in 1823 did Berzelius isolate fluorine, which had been tentatively identified sixteen years earlier.
69. For instance by I. Guareschi in his *Discorso storico-critico* (Turin, UTET, 1911) and J. R. Partington, *History of Chemistry*, vol. 4, pp. 4, 215. Recently S. H. Mauskopf, *Isis,* **60** (1969), 65, gave credit for this to Gaudin who suggested SiO_2 only in 1833, *Ann. Chim.* 2, **52** (1833), 126.
70. *J. Phys.* **78** (1814), 151.
71. According to current figures, Avogadro is incorrect for P, F, B, Au, Pb, Cu, Sn, Sb, As, K, Na, Ba, Al, Sr; Berzelius is incorrect for P, Hg, Mn, Ag, Cu, Sb, As, K, Na, Si, Sr, Mg.
72. Reported by Dulong in *Ann. Chim.* **86** (1813), 37, although the actual discovery had taken place in 1811.
73. Avogadro assumed that the density (relative to air) of gaseous carbon is 0.832. By adding twice the density of hydrogen (2×0.0732) the result was 0.9784, while the experimental value was 0.978.
74. 'Mémoire sur la combinaison . . . ', *Mem. Phys. Chim. Soc. Arcueil,* **2** (1809), 233 (*Alembic Club Reprint*, trans., pp. 10–11).

75. Avogadro confuses ammonium bicarbonate with the neutral salt (as the analysis indicates), an error made also by Gay-Lussac. He does not mention, however, Gay-Lussac's observation that 'the neutral carbonate of ammonia can only exist through the medium of water', (*Alembic Club Reprint*, trans., p. 11).
76. 'Essay on the combinations . . . (*Alembic Club Reprint* trans , p. 11.)
77. *J. Phys.* **69** (1809), 142.
78. Only recently J. K. Bonner in his dissertation ('Amedeo Avogadro: a reassessment of his research and its place in early nineteenth century science'), Johns Hopkins University, Xerox Univ. Microfilms, 1974, p. 130, has discussed it in great detail.
79. For instance, sulfur would show its acidic nature, were it not for its great chemical cohesion.
80. *J. Phys.* **78** (1814), 153.
81. Ibid., p. 154.
82. 'Gravitational' is used here by inference, since Avogadro does not employ this term. He says ' . . . quoique leur attraction soit moindre entre elles molécules en raison de la petitesse de la masse . . . ' *J. Phys.* **78** (1814), 153.
83. *J. Phys.* **78** (1814), 154.
84. 'Lettre de M. Ampère à M. le comte Berthollet . . . ', *Ann. Chim. Phys.* **90** (1814), 43–86, 1st ser.
85. See p. 13 of 'Nouvelles considérations', *Mem. R. Acc. Sci. Torino*, **26** (1821), 1–162.
86. See *Fisica dei corpi ponderabili* (Torino: Stamp. Reale, 1837–41), vol. I, chap. 4, 305.
87. *J. Phys.* **78** (1814), 155.
88. Three articles submitted by Davy to Gay-Lussac and Thenard in 1810 for publication in *Annales de Chimie* were not accepted. As a result Davy asked de La Métherie to publish them in *J. Phys.*, appealing to him as a 'lover of truth and justice'. J. Z. Fullner, *Sir H. Davy's Published Works* (Harvard Press, 1969), p. 60.
89. T. M. Lowry, *Historical Introduction to Chemistry* (Macmillan, 1916), p. 343.
90. For N. G. Coley ('The Molecular Hypothesis', Diss., Leicester University 1964, p. 5) Berzelius almost certainly had never heard of Avogadro's hypothesis. This implies that the Swede had not read the *Journal de Physique*, which is implausible, since some of his essays were published there, one in 1811.
91. This forms the subject of chapter 5 of this work. Nernst in 1893, Grimaux in 1884, P. Muir in 1883, Roscoe in 1898, all came forward in full support of Avogadro's priority over Ampère.
92. See, for instance, U. J. Kay-Shuttleworth, *First Principles of Modern Chemistry*, (London, 1868). '. . . the most valuable evidence of its atomic constitution is afforded by the remarkable law, the earliest conception of which is attributable to Gay-Lussac, though it was first enunciated by Ampère, namely that all gases and vapours contain the same number of molecules within the same volume'. H. Debus in *Uber Einige Fundamental Satze der Chemie* (Kassel, 1894), considers Dalton, in 1801, the author of the equal volume–equal number generalization. Guareschi in his *Discorso storico-critico*, p. 39, recalls that the molecular hypothesis was also attributed to Laplace, Lavoisier, Bernoulli, and Kant. For this see *La Philosophie de la Nature dans Kant*, by C. Andler and Chavannes. In 1833 Gaudin credited Ampère, although he knew that Ampère had recognized Avogadro's priority. D. Knight's view, *Atoms and Elements*, (Hutchinson, 1967), p. 90 that Avogadro was an 'established scientist', does not seem valid, if referred to 1811.
93. This subject is examined in detail in chapter 4. G. Buchdahl in 'Sources of Scepticism

in Atomic Theory', *Brit. J. Phil. Sci.* **10** (1959), 126, mentions Berzelius' criticism as an example of an hypothesis 'impugned for an apparent lack of plausibility'.

94. See A. E. McKenzie, *The Major Achievements of Science*, vol. I, p. 142.
95. F. Greenaway, *John Dalton and the Atom*, Cornell Univ. Press 1966, p. 174, raises this point which is shared also by L. K. Nash, *Atomic-Molecular Theory*, Harvard Case Histories, vol. I (Harvard, 1957), 217–321.
96. Coley's view that Avogadro's concepts were clear beyond doubt, may be acceptable only to the reader who knows in advance what the author wishes to express. I am wondering if this was true at the time when the 'Essai d'une manière' appeared. F. Greenaway feels that Avogadro's obscurity was the most important reason for his neglect. (See also fn. 90)
97. Avogadro's articles in the *Annales de Chimie* (the first appeared in 1813) dealt mainly with physical subjects.
98. This view is supported by B. Jaffe, *Crucibles* (Schuster, 1930), p. 165. Iodine was actually isolated by Courtois at the end of 1811.
99. For instance, Avogadro, on the basis of data reported for their oxides, gave to copper, iron, and silver values double those eventually accepted. On the other hand, he was amazingly close to the correct figures for lead and mercury. This will be reviewed in detail in examining the 'Mémoire sur les masses relatives' (1814).
100. See E. von Mayer, *History of Chemistry* (Macmillan, 1906), p. 225. According to N. G. Coley, Diss., 1964, Appendix V, this can be considered an example of Avogadro's lack of understanding of the chemical tradition of his time.
101. *Avogadro and Dalton and the Standing in Chemistry of Their Hypotheses* (Edinburgh: W. F. Clay, 1904).
102. Icilio Guareschi (1847–1918) studied under Francesco Selmi, founder of the inorganic colloid chemistry, and became professor of pharmaceutical chemistry at the University of Turin in 1879. His research in organic chemistry ranged from thioaldehydes to asparagine, pyridine, and naphthalene derivatives. He is the most prolific and best known of the Italian historians of chemistry.
103. Published in 1911 as part of *Opere scelte di A. Avogadro* (Turin: Reale Accad. Scienze).
104. Graebe, then seventy-two, had retired after teaching at Zurich, Geneva, and Frankfurt. He was an organic chemist and is well known for his research in derivatives of anthracene, phenanthrene, pyrene, and crysene. In 1869 he synthesized alizarine for the first time with Liebermann. 'Der Entwicklungsgang der Avogadroschen Theorie', *J. Prakt. Chemie,* 87 (1913), 145, is a detailed chronology of events.
105. Wurtz, *Histoire des doctrines chimiques depuis Lavoisier* (Paris, 1868).
106. *Comptes Rendus*, **84** (1877), 977, 1183, 1262, 1347, 1349.
107. See *La Théorie Atomique* (Paris, 1879).
108. Published between 1843 and 1847. In 1863-65 H. Kopp was the author of the generalization relating molecular and atomic heats.
109. See *Die Entwickelung der Chemie in der Neuren Zeit* (Munchen, 1973), p. 353.
110. A good number of examples could be taken from the history of the development of scientific ideas to prove just the opposite of Kopp's contention. To remain within the domain of this study, one may recall that Gay-Lussac's first important paper in which he enunciated the law correlating temperature to the volumes of gases, appeared in 1802, when he was only twenty-four and little known. As for the conjectural approach to the definition of a theory, it has been said recently: 'The origin of the new scientific ideas are attributed to induction from factual observations whereas, in many cases they really arose from flights of speculative fancy entirely worthy of the epithet poetic.' See L. P. Williams, 'The Physical Sciences in the 19th Century', *Hist. Sci.* **1** (1962), 9.

111. Published in Leipzig, 1889. A very successful book with three editions in Germany alone. It was translated into several languages.
112. 'As will presently be seen, the fruits of indiscriminate attention to figures or of neglect of principles are inconsistency, obscurity, disorder and confusion in the history of chemistry.' Taken from A. Meldrum, *Avogadro and Dalton* (Edinburgh, 1904), p. 84 ff.
113. *Nature*, 178 (1956), 8.
114. In his speech before the Academy of XL in Rome, 6 June 1956.
115. Vol. 2, (Turin, Stamperia Reale, 1837), 843 ff. See J. R. Partington *History of Chemistry*, vol. 4, Macmillan, 1964, p. 217.
116. N. W. Fisher, 'Avogadro's Hypothesis: Why Did Chemists Ignore it?' read at the British Society for the History of Science, Lancaster, July 1973. Published in *Hist. Sci.* **20**, (1982), 77 and 212.
117. *Nuovo Cimento*, 5 (1856), 473. Carlo Matteucci (1811–68), professor at Pisa after 1841, a well-recognized physicist, author of numerous works on the relation between electrical phenomena and muscular activity, recipient of the Copley Medal in 1848. He founded *Nuovo Cimento* in 1854 together with R. Piria. See note 130, below.
118. See Guareschi, *Amedeo Avogadro e la teoria molecolare* (Torino: UTET, 1901).
119. *Nuovo Cimento*, 6 (1857), 24. Title is 'The simple bodies in the free state are made of diatomic molecules; when they mutually react they do not combine, but mutually decompose'.
120. Raffaele Piria (1813–65) was born in Calabria and studied in France under Dumas and Boussingault. He left Dumas in 1835, did research alone, mostly on the structure of salicine, moved to Pisa in 1842 and for political reasons had to leave; was then appointed in 1856 to the chair of organic chemistry at Turin.
121. See M. Giua, *Storia della Chimica* (Turin: Chiantore, 1946), p. 143.
122. Stanislao Cannizzaro, eleven years younger than Piria, had been his assistant at the University of Pisa from 1845 to 1847. M. Melloni, the physicist well known for his studies on the radiant heat, introduced Cannizzaro to Piria who convinced the nineteen year-old Sicilian to study chemistry.
123. 'Sunto di un corso di filosofia chimica fatto nella R. Universita di Genova', *Nuovo Cimento*, 7 (1858), 301.
124. *Fisica dei corpi ponderabili* (Turin: Stamperia Reale, 1937).
125. Probably there were about one hundred Italian chemists and physicists involved in academic work at the time. Italy had thirteen universities in the Napoleonic era, seven in the north, four in the center, two in the south.
126. Cannizzaro himself, at the age of twenty-two, took part in the Sicilian revolution (1848) and, while a fugitive, was sentenced to death by the Bourbon King, then ruling the southern part of Italy. Matteucci, Mossotti, and Melloni were among the young Italian scientists of their generation who found themselves involved in the revolutionary movements of the 1830s and 1840s, and paid the price of exile for their liberal ideas.
127. Avogadro participated in this in 1840.
128. See D. M. Smith, *The Making of Italy* (New York: Harper and Row, 1968), p. 56.
129. Ottaviano Mossotti was born in Novara in 1791 and died in Pisa in 1863. He spent several years in political exile, first in Switzerland and then in England, where he became a friend of T. Young. Later on he moved to Argentina and returned to Italy to teach only in 1835. In 1840 he was appointed to the chair of Pisa. A large portion of his work is devoted to astronomy; less known is his research in optics and on the polarization of dielectrics. His atomic theory published in Britain in 1837 created a great deal of interest, especially in Faraday.

130. In 1838 Matteucci established the existence of a relationship between the amount of electricity passing through a solution of an electrolyte and the amount of electrolyte decomposed. This he accomplished, independently of Faraday, while he was still living in Florence.
131. Mossotti's paper, as reported by Partington, *History of Chemistry*, vol. 4, p. 127, is in *Bibl. Univ. de Genève*, **6** (1847), 193.
132. See *J. Phys. Chimie*, **65** (1807), 130.
133. In the short 'Note sur la nature de la charge éléctrique', *Arch. Electric: Bibl. Univ. Genève*, **2** (1842), 102, which has been reviewed in chapter 2 of the present work.

CHAPTER FOUR

THE MOLECULAR HYPOTHESIS: FROM 1811 TO CANNIZZARO'S 'SUNTO' IN 1858

In the two decades which followed the 'Essai d'une manière', Avogadro published three more essays dealing with the molecular hypothesis: in 1814 the 'Mémoire sur les masses relatives' appeared in the *Journal de Physique* of de La Methérie and the other two in the official organ of the Royal Academy of Sciences of Turin. One may assume that at least the two articles in the Parisian journal should have reached a rather large audience outside Piedmont. Due to the peculiar position of this region, then a Department of the French Empire, there was never a problem of communication between Paris and Turin even during the Napoleonic Wars. As for the 'Nouvelles considérations' and the 'Mémoire sur la manière', they both were published after Waterloo when Europe had been at peace for many years and circulation of news and mail no longer presented a problem.

Quite likely, both the 'Essai d'une manière' and the 'Mémoire sur les masses relatives' came to the attention of that small group of individuals who, in those years, were inquiring into the intimate constitution of matter. The solution to the problem of finding a link between atomism and volumism appeared particularly puzzling. A corpuscular system, although rejected mainly on empirical grounds, was valid; on the other hand, very few people, and Dalton was among them, could challenge the reality and the meaning of the conclusions reached by Gay-Lussac in his 'Mémoire sur la combinaison'. The way to relate these two pivotal standpoints remained elusive. The answer provided by the unknown professor of physics at the Collegio of Vercelli was simply not acceptable: to some, probably, it seemed too easy a way out; to others, it appeared as a visionary solution introducing incompatible and awkward notions. Definitely, the disregard to which the molecular hypothesis fell prey cannot be explained by appealing to the calamities of ignorance,[1] for, after all, the men involved in the issue were highly talented and eminently qualified. Nevertheless, the history of the development of scientific doctrines contains quite a few examples of minds which were brilliant and arrogant at the same time. If to examine new ideas in an open and unbiased way, and to avoid their rejection just because someone else has supported them; if to doubt one's own views, to assess one's own limits, to accept criticism and exercise prudence in passing judgment on another's opinion;

if these represent the traits of the true philosopher, then one must conclude that in the early part of the nineteenth century only a few among those distinctively engaged in the physical sciences could be properly considered so.

In the years following the publication of the molecular hypothesis the scientific community in Europe was still so small that no more than a handful of individuals could have grasped the significance of the 'Essai d'une manière'. Their attitude and reaction to Avogadro's assumption are necessarily important and should be examined in detail.

When the fundamental contribution of the Piedmontese physicist appeared, the atomic theory was three years old; it had been received by the French with varying degrees of skepticism and with criticism in Britain.[2] As for Gay-Lussac's law of volumes, although generally accepted by his countrymen and recognized early by Berzelius, it failed to convince Dalton and the atomists. From the debates which at times filled page after page of the journals on both sides of the Channel, it is apparent that there was plenty of room for dissent. The influence of Berthollet was still so overwhelming that Dulong, in writing to Berzelius in 1820[3], felt more inclined to hold Laplace responsible for the French opposition to the atomic theory than to antagonize Arcueil's grand old man and leading opponent of Dalton. A similar attitude was taken in those years by Gay-Lussac and Thenard.[4]

In Britain, Dalton's corpuscular doctrine was never fully accepted by the most influential and renowned of his contemporaries. Humphry Davy not only objected to the validity of some of the numerical results published in the *New System* – for instance, the atomic weight of nitrogen – but even raised doubts about the originality of the ideas expressed there. His feelings towards Dalton, articulated just a month before Avogadro's Hypothesis appeared in the *Journal de Physique*, were presented thus:

> I shall not enter any further at present into examination of the opinions, results and conclusions of my learned friend: I am, however, obliged to dissent from most of them and to protest against the interpretation that he has been pleased to make of my experiments; and I trust to his judgment and candor for a correction of his views.[5]

Given his mistrust of any speculation set forth just to explain observable facts, it is not difficult to imagine what would have been Davy's reaction to the molecular hypothesis had he known of it. After emphasizing once again his aversion to any unproven conjecture, he says: 'Mr. Dalton has arranged, combined, weighed, measured his atoms, but it is not, I conceive, on any speculations upon the ultimate particles of matter that the true theory of definite proportions must ultimately rest.'[6] In 1826 the President of the Royal Society,[7] although recognizing the merits of its author, could not yet express confidence in the atomic doctrine, whose full significance – it should be recalled – was still forty years later a subject of debate in the British Isles.[8]

Berzelius

Most significant, however, remained the position of the man who in those years, among the continental chemists, came closest to Avogadro in his attempt to associate the corpuscular system with the law of combining volumes. For his early recognition of the validity of the latter and, at the same time, for his appreciation of the former, Berzelius appeared as particularly qualified to assess the full implications of the molecular hypothesis.

Unlike Davy, the Swedish chemist had in 1811 drawn a very favorable conclusion from his first contact with the Daltonian theory. As he wrote then, it represented 'the greatest advance that chemistry had ever yet made in its development'.[9] It was a conditional statement, however, because Berzelius could not admit the validity of such a remarkable conjecture based only on very limited experimental evidence.[10] So much proof had accumulated in the following year that Berzelius could write to Dalton, 'All the results so far obtained have contributed to justify this hypothesis.'[11] On the basis of further experimental results, however, he dramatically reversed his position in 1813, when he admitted that the atomic doctrine 'can neither be adopted nor considered as true'.[12] He also recognized the possibility, through Gay-Lussac's generalization, of applying the law of definite proportions to the gaseous combinations.

This was the time when for the Swedish chemist the theory of volumes assumed an even greater authenticity than the Daltonian system, because it could be proven by experiment. This was also the occasion when the equivalence atoms–volumes was set. By establishing a close relationship between the law of definite proportions and that of combining volumes, Berzelius moved quite close to the main premise of Avogadro's 1811 essay. However, while he realized the advantage offered by the notion of divisible volumes over that of indivisible atoms, it did not occur to him that such a notion could be the correct answer to the well-known objection raised by Dalton to the equal volumes–equal numbers generalization.

Early in 1814 came Dalton's often quoted reaction to these considerations: how could the atomic theory be helped by substituting one term – volume – with another, atom? The following year, in his 'Address to Chemists'[13] Berzelius, after expressing his surprise at Dalton's criticisms of some alleged inaccuracies in a previously reported analysis of oxalic acid, set his view straight: first of all, he had no intention of refuting the corpuscular theory, but only of confirming it through experimental evidence; then, that the volume theory was not intended to replace the atomic theory, since for him 'both are absolutely the same thing' and finally that the only difference between the two was represented by the 'state of aggregation of the elements'.

On Dalton's side there was undoubtedly a confusion *in verbis*, because the point raised by the Swedish chemist seemed quite unequivocal: volumes applied to gases and atoms to solids, and there was no reciprocal substitution of these terms. One may then sympathize with Berzelius' reaction when the 'Annual Review' of the most important events in the physical sciences appeared in the January 1815 issue of the *Annals of Philosophy.*[14] In referring to the volume theory, T. Thomson, after paraphrasing Dalton's opinion, concluded with the ironical comment that the standing of the Stockholm scientist in those years was such that 'he may be indulged in any innocent whim which produces no deterioration'.[15] Thus, four years after the 'Essai d'une manière', not only had Dalton maintained his original critical position toward Gay-Lussac, but also his dialogue with Berzelius was becoming more confused and less fruitful.

As has been observed, on the basis of the very simple conclusions drawn from Gay-Lussac's 1809 essay, the author of the atomic doctrine found himself facing a dilemma: since the atoms could not be divided, the only alternative left to explain the simple combining volumes' ratio was to assume as incorrect the experimental data proposed by the French. From those then involved in researches on gases came reports so discordant and results so confusing, especially when involving compounds of nitrogen with oxygen,[16] that it seemed rather easy for the Manchester physicist to challenge the validity of Gay-Lussac's work. Naturally, he accepted the evidence of the figures when they did not conflict with his fundamental assumption; in an 1817 paper[17] he agreed with Davy and Gay-Lussac regarding the even ratio displayed by the progressive amounts of oxygen combining with a set amount of nitrogen; but he certainly differed from their conclusions in the nitrogen/oxygen relative ratios observed for the same group of compounds.

While Dalton, in order to support the validity of his theory, chose to reject the law of combining gases, Berzelius, after the formulation of his dualistic system, faced a no less thorny predicament. A dualistic interpretation could be useful for most inorganic compositions, but certainly it did not help to explain how gaseous elements combine with each other in the simple ratios described by Gay-Lussac. The polyatomic nature of nitrogen, hydrogen, and other elemental gases might be the solution;[18] however, such complex molecules, formed by identical atoms, seemed decisively at odds with the dualistic theory. Faced then with the alternative of modifying his system, if only for a few elements, or of retaining, even for the elemental gases, a monoatomic structure, Berzelius decided in favor of the latter course.

It has been remarked, quite correctly, that by doing so he incurred a conceptual inconsistency.[19] He found it convenient to use Gay-Lussac's generalization only for the gaseous compounds which could be interpreted in dualistic terms, while in fact it covered both gaseous elements and compounds.

Before 1820 Berzelius came within reach of the notion of polyatomic particles in elemental gases. By suggesting the advantages offered by fractionable volumes over indivisible atoms, he probably realized that for gaseous elements, at least, it would eventually become necessary to introduce a new concept. He, therefore, used the term 'elementary volume' to indicate the smaller part of an elemental gas and this, unlike the Daltonian particles, could be divided. Then, the emerging conflict with his own dualistic theory did not allow any further elaboration of the subject. It may be interesting, however, to observe that later in his life he took a much less rigid position on the subject. It is only a matter of *nuances*, but even these should not be neglected in examining the writings of a man who was always very careful in expressing his opinions.

In the fifth edition of his classic *Lehrbuch*[20] Berzelius devoted only a few paragraphs – out of more than a hundred pages in the introductory section – to the method of correlating the ratios by weight of the ultimate particles of gaseous elements and compounds to the ratios of their volumes. Though generally unnoticed, these considerations are important for the history of the molecular hypothesis because they reveal that, despite Dumas' faulty interpretation of it, Berzelius had not entirely abandoned the key notion of a molecular division.

After recalling how he had arranged the gases in three groups according to their physical behavior[21] he offered his views. First of all, he said, gases are formed by mutually repelling compound atoms or simple atoms; this mutual repulsion, when not limited by the walls of the containers, is balanced by the particles' weight; for this reason, the atmospheric gases, although rarified at high altitudes, cannot escape the earth's gravitation.

Next he gave his interpretation, in terms of the atomic theory, of the gases thus classified:

> It is necessary to admit that the permanent gases of elementary bodies – oxygen gas, hydrogen gas, and nitrogen gas – contain for the same volume the same number of atoms. . . . Moreover, one admits with as much probability that the coercible gases of simple bodies enclose in equal volumes the same number of atoms as the permanent gases; this law, however, does not apply to nonpermanent gases of simple bodies.[22]

Berzelius had thus excluded from the equal volumes–equal numbers generalization not only the molecules of the compound gases, regardless of the nature of their components, but also the molecules of both permanent and coercible elemental gases as well as the atoms of the nonpermanent elemental gases.

These conclusions would seem in agreement with his earlier interpretation of the same facts if they had not been preceded by a statement which, although seemingly clear, is in fact puzzling in its implications and rather inconsistent with the Swedish chemist's approach to the problem: 'The number of simple or compound atoms contained in different gases, in a given volume and at equal pressure and temperature, is the same or rather is for some a multiple or a submultiple, by an integer, of the number contained in others.'[23] This is an important assertion tempered, in the alternative offered, by the admission that a ratio different from the unity may be observed between the numbers of the particles contained in equal volumes of combining and combined gases.

Berzelius then offered several examples of gas combinations, all included in his generalization just mentioned. They can be substantially reduced to two cases: the first is when one volume of a compound gas is obtained from two or more volumes of elemental gases. This also represents the combination of two or more simple atoms to form one compound atom; the result is that, under the same conditions of pressure and temperature, the number of compound atoms (the term molecule is never used on this occasion[24]) in a given volume of combined gas is equal to the number of simple atoms present in each volume of combining elemental gas.

The second situation is observed when two volumes are formed from two or more volumes of elemental gases. Even in this case two or more simple atoms yield a compound atom, with the difference – Berzelius emphasized – that each volume of combined gas now contains only half as many compound atoms as the single atoms present in each volume of the combining gas. It may be worthwhile to recall that Avogadro, in the second section of his 'Essai d'une manière', had arrived at a similar position, without drawing the same conclusions.[25] His assumptions were: (a) a molecule composed of two or more elementary molecules should have its mass equal to the sum of their masses, and (b) if the elementary molecule of a substance combines with two or more elementary molecules of another, the number of compound molecules should remain the same as the number of elementary molecules of the first substance.

The results offered by Gay-Lussac indicated, however, that when a gas unites with two or more times its volume of another gas, the gaseous compound thus obtained has a volume different from that of the first gas. A typical example is represented by water whose volume is double that of the oxygen entering into combination. The problem for Avogadro at this point was to reconcile the experimental evidence with the equal volumes–equal numbers generalization. He overcame the obstacle by assuming the existence of composite molecules, whereas Berzelius could find no acceptable explanation, at least for some gaseous combinations, except to halve the number of compound atoms.

It is generally recognized by most observers that Berzelius' attitude toward the notion of complex elementary molecules remained unaltered and that, in view of his influence, the acceptance of the molecular hypothesis was thus considerably delayed. However, in the fifth edition of *Lehrbuch* there are a few sentences which seem to indicate that he had somehow softened his earlier uncompromising conceptual position in relation to Avogadro's hypothesis. After defining the problem as 'important' and one that will 'never' be definitely solved, he asks: are the mutually repelling particles present in the atoms of elemental gases or in the groups of atoms?[26] To this, the answer was:

> Ceux qui aiment mieux représenter les atomes par groups trouvent une explication facile, en admettant que, dans les combinaisons, les atomes des molécules d'un corps simple s'échangent contre des atomes des molécules de l'autre, de manière que dans le composé gazeiforme obtenu, le nombre de molécules reste le même pour un volume donné.[27]

Immediately after this, and after dismissing as obviously contradictory Dumas' suggestion of considering as divisible even the atoms when they combine,[28] Berzelius pointed out that the 'Doctrine des atomes groupés' – as he defined the polyatomic molecules – may be applied only when a combination occurs involving equal numbers of atoms of each element.

When this is not the case, then one group of atoms 'devrait perdre par l'exchange plus qu'il ne recevrait par voie de substitution', as when one atom of an element combines with two or three of another. He thus reached the conclusion that, for instance, the combination of nitrogen and oxygen – where elementary atoms exchange with each other and, therefore, the number of the molecules of the compound gas remains the same for a given volume – is acceptable in the formation of nitrous gas. On the other hand, ratios such as those observed between nitrogen and hydrogen when forming ammonia can only be explained by admitting again that a smaller number of molecules is contained in a given volume of the compound gas.

There is no doubt, then, that by this time Berzelius had recognized the validity of an equal volumes–equal numbers generalization, at least for some compound gases; moreover, the existence of groups of 'homogeneous' atoms (he never calls them molecules) in elemental gases is no longer discarded by him as a 'fantasy game'.[29] One may detect a departure from his previously accepted view, even if attenuated by a major restriction.

What was behind Berzelius' shift toward the molecular hypothesis? Of the four men, who until 1834[30] had turned their attention toward establishing a possible link between atomism and volumism, Avogadro had been ignored and Dumas had been criticized by Berzelius, who, on the other hand, took only a very moderate stand in relation to Ampère and Gaudin. In 1834, in commenting on a series of experimental data reported by Dumas in his attempts to determine the atomic weights of the elements through their vapor densities, Berzelius said that a hypothesis on the cause of the multiple proportions relating by integers the gaseous volumes with the number of the atoms contained therein might be 'premature'.[31] He also pointed out that, especially for nonpermanent elemental gases, their specific weights were not 'necessarily' proportional to their atomic weights.

The following year Berzelius, rather inaccurately, cited[32] Ampère as the author of the 'proposition' assuming that in all gases the distance between atoms is the same, but commented no further. On the other hand, he extensively reviewed a very profound and comprehensive note written by Ampère in 1835, a few months before the latter's death.[33] After distinguishing the 'particles' at a higher level of aggregation from the molecules which in turn are formed by a group of atoms, Ampère assumed that mechanical forces alone are responsible for the separation of such particles. Once they have been detached from each other, then only chemical forces can act upon the molecules and cause their division into the component atoms. A very clear example follows of what happens when a mixture of two volumes of hydrogen with one of oxygen explodes to form two volumes of water vapor. Every molecule of oxygen, when subjected to the particular conditions of the reaction, is divided in two, and the atoms of each half unite with those of one hydrogen molecule to yield one molecule of water. One should not neglect to mention the importance of the fact that both the notion of the molecular partition so unambiguously stated and its application to gaseous combinations were reported verbatim in Berzelius' *Jahresberichte*. His comments should be recorded: 'Dieser Versuch enthalt eine so klare und nüchterne Corpuscular-Ansicht, dass ich ihn mit seinen eigenen Worten anführen zu mussen glaube'.[34]

Also quoted in the same context was the equal volumes–equal numbers of molecules generalization as expressed by Ampère who recalled his twenty year-old essay on the same subject.[35] Yet this hypothesis was still thought premature in 1834. Berzelius, who shortly afterwards commented that only time would establish its validity or not, was to give it serious consideration three years later. A possible explanation of this development is offered by the combined emergence of two concepts, which in the history of the molecular hypothesis is indeed unprecedented: the existence in elemental gases of two different orders of particles and their behavior when they chemically combine; not only were they proposed in a lucid and also reasonable form,[36] but their proponent was by then one of Europe's most renowned physicists.

It seems evident that Berzelius, although reluctant to take a definite position, was, above all in those years, keenly looking for an acceptable conceptual system which could again harmonize Gay-Lussac's findings with Dalton's atomic speculation. He apparently found something unequivocal in a paper which appeared in 1833 in the *Annales de Chimie*. It was written by a Frenchman, Marc A. Gaudin, possibly as little known in his own country as Avogadro in his,[37] and had the ambitious purpose of attempting to clear 'les points de la théorie atomique qui sont encore indécis'. In his discussion of this essay two years later,[38] Berzelius, surprisingly so in view of his generally critical reaction to such conjectural works, passed a rather mild judgment on Gaudin's ideas; furthermore, he did not discard the interpretation offered by the Frenchman for the combinations of elemental gases when they take place without changes in volume: a corresponding number of atoms of an element is 'exchanged' – he argued – with the atoms of the other element, so that the number of the molecules of the compound becomes equal to the number of molecules of the element.[39]

Graebe correctly observed[40] that the undeviating position taken by Berzelius for almost thirty years on the constitution of gases was somehow mitigated as if he were not completely sure of the validity of his arguments. In 1842, again in the fifth edition of his *Lehrbuch*, he posed a question revealing his doubts: an important question, he stressed, 'gewiss niemals zu entschieden sein wird'.[41] Are the smallest particles which in gaseous elements mutually repel under the influx of heat, atoms, or groups of atoms? To this question, Berzelius, who – according to Graebe – 'saw difficulty where it was not the case' replied that he cared less for the hypothesis of the atomic groups than for that of individual atoms. Or, in other words, while he did not approve of it, he did not immediately reject the concept of polyatomic molecules. This attitude its perfectly understandable if one remembers Berzelius' instinctive reluctance to change his view on an issue as momentous as the existence of a new order of particles. He was not that kind of man.[42]

Gifted with a highly organized mind and a deeply methodical nature, Berzelius had reached a superb skill in both preparative and analytical procedures. While his unprecedented precision and systematic experimental approach certainly represented a turning point in the development of chemistry, he unfortunately had not been generously provided with one of the most characteristic traits of genius: unlike Dalton, Faraday, and Avogadro himself, just to mention a few among his contemporaries, he markedly lacked imagination; he followed a very systematic approach in his research while leaving little or no room for conjecture. 'Erfahrung' became his key word and everything not supported by experimental 'gründlichkeit' was either arbitrary assumption or fanciful speculation. He felt hypotheses should be avoided and only theories supported by facts (and not the other way around) need be taken into consideration; along with this strict empiricism there followed, as his fame and authority expanded, a typically Manicheistic attitude in his relations with other scientists.[43] In April 1838 Berzelius wrote to Liebig,[44] 'In dealing with scientific topics one cannot have either friends or foes. What is wrong must be fought regardless of who is wrong. Points of view are not people, and we can reject them without arguing with their author.'[45] Although it is widely recognized that the molecular hypothesis was not fully accepted by Berzelius, it is seldom mentioned that he never directly criticized or even mentioned Avogadro in relation to this fundamental doctrine. This attitude is quite puzzling and the few explanations proposed thus far do not survive close examination. It has been said, for instance, that due to the war conditions, the Swedish chemist could not receive the French publications.[46] Berzelius seemed, however, very familiar with the work of Gay-Lussac, Thenard, and Berthollet in those same years.

Another point raised is that since both the 'Essai d'une manière' and the

'Mémoire sur les masses' were published in a non-chemical journal, they did not attract his attention.[47] To this it can be answered that the Swede was a very attentive and minute reader and that, as editor of the *Jahresberichte*, he perused a large number of publications which were not strictly devoted to chemical research. On the other hand, the criterion he often applied in choosing works to review is debatable. For instance, instead of examining Avogadro's significant and extensive 'Nouvelles considérations' in the second volume of the *Jahresberichte* (1823), he commented on a rather trivial article on a new mineral water spring found in a Piedmontese village (not too far from Turin).[48] This example, and there are many others, negates the opinion that Berzelius' journal reviewed only major topics. A more acceptable explanation is that the Memoirs of the Royal Academy of Sciences of Turin were not among the journals which regularly came to his attention, and this appears confirmed by the fact that all subsequent articles by Avogadro published in Italian periodicals and reported in the *Jahresberichte* were abstracted in French journals.

Altogether, five essays by the author of the molecular hypothesis were reviewed in the *Jahresberichte* during Berzelius' lifetime. The first one in 1825. 'Mémoire sur la construction d'un voltimétre multiplicateur'[49] was the object of only a short comment.

Almost a decade went by before another paper attracted Berzelius' attention; it covered the determination of vapor tensions of mercury over a wide range of temperatures (from 100° up to 360°).[50] This time, the experimental procedure described did not escape the criticism of the reviewer. After suggesting that corrections should be introduced in order to take care of factors which obviously had eluded Avogadro's attention, Berzelius comments rather sharply on the peculiarity of the fact that, despite these faults in the procedure, the results reported 'nicht mit sehr grossen Fehlern behaftet waren'.

In 1836 a very extensive 'Memoir' published two years after Avogadro's reappointment to the chair in Turin was summarized in the *Annales de Chimie et Physique*;[51] the topic, again, is the relative position of the elements in the electrochemical series. The reviewer quickly disposed of the subject, reporting little of the data given in the paper and reaching the conclusion, on the basis of the ratio observed between the electronegativities of oxygen and chlorine, that not all the evidence had been taken into consideration by the author in conducting these determinations.[52]

Another essay, this time by Avogadro and his assistant G. D. Botto, appeared in 1839.[53] It is rather thoroughly discussed in the 1841 *Jahresberichte*. The point raised by the article is that, when a current is passed through a liquid, it always exerts a chemical effect which may be occasionally concealed. Berzelius, in summarizing this and other conclusions reached by the Piedmontese physicists, appeared very restrained in his comments; his only remark concerned the neglect into which the then fourteen year-old Ohm's Law[54] had fallen, whose application, he adds, would have explained some of the considerations raised by Avogadro and Botto in their essay.

Between 1846 and 1852 Avogadro published in Turin a series of four memoirs on the atomic volumes and their relationship with the physical characteristics of the chemical elements. They were all reported in the *Jahresberichte*, the first one in 1847,[55] the year before Berzelius' death, while the others appeared when Kopp took over as editor. This review is recalled by Guareschi and occasionally by other historians to prove the consideration of Berzelius for the author of the molecular hypothesis. The opening remark, 'Avogadro hat darüber eine sehr lesenswerthe Abhandlung mitgetheilt', is actually quite remarkable if one considers that in this memoir he again restated the equal volumes–equal numbers of molecules assumption which had earlier been challenged by the reviewer. Moreover, the purely speculative tone of a work devoted to finding a link among such physical properties as the molecular mass, the affinity for heat and the relative electrochemical activity did not harmonize with the basic conceptual attitude of Berzelius.

The formal relationship between Avogadro and Berzelius for a span of over thirty years is all here, in these five reviews published in the *Jahresberichte*.[56] There was, however, very little, if any, personal relationship between them. Among the men involved in scientific speculations at the time, Avogadro was one of the few with whom Berzelius had no correspondence. Moreover, in the course of his journeys the Swede never visited Italy, and so there was no opportunity for even a brief direct contact.

According to Soderbaum[57] who carefully studied this subject, Berzelius in the course of his seventeen-year epistolary with Davy, not once mentioned the molecular hypothesis and its author. On the other hand, there is plenty of evidence, through his profuse quotations, that Avogadro was well acquainted with Berzelius' work since the earliest years,[58] and that he constantly held the Stockholm scientist in high esteem. A close examination of his writings discloses, furthermore, that through 1823 Avogadro seemed quite confident that his main assumption might be accepted at least by the French, and that he earnestly hoped to convince Berzelius. Then, suddenly in 1828, when the seventh volume of the *Jahresberichte* published the sharp criticism of Dumas' interpretation of the molecular hypothesis as proposed by Ampère, many of his expectations were probably shattered. Although Avogadro insisted even in the late papers[59] on expressing the feeling that his fundamental idea had been accepted – either in an explicit or an implicit form – by his contemporaries, the fact remains that quite significantly he never published any major chemical memoir[60] after Berzelius' position became known in 1828.

The Molecular Hypothesis in Britain

In Britain, a general review of the results obtained in the physical sciences during the previous decade appeared in the *Annals of Philosophy* in January 1815. Written by Thomas Thomson, it revealed the unequivocal Daltonian views of its author. While Wollaston, Gay-Lussac, Berzelius and, needless to say, Dalton were recalled for their contributions to the corpuscular doctrine, the 'Essai d'une manière' and its author were ignored. Again, poor communication with the Continent seems to be the first plausible explanation for this attitude. It should be recalled, however, that after the abdication of Napoleon at Fontainebleau (April 1814), free circulation of news from France to Britain resumed for more than a year. By keeping in mind that Avogadro's first two major essays appeared in a French journal, it is legitimate to infer that somebody in Britain may have noticed them. Thomson, in the course of his review, did not neglect to mention – quite surprisingly – a rather insignificant work published by Morichini, then professor in Rome;[61] this seems to suggest that reports of scientific proceedings south of the Alps were somehow reaching the editor of the *Annals of Philosophy*.

A more acceptable interpretation for the lack of interest which greeted the molecular hypothesis in Britain is offered by considering that, while for the atomists the theory of volumes – even after its interpretation by Berzelius – appeared groundless, for Davy and Wollaston it was not beneficial to introduce into their systems the concepts proposed by the 'Essai d'une manière'. Since, in principle, they were distrustful of a corpuscular vision of the matter, they could not *a fortiori* consider a structure as further complicated as that proposed by the Piedmontese with his composite molecules.

One had to wait until 1832 to see Avogadro's hypothesis revived in Britain, namely by M. Donovan in his *Elements of Chemistry*.[62] A fervent supporter of atomism and of the caloric theory, he introduced the equal volumes–equal numbers hypothesis as an explanation for the unusual volume ratios observed in the chemical combinations of gases[63] when compared with the ratios by weight found for two nongaseous bodies. Avogadro was credited as the first one to have proposed this hypothesis, and Ampère was also mentioned. However – and this does not cause surprise, because many among Donovan's more celebrated contemporaries had stumbled on the same error – the crucial distinction between the concept of atoms and that of molecules is completely missed; no wonder, then, that in the absence of such a premise, the assumption set forth by Avogadro could not be understood.

From the *Elements of Chemistry* it is clearly understood that its author not only misrepresented the concepts as originally expressed in the 'Essai d'une manière', but also used the wrong terminology:[64] thus, 'atoms' or 'ultimate particles' are used interchangeably, while – it should be recalled – an elaborate differentiation had been adopted in 1811 by Avogadro. Donovan certainly excelled in the simplification of someone else's ideas, a process which is not always beneficial.

Donovan proceeded to prove the fallacy of the hypothesis that equal volumes of all gases contain under like temperature and pressure the same number of atoms.[65] First of all, he recalled the case of three compounds of carbon with hydrogen, all having the same relative ratio of combining volumes, but in progressively increasing amount and all yielding one volume of compound; for him this was sufficient proof that the number of atoms in these three gases could not be the same. But, of course, it was easy to prove Avogadro wrong by twisting his concepts.

Quite typical was also the second argument raised by Donovan. It was based upon the old Daltonian interpretation that when two combinations take place between the same elements, one must be presumed to be binary and the other ternary; since this is the case for hydrogen and oxygen, then water must be binary or, in other words, in two volumes of hydrogen the same number of atoms sould be present as in one volume of oxygen. Needless to say, such reasoning – an essential part of Dalton's system – could not be challenged by one of his supporters; the conclusion then is that, since *tertium non sinitur*, Avogadro had to be mistaken.

To Donovan, furthermore, the equal volumes–equal numbers assumption seemed an insufficient explanation for the unique contraction observed in a number of gaseous compounds at the moment of their formation; contraction which, in agreement with Gay-Lussac's Law mentioned at this point, is in simple relation to the volume of one of the combining gases. In the case, for example, of nitrogen protoxide, two volumes of this gas, rather than three, are formed from the combination of one volume of oxygen and two of nitrogen. How, Donovan asks, is it possible to explain that 'one-third of the whole bulk is lost by approximation of the constituent atoms'?[66]

One may conclude at this point that Donovan never had the opportunity of reading the 'Essai d'une manière';[67] otherwise, he would have known that the obstacle represented by compounds such as ammonia, nitrous oxide, water, etc. had been overcome by Avogadro's conception of molecular division. Two years after being treated by Donovan in such an inauspicious way, the molecular hypothesis was reproposed in Britain by William Prout. Graduating from Edinburgh in the same year as the publication of the 'Essai d'une manière', he had already earned distinction for his proposal in 1815 relating all the atomic weights to the atomic weight of hydrogen. Prout's, though following Avogadro's, were among the first attempts to apply the law of combining volumes to the determination of the specific gravities[68] of the elements, and to relate them to the atomic weights. At that time he had independently arrived at the same conclusions as Avogadro; however, although Prout claimed to have understood the relation between the doctrine of volumes and that of atoms, it is not clear whether he adopted the same line of reasoning followed by Avogadro. He inferred that a direct relation could be established between the masses of the ultimate particles and their densities, but certainly he was quite implicit in his approach. Was his just an intuition or, if not, then why did he step right away *in medias res*, without further elaboration on the necessary and, in those years, scarcely understood premises of his assumption?[69]

Despite his later assertions, Prout never mentioned in 1815 the equivalence between number of molecules and volumes under the same conditions of pressure and temperature. From some of his examples it appears that the fundamental notion of polyatomic particles in elementary gases was absent. A typical interpretation, thus, was that of ammonia, 'composed of one atom of azote and only 1.5 of hydrogen, which are condensed into two volumes, equal therefore to one atom'.[70] This seemed in line with Dumas' 1826 view, but definitely had no resemblance to Avogadro's explanation of the same facts.[71]

From an historical viewpoint one may wonder what would have been the aftermath of the molecular hypothesis, at least in Britain, had Prout's contemporaries actually understood his ideas concerning the relation between densities and atomic weights.[72] A legitimate answer is that those ideas had only a slight chance of acceptance; first of all, they supported an inconceivable division of atoms and an outdated concept of air as a chemical compound; they were extended prematurely (as, of course, was the case with Avogadro) to elements which are non-gaseous under ordinary conditions, and finally their speculative basis seemed far from clear. As a result, Prout's short essays of 1815 and 1816 became eventually well known for the integral multiple-weights hypothesis they exposed, while their attempts to correlate the law of combining volumes with the corpuscular doctrine receded into the background.

Almost twenty years later, in his *Chemistry, Meteorology, and Function of Digestion,*[73] Prout claimed that he had arrived independently at the assumption made by Dumas, Ampère, and Avogadro: that under the same conditions, all gaseous bodies contain an equal number of molecules. This assertion appeared as a footnote at the end of the section devoted to gaseous substances.[74] Those same paragraphs contained a series of considerations aimed at proving that all the laws governing the behavior of such substances – Boyle's, Charles', and Dalton–Graham's – and the experimental evidence[75] supported beyond any doubt the equal volumes–equal numbers hypothesis.

Chronologically speaking, it is correct to believe that Prout formulated his ideas sometime in 1815, before the publication of his anonymous 'Relation between specific gravities of bodies'. More questionable was his avowal[76] of ignorance about both the Avogadro and Ampère papers, which preceded his own by four years and one year, respectively. It is very likely, on the other hand, that he became only indirectly familiar with both works through Donovan's *Treatise on Chemistry*[77] which quoted the same 1827 paper by Avogadro. If one assumes that Prout did not bother to examine the original references and their dates of publication, he may have then remained under the erroneous impression that his views had not been anticipated. He admitted, however, that they were 'most nearly' similar to those of Dumas as reported by J. W. Johnston in the *Report of the British Association*; needless to say, Dumas in 1826 had based his conclusions on the Avogadro–Ampère assumption.

There is another aspect of Prout's speculative system which, because of its relation of Avogadro's hypothesis, should be recalled here. A close examination of the third chapter of Prout's *Chemistry* reveals that his ideas on the nature of the ultimate components of matter were rather ill-defined. In 1815 he had started by considering atoms as divisible, and now quite clearly he indicated that the molecules – a term which he used generally to denote atoms[78] – are not capable of further division. This view is supported throughout the extensive discussion of the type of forces influencing the ultimate particles of matter. There is either aggregation or repulsion between molecules which may be similar (or homogeneous) and dissimilar (or heterogeneous). In the case of the latter[79] there are chemical forces involved which may cause their mutual aggregation or repulsion, according to the relative position of the particles. On the other hand, the homogeneous molecules (compound molecules) are subject either to cohesive forces (responsible, for instance, for the transition from water to ice) or to self-repulsive forces prevailing in gaseous bodies. There is an equal number of these homogeneous or self-repulsive molecules in equal volumes of all gases at the same temperature and volume, Prout concluded shortly afterwards.[80]

What about the dissimilar molecules when in the gaseous state? In discussing the application of the Graham–Dalton Law to his generalization, Prout indicated that two molecules of different matter remained passive and had the tendency neither to cohere nor to repel each other. The question remains whether he limited the equal volumes–equal numbers hypothesis to atoms or to molecules in their current meaning.

A possible answer was given by Prout himself in a subsequent section of the *Chemistry* when he specifically asserted, after repeating that the number of self-repulsive molecules is the same for equal volumes, that every self-repulsive molecule (of oxygen, in the case quoted) had been divided in two during the formation of water and 'consequently must have originally consisted of at least two elementary molecules . . . somehow or otherwise associated as to have formed only one self-repulsive molecule'.[81] This apparently was the conclusion of a line of reasoning which moved from indivisible unqualified molecules to self-attractive indivisible molecules and, eventually, to simple self-repulsive but divisible particles. Prout, however, did not rationalize how those two 'elementary' particles of the same matter could stay together in the first place. 'Somehow', he said without further elaboration, 'the push-and-pull approach could explain some but not all the facts.'

There is a lack of coherence emerging from Prout's interpretation of a molecular framework which makes it far less convincing than that suggested by Avogadro almost a quarter of a century earlier, and also much less lucid than that proposed by Gaudin just one year prior to the publication of *Chemistry*. Surprisingly, this first attempt in Britain to find a correlation between volumism and atomism escaped the attention of Carl Graebe;[82] in his generally very accurate 'Entwicklungsgang der Avogadroschen Theorie' (1913) he never mentions Prout.

The short anonymous notes published in the November 1815 and February 1816 issues of the *Annals of Philosophy* which suggested a relation between specific gravities of bodies in the gaseous state and their atomic weights, were a few months later summarized in the *Annales de Chimie et Physique.*[83] In 1820, while drafting his major chemical essay,[84] and reviewing the relevant works subsequent to 1811, Avogadro also read the abstract of Prout's paper in the French journal. Later, in the introductory section of his memoir, he dismisses in a few sentences the conclusions reached by the British physician. In so doing he took, it should be recalled, a position close to that of Berzelius.

When his own name is quoted by Prout thirteen years later in *Chemistry*, Avogadro was unaware of this citation.[85] Prout's book actually received very little attention on the Continent; even in his own country his scientific reputation steadily declined, and when he died in 1850 the Royal Society ignored the event.[86]

The Molecular Hypothesis In France

With no other group of individuals, nor with any other European school of thought, were the origins and the development of Avogadro's speculation more closely linked than with those of the French scientists and the system of doctrines they had formulated in the first two decades of the nineteenth century. His intellectual formation was essentially French. While this represented the trend of the times in many countries all across the Continent, it seemed particularly true for Piedmont, where for historical and traditional reasons, France had exerted its cultural influence even during the eighteenth century. The events which followed the Revolution in 1789 and above all the subsequent Napoleonic annexation greatly enhanced this situation.

It should be further remembered that the years of Avogadro's critical conversion from the practice of jurisprudence to the study of natural philosophy coincided with the second invasion of his country by the armies of the Republic. When the French occupation came to an end seventeen years later, after the armistice of Schiarino, he was by then thirty-eight; his education and imagination had been inspired by the ideas of the men who had been part of the circles of Arcueil and Auteuil, by members of the *Acadèmie* and by teachers at both the *École Polytechnique* and the *Collège de France*. Not only was the language spoken by these men a second mother tongue for him, but through that language he also became aware of the works of Dalton, Davy, Wollaston, Berzelius, and Mitscherlich.

Although it is legitimate to assume that, to some extent, the molecular hypothesis was 'thought'[87] as well as written in French, it should also be added that Avogadro elaborated his major doctrine in a rather independent way. There was no personal constraint upon him, no allegiance to any school. In his position as an outsider, he could accept whatever appeared coherent with the framework of his ideas and set aside whatever he felt to be superseded and inconsistent.

It may be appropriate to recall here that in the late eighteenth and early nineteenth centuries the prevailing scientific tradition in France was that which had been defined as 'mechanico-molecular' and, less generically, as 'Laplacian' in honor of its foremost exponent. Within this tradition which, among other concepts, supported the existence of molecular forces and considered chemistry as a branch of physics, one finds not only Laplace but his close friend Berthollet and many of their younger disciples, like Gay-Lussac, Biot, Thenard, and Poisson. They met often in Arcueil, they exchanged ideas and their views, published in the most authoritative journals, were seldom challenged, at least in France.

Only two major names on the French scientific scene of the time seemed outside the Laplacian circle: Ampère and Fourier, although the former until 1815 still maintained a position, in relation to the particulate structure of matter, not too dissimilar from that of the Arcueil group.[88] As for Fourier, a prominent theoretical physicist, he took a critical stand on the dynamical interpretation of all physical facts generally shared by Laplace, Berthollet, and their followers.[89]

There is no doubt that Avogadro fully absorbed the doctrines of the French mechanico-molecular tradition. This is clear from his early writings in 1809 through his *Fisica dei corpi ponderabili* in the late 1830s. That he remained an adherent of the Laplacian position even at the time when this scientific philosophy became obsolete in France and elsewhere is also an irrefutable fact.[90] One may argue about the reasons behind this intellectual conservatism of the Piedmontese. Physical distance and lack of personal contacts certainly are important. Very likely age also had its effect: he was in his late fifties when his scientific *Weltsanschauung* began to crumble, and he most probably realized it. As transpires from *Fisica dei corpi ponderabili*, it is also possible that he found it difficult to reconciliate his views on molecular constitution with the new wave theory of heat; and when later on after so many of his conjectures had been demolished by Dulong's experimental work, he did not have the confidence to further modify the conceptual framework upon which his 1811 gas hypothesis rested.

There are areas where the men within the Laplace circle had more effect on Avogadro's scientific philosophy, but to reach any definite and clear assessment of their influence, however, seems rather difficult. For instance, to link him too closely to Berthollet would be to overstate the case.[91] Yet there are elements in Berthollet's system which certainly inspired Avogadro as early as 1808, in the years immediately following the appearance of *Essai de Statique Chimique.*[92] Among these, the most important is, perhaps, that dealing with the presence of antagonistic forces, such as acids and bases, in all chemical phenomena. This subject was examined and developed in Avogadro's 1809 'Idées sur l'acidité'[93] where he proposed a scale of elements according to their acid and basic nature. The importance of this scale broadly sketched in 1809 and which had only been hinted at by Berthollet in his treatise[94] lies in Avogadro's recognition of oxygen as not necessarily the most acidic element. Avogadro emphasized that the term acid denotes only an accidental property and should not be used without qualification.

In the same essay he correlated his scale of relative acidity and alkalinity to Davy who assumed a link between electromotive power and chemical antagonism, thus suggesting that through 'electrical heterogeneity' of the various elements one could measure their chemical affinity. This entire concept certainly originates with Avogadro. To represent the relative position of an element in his scale of acidity and alkalinity, he, on the same occasion, introduced the term oxygenicity, a term which, although in a modified form, remained for a long time in his thought. Thus, while the concepts of affinity and antagonism were present in Berthollet's *Essai de Statique*, Avogadro, who was certainly not the only one to adopt them at the time, went beyond them in 1809.

In the 'Essai d'une manière' (1811) he returned to his ideas of neutrality, acidity, and alkalinity only in the concluding section of his paper, yet oxygenicity, together with the proportions by weight of the combining substances, appears as an important parameter in Avogadro's system. Although he briefly mentioned Berthollet's notion of variable proportions in chemical combinations in the 'Essai d'une manière', he allowed this concept of Berthollet's system to recede into the background. In three different parts of his 1811 paper, Avogadro unequivocally stated that acids and alkalis combine by a definite number of molecules in simple ratios to reach the neutrality.[95] On the other hand, he emphasized that his ideas of 1809 could be 'reconciled' with those put forward on the same subject by Laplace and Haüy.[96]

If one, then, can easily detect in Gay-Lussac's 'Essai sur la combinaison' an attempt to pass beyond the position of his mentor at Arcueil with regard to the variable proportions, quite evident is Avogadro's rift with Berthollet, in the two major chemical papers of 1814 and 1821. In the 'Nouvelles considérations' (which appeared the year before Berthollet's death) he referred to his own contribution to the 'combined theory of the simple ratios by volumes with the theory of fixed proportions or atomistic . . .',[97] after stressing the importance of densities and volumes as the only non-arbitrary terms of reference in determining the molecular weights and the combining ratios. In the introductory section of the 'Nouvelles considérations' he mentioned, after pointing out that his own molecular hypothesis lies *within the realm of physics*, several chemists: Berzelius, Dalton, Gay-Lussac, Thenard, Davy, Thomson, Vauquelin, and Wollaston, but Berthollet's name never appears.

Avogadro's faith in the laws of definite and multiple proportions cannot be disputed, as disclosed by his assumption in the 'Nouvelles considérations' that even the composition of elaborate organic substances could be construed as being in harmony with those laws.

A further differentiation between the thinking of Gay-Lussac, Berthollet, and Avogadro resides in their attitudes towards atomism. Gay-Lussac, again out of personal deference to Berthollet, firmly denied any correlation between his own investigations and the ideas set forth by Dalton; he stressed in his famous 1809 essay that Dalton's theory was not 'entirely exact'. Later on, at the conclusion of the same paper, Gay-Lussac, after acknowledging that not only Thomson's and Wollaston's but also his own results showed the validity of the law of multiple proportions, argued at the same time that that did not necessarily mean Berthollet's views were fallacious. And he proceeded to explain how the two positions could be reconciled.[98]

Avogadro, on the other hand, pointed out in the 1811 essay the 'arbitrariness' intrinsic in the Daltonian method of establishing the number of molecules present in compounds. He also stressed[99] that totally absent in the system set forth by Dalton – and he could have added in Gay-Lussac as well – was the concept of molecular division occurring upon formation of gaseous compounds.

Several other conclusions reached by Dalton were criticized in the course of the same essay, but Avogadro concluded that there were also 'many points of agreement' with him.[100] Furthermore, although he refrained from using the term atom until later in his career, no record can be found in his writings of any criticism of the essence of the atomic doctrine. Here, definitely, he took a stand different from that of the Laplacian school. Finally, his praise of the work and gratification at observing how some of his conclusions paralleled those reached by Davy in *Elements of Chemical Philosophy*,[101] put Avogadro – as earlier mentioned – at odds with the exponents of the French chemical school who in those same years had been involved in rather fierce polemics with Davy.

While the influence of Berthollet on Avogadro, after 1813, remained in the background, and while Gay-Lussac's observations on gas combinations still had an unquestionable effect on the formulation of the molecular hypothesis, the man who, among the French natural philosophers of the time, probably most inspired Avogadro to elaborate on his molecular system was Siméon Denis Poisson. A former student of the *École Polytechnique*, assistant to Biot, a member of the Arcueil circle, a full-fledged Laplacian, he had since 1808 embarked on applying mathematics to molecular forces in general. Like Avogadro, Poisson interpreted the structure and the behavior of submicroscopic matter in merely mechanical terms.[102] Magnetism, electricity, capillarity, heat, as well as elasticity, all could be dealt with at the level of the constituent particles.

Poisson and Avogadro, both with highly speculative minds, also shared the characteristic of being very poor experimentalists. In view of the fact that the gas hypothesis was formulated by Avogadro in the latter part of 1810, Poisson's work which had just begun at the time definitely had no effect on it. However, in the following twenty years, the appearance, among other writings, of both the *Traité de Mécanique* and of *Théorie nouvelle de l'action capillaire* gave to the Piedmontese wider foundation for his system of molecular constitution of bodies. This Avogadro unequivocally recognized and gave Poisson proper credit.[103]

How did French science react to Avogadro's fundamental assumption on gases, so inspired by the ideas of some of its most reputable representatives? The essay asserting the molecular hypothesis appeared in the summer of 1811 in the *Journal de Physique et Chimie*, edited by J. C. de La Métherie, which occasionally contained articles contributed by renowned men of science from all over Europe.[104] Almost every important chemical researcher then working in France was quoted in the essay, yet not one of them reacted to or expressed in writing an opinion on the postulate set forth by the scarcely known physics teacher from Vercelli. It should be recalled that, although this was Avogadro's fourth article to be published in the *Journal de Physique*, it was the first one essentially devoted to chemical investigation.[105]

Any attempt to identify the reasons behind this total absence of reaction in France in the years after the publication of the 'Essai d'une manière' should take into special consideration the following elements which had nothing to do with the intrinsic value and meaning of the ideas debated there: the lack of support for its author since he resided outside the Parisian scientific patronage; moreover, French science, at its peak then, had begun to suffer from chauvinistic tendencies which unfortunately lasted for over half a century,[106] and were certainly not conducive to the development of an objective evaluation of any foreign contribution.

A further explanation for the lack of an affirmative attitude towards

Avogadro's hypothesis can be found within the exponents of French chemistry at that time. Although he did not mention Avogadro's generalization, one may still feel its relevance in Claude Berthollet's comment on the Daltonian theory of atoms:

> It was better that chemists should repeat and extend experiments relating to constant and multiple proportions than that they should devote themselves to hypothetical speculations on the number, arrangement and figure of atoms which escape all experience.[107]

Among these hypothetical speculations, that proposed by Avogadro is typically the one concerning the number of particles. In those years it was difficult enough for the French natural philosophers – and not only for them – to comprehend and appreciate the corpuscular system just elaborated in Britain, without facing the further speculative effort required for 'counting' or 'shaping' the ultimate components of matter.

The problems of the forms and number of these indivisible entities had been undertaken by Dalton, but his solution to the latter one did not agree with the generalization offered by Avogadro. His findings, moreover, seemed contradicted – as the often quoted example given by Dalton himself indicated – by the experimental evidence.[108]

On the basis of the skeptical attitude towards the atomic theory then prevailing in France, one may understand the greater reluctance to take into consideration an hypothesis which introduced a concept, such as the aggregation of like particles, certainly awkward, hard to comprehend and unsupported by direct evidence. In 1811, among the French Academicians, Gay-Lussac seemed the most qualified to grasp the meaning of the ideas expressed by Avogadro in the 'Essai d'une manière'; after all, they represented a corollary to the conclusions he himself had reached only a few years earlier on the regular combining ratios of gaseous substances. To the obvious question of why he had not taken into consideration the equal volumes–equal numbers generalization and if he had, why had he not proposed it himself, one possible answer is given by recalling his reluctance to accept even the Daltonian atoms. His position towards the corpuscular theory remained one of prevailing doubt for almost two decades.

Among those who further elaborated Gay-Lussac's conclusions on gas behavior, Avogadro himself, unlike Berzelius, did not accept atomism *in toto*. At the end of the 1811 essay he explained that the deductions drawn from his assumption are 'un nouveau moyen de précision' over Dalton's system.[109] If one considers the serious criticisms raised earlier by Avogadro in the 'Essai d'une manière' on some of the results set forth by Dalton, this final comment represents an understatement typical of the personality of its author.

Ampère

In the April 1814 issue of the *Annales de Chimie* a note appeared in the form of a letter addressed by Ampère to Claude Berthollet. This event – it should be stressed – took place three years after Avogadro's fundamental paper and three months after its sequel, the 'Mémoire sur les masses', had been published in the *Journal de Physique*. For the historical importance of the letter of Ampère, no analysis of the events related to the molecular hypothesis may be conducted without a detailed evaluation of its content.

What was the standing of Ampère in France at the time? Self-educated and after a short teaching career, he had been appointed professor of calculus at the École Polytechnique in 1809; although not yet at the peak of his fame,[110] he already enjoyed such reputation as to justify his admission to the Institute five years later. He was then thirty-nine and a voracious reader with a highly versatile mind. His cultural interests ranged from classical languages to botany, from poetry to chemistry and mineralogy.

One cannot disclaim that Ampère's familiarity with Daniel Bernoulli's *Hydrodynamica*[111] suggested to him further study on the behavior of molecules of fluids. He was especially intrigued by the problem of the distance between these particles and its relation to such agents as pressure and temperature; it is reasonable to assume that he perceived in Gay-Lussac's conclusions on combining gases and in the law of Mariotte the confirmation of the conjecture hinted at in the *Hydrodynamica*, namely that the particles of all the gases, under the same physical conditions, are equidistant.

Due to the wide variety of subjects touched upon – clearly revealing the eclecticism of its author – the note addressed by Ampère to Berthollet represents a remarkable piece of scientific literature; although its immediate impact was negligible, it had far-reaching consequences on the further development and eventual acceptance of Avogadro's hypothesis. Unfortunately, by introducing the existence of a tetratomic structure of molecules within an intricate geometric framework, the main assumption became clouded. This, the equal volumes–equal numbers generalization, is expressed in terms which are much less clear in their form than those used by Avogadro.[112] As for the consideration made on gaseous compounds, it is true that Ampère relates

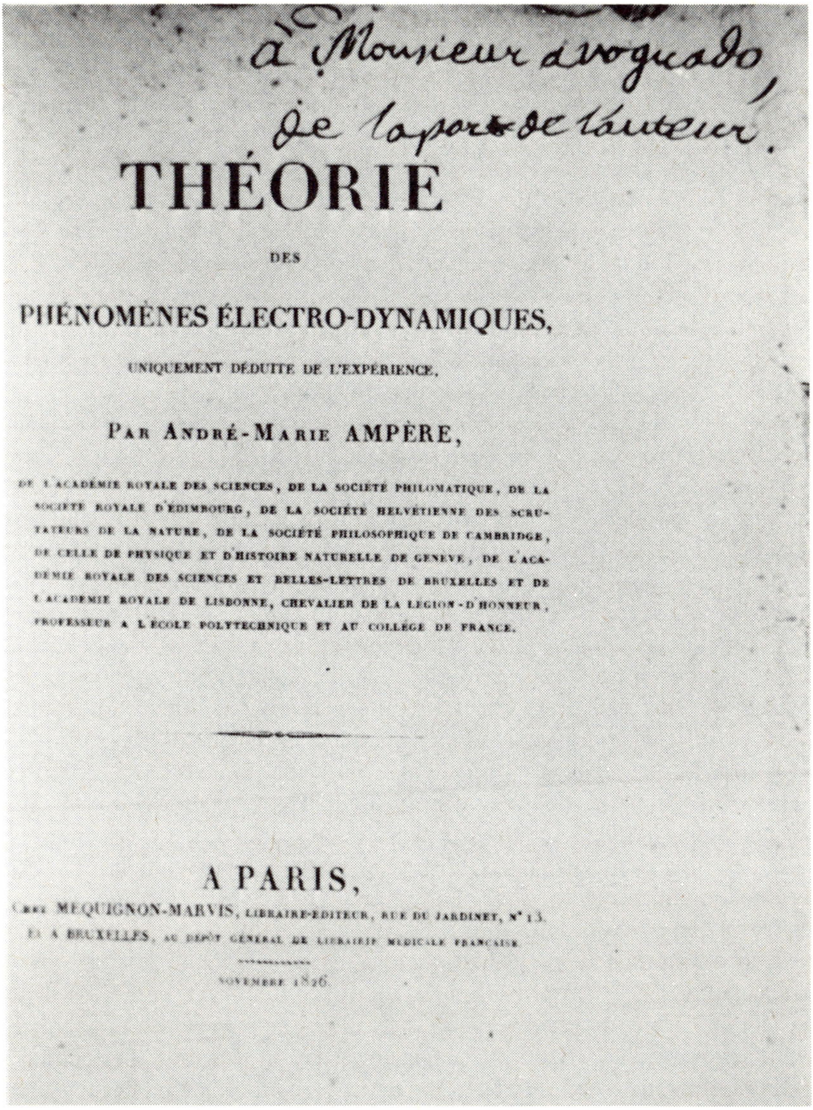

à Monsieur Avogrado, de la part de l'auteur.

THÉORIE

DES

PHÉNOMÈNES ÉLECTRO-DYNAMIQUES,

UNIQUEMENT DÉDUITE DE L'EXPÉRIENCE,

PAR ANDRÉ-MARIE AMPÈRE,

DE L'ACADÉMIE ROYALE DES SCIENCES, DE LA SOCIÉTÉ PHILOMATIQUE, DE LA SOCIÉTÉ ROYALE D'EDIMBOURG, DE LA SOCIÉTÉ HELVÉTIENNE DES SCRUTATEURS DE LA NATURE, DE LA SOCIÉTÉ PHILOSOPHIQUE DE CAMBRIDGE, DE CELLE DE PHYSIQUE ET D'HISTOIRE NATURELLE DE GENÈVE, DE L'ACADÉMIE ROYALE DES SCIENCES ET BELLES-LETTRES DE BRUXELLES ET DE L'ACADÉMIE ROYALE DE LISBONNE, CHEVALIER DE LA LÉGION-D'HONNEUR, PROFESSEUR A L'ÉCOLE POLYTECHNIQUE ET AU COLLÉGE DE FRANCE.

A PARIS,

CHEZ MEQUIGNON-MARVIS, LIBRAIRE-ÉDITEUR, RUE DU JARDINET, N° 13.

ET A BRUXELLES, AU DÉPÔT GÉNÉRAL DE LIBRAIRIE MÉDICALE FRANÇAISE

NOVEMBRE 1826.

Plate 8. Copy of A. M. Ampère's *Théorie des phénomènes électrodynamiques* (1826), dedicated to Amedeo Avogadro. At the time, his chair at Turin had been abolished, and he was living under forced retirement. Note the misspelling of Avogadro as *Avogrado* by Ampère.

volumes to 'particules', but he does not explain why, for instance, from the combination of one 'particule' of nitrogen with three of hydrogen, only two 'particules' of ammonia are formed. No attempt is made to explain – as Avogadro did – the very intriguing question of the number of 'particules' of a compound gas in the case of the contracting volumes of its components.

Before proceeding to a more detailed examination of Ampère's letter, there is another aspect of this episode which still requires some inquiry: the apparently peculiar conduct of the French physicist towards the man who had preceded him in the formulation of a molecular generalization. His behavior is, indeed, very difficult to explain, unless it is seen within the prevailing attitude of the time when regular scientific communication, although on its way to becoming an essential institution, was not yet considered a usual implement for the natural philosopher.

The following are the facts which, from the point of view of a twentieth-century observer, cannot be easily explained: first of all, Avogadro's 'Essai d'une manière' was published in French in a French journal mainly devoted to physical problems and well known at the time. In the second place, as Ampère himself emphasized, Gay-Lussac's work on combining volumes which had attracted his attention had induced him to find a physical interpretation for those same facts. It is also very clear that his letter to Berthollet, certainly not the result of a *foudroyant* insight, represented an abstract in anticipation of a comprehensive theory covering also the composition of nongaseous substances. This leads to the question of how deeply was Ampère involved in this particular endeavor. In a letter to the Genevian Jacques Roux (a member of a small group of Ampère's friends in Lyon) dated 11 March 1814, Ampère defines his 'discovery' of the Law of Nature enabling to predict *a priori* the set combining ratios of bodies, the 'most important thing I ever conceived in all my life after what I have done last summer in metaphysics'.[113] Undoubtedly, Ampère was aware of the importance of his speculation in this field of inquiry.

It is worthwhile to recall here the answer given not too long ago by a French eulogist:[114] 'L'esprit d'Ampère était trop mobile et, pour désir de la perfection, à force de retouches successives, il arrivait à ne publier que tardivement ses recherches.' After explaining the delay in this way, the following clearly reveals that, a century later, the old chauvinism is not yet dead:

> Ampère avait donc gardé inédit son travail commencé de 1809 et, lorsque il se décida, sur les instances de Berthollet, à en donner le résumé en 1814, il ignorait l'existence du mémoire d'Avogadro qui accompagné de considérations

inexactes, avait été à peine remarqué ou considerée seulement comme une théorie de plus ajoutèe à tant d'autres.

If one accepts this explanation which is quite open to criticism[115] then one may also understand why Ampère had ignored Avogadro's earlier memoir: it was, after all, just another theory and, furthermore, it contained 'inaccurate considerations'.

Another intepretation, probably more legitimate, is that Ampère was not only completely unaware of Avogadro's essay, but perhaps even after it was reported to him, he still did not care to study it. Otherwise, he would not have dismissed so lightly a conceptual system much clearer, more complete, and systematic than his own. The alternative – that he proceeded anyway to elaborate his ideas despite his knowledge of the content of the 'Essai d'une manière' – cannot be taken into consideration if one accepts at full value Ampère's assertion, stated in a footnote in the letter to Berthollet, that he became familiar with Avogadro's essay only after he had completed his writing. Apparently since 1809 Ampère had drawn the outline of the gas hypothesis and only in 1814, upon Berthollet's influence, did he decide to publish it. Avogadro and Ampère arrived independently at the same conclusion.[116] Later in the 1820s Ampère dedicated a copy of his 'Théorie des phénomènes électrodynamiques' to Avogadro. On this occasion he again misspelled his name as 'Avogrado';[117] a detail which suggests that there was no real acquaintance between them, and that Ampère may have forwarded a copy of his work only upon the request of Avogadro himself.

As for the full-sized work which had been anticipated, it was apparently never published. Did Ampère then have such slight interest in further evolving his own hypothesis? According to some views, he seemed too involved in other speculations to devote more time to the molecular issue. It is more likely that he had serious doubts about the validity of some of the ideas he had expressed or about the timing of their submission for consideration. There is also the possibility that, aware of Avogadro's subsequent publication on the same subject,[118] Ampère discontinued his investigations for the time being. He returned to this topic almost a quarter of a century later, in the 'Note sur la chaleur', this time introducing a clear distinction between 'particules' and 'molécules', unfortunately neglected in his earlier paper. His interpretation was then much freer of ambiguities and in harmony with the ideas expressed by Avogadro. The latter again was not mentioned by Ampère in his 'Note' nor, as a matter of fact, were others who in the preceding twenty years had devoted their attention to the intimate constitution of matter.

In agreement with his personality, Avogadro reacted to Ampère's note of 1814 by simply taking the latter's statement at face value and properly avoiding any pointless polemical debate with him. In 1816 in the 'Memoria sul calore specifico dei gas composti'[119] he mentions Ampère's letter for the first time. He states, and without further comment, that the Frenchman 'has adopted my hypothesis'.

From the historical point of view, the appearance of the 'Letter to Berthollet' is of great importance. As mentioned earlier, the repute of its author and the fact that it had been addressed to one of the most influential scientists of the time eventually gave to this document a weight far exceeding its intrinsic value. It was not accepted immediately because of the prevailing attitude of those years which was so minimally receptive to pure speculations; sheer intuition unsupported by experimental evidence had limited appeal. After the appearance of atomism which still had more critics than supporters, there was simply no room for further corpuscular interpretation of the matter despite the fact that many gaseous combinations could not be explained solely in Daltonian terms.

To those who have observed that Avogadro's hypothesis was too permeated with physical concepts to be fully appreciated by the chemists of his time, it is easy to answer that the ideas set forth by Ampère in the early decades of the century were even more remote from the practical problems intriguing men such as Davy, Berzelius, Gay-Lussac, and Thenard. In its relation to the belated acceptance of Avogadro's generalization, the note of Ampère muddled the issue of the correct priority of the two theories to the extent that, even at the present time, for some authors Ampère and Avogadro are considered co-proposers of the molecular hypothesis. More important, however, is the fact that when the ideas proposed by Ampère were finally revived in 1826 by J. B. Dumas, the then-young pupil of Thenard, instead of clarifying them, added further confusion to a scheme which in itself was already anything but lucid.

A comparison of the two essays clearly indicates that while Avogadro elaborated his hypothesis as a synthesis between Dalton's quantitative observations on the combining weights and those of Gay-Lussac on the combining gases, Ampère based his line of reasoning only on the conclusions reached by the latter. It is likely that he was aware of Dalton's work, at least indirectly,[120] but he chose to ignore it in his letter to Berthollet.

Ampère's attention was centered upon the mutual position of the molecules (atoms) and the shape of the particles which they form. To each particle was assigned a geometrical form which defined the position of the molecules, whose number was set as four, but could be much higher. Due to their 'infinite' mutual distance, the molecules were enclosed in a space much larger than the volumes they occupy. Most elemental particles were represented by tetrahedrons, by combining with each other they yielded more complex geometrical shapes, such as a parallelepiped, a prism hexagonal, and a rhomboid dodecahedron corresponding to particules containing eight, twelve, and fourteen molecules, respectively.

Although this intepretation applied to both solid and gaseous substances, Ampère found that it was possible to use much simpler considerations in dealing with the latter. As Gay-Lussac had pointed out, there is neither cohesion nor attraction among the particules[121] of gases in general, and in these conditions they 'obey simple and regular laws'. For Ampère then – he arrived at his conclusion in a more straightforward and simpler way than Avogadro – the mutual distance among gaseous particules was determined only by the temperature and pressure to which they were subjected. This meant that under equal volumes, particules of any gas (simple or compound) were equidistant. Or, as Ampère concluded, 'the number of particules was proportional to the volumes of gases'. A number of examples – based upon the quantitative observations reported by Gay-Lussac – were then offered to prove the validity of the aforementioned relationship between volumes and numbers of particules.

However, it should be said at this point that the approach followed here by Ampère was quite inadequate to explain the behavior of gaseous combinations whenever odd ratios of component gases were involved. Such was the case, for instance, of water and ammonia. He stated only that since water vapor was formed by an equal volume of hydrogen and half its volume of oxygen, then its particle was composed by a whole particule of hydrogen and a half particule of oxygen. This could hardly be considered an explanation for the obstacle represented by the contractions in volumes observed in some chemical reactions among gases, an obstacle that Gay-Lussac had, without success, attempted to overcome in his 1809 memoir. If one recalls the objections raised by Dalton, among others, to the molecular generalization for its inability to solve these contradictions,[122] then it is easier to understand Ampère's failure to have his ideas accepted at the time. On the other hand, Avogadro not only realized that there existed a problem of reconciling the experimental evidence with the equal volumes–equal numbers generalization, but he also offered a solution for it which is still valid now.

Ampère had suggested that one volume of ammonia was formed by the combination of a half volume of nitrogen and one and a half volumes of hydrogen;[123] in agreement with this hypothesis, one particule of ammonia should result from the combination of a half particule of nitrogen and one and a half particules of hydrogen which, of course, is a *contradictio in terminis*. The reply to this question, left unanswered by Ampère, came only through Avogadro's assumption that the compound molecules, once formed, would undergo a division leading to bi-, tri-, and tetratomic molecules in agreement with the number of elementary molecules entering into combination.

It is evident then that the notion of molecular division, whenever gases combine chemically, was not present in Ampère's letter to Berthollet. Here lies the most striking difference between his interpretation and that offered by the Piedmontese.

Despite its inadequacies, the equal volumes-equal numbers generalization, as proposed by Ampère in his letter to Berthollet – it may be compared to a gem embedded in a vast amorphous matrix – overshadowed the more comprehensive hypothesis proposed earlier by Avogadro. In France, beginning with Dumas and Gaudin through Wurtz, Hoefer, Delacre, and recently Taton, the work of Avogadro was either ignored or thought of as being on the same level with that of Ampère. The notable exception was Grimaux and earlier Baudrimont who recognized the priority of the 'Essai d'une manière'.[124] In the British Isles both Prout and Henry mentioned Avogadro together with Ampère, while Donovan referred only to the former.[125]

Dumas

In the *Annales de Chimie et Physique*, twelve years after Ampère's letter, a memoir with a very unassuming title appeared; submitted by a young researcher then working under Thenard at the École Polytechnique, it dealt with the very first attempt to extensively apply the molecular hypothesis to the determination of the atomic weights, moving from the experimental finding of the vapor densities of a large number of elements. The impact of this early work of Jean B. Dumas cannot be overlooked; although only twenty-six at the time, he was already a member of the French scientific oligarchy. Through Thenard he had easy access to Gay-Lussac, Vauquelin, Dulong, and Chevreul, among others.[126] Eventually, he would become and remain the authoritative spokesman of the French school for a whole generation of chemists.

In his 1826 'Mémoire sur quelques points de la Théorie chimique',[127] Dumas, although accepting in principle the assumption of the equal volumes–equal numbers and making this the basis for his experimental work, seemed to have compromised Ampère's and Avogadro's ideas. However, he gave no credit to Avogadro for his earlier and more lucid intepretation of the same facts. Graebe supports the view that Dumas was personally influenced by Ampère when he began his investigations of the atomic theory.[128] According to the same source, he in fact also owed to Ampère his appointment as successor of Robiquet at the Atheneum.

A parallel reading of the first pages of Dumas' memoir, of Ampère's letter to Berthollet, and of Avogadro's 1811 essay can be quite revealing; Dumas offers this interpretation of the molecular hypothesis: 'que dans tous les fluides elastiques sous les mêmes conditions les molécules se trouvent placées à égale distance, c'est-à-dire qu'elles sont en même nombre'.[129] Avogadro suggested, 'les nombres des molécules integrantes dans les gaz quelconques est toujours proportional aux volumes', but he is even clearer in his 1814 text 'que les volumes égaux des substances gazeuses, à pression et températures égales, représentent des nombres egaux des molécules',[130] with the result that 'les densités des differens gaz sont la mesure des masses des molécules propres à ces gaz', and finally 'que les rapports des volumes dans les combinaisons ne sont que les rapports entre les nombres des molécules qui se combinent pour former des molécules composées'.

As for Ampére, his conclusions 'que à des pressions et des températures égales, les particules de tous les gaz soit simples, soit composées, sont placés à la même distance les une des autres', and therefore, that 'le nombre des particules est dans cette supposition proportionel aux volumes des gaz',[131] are less lucidly expressed than those of the Piedmontese physicist.

Dumas at first ran into a series of factual inaccuracies – such as that all physicists of the time were in agreement with the equal volumes–equal numbers hypothesis, that nobody had yet considered the experimental application of this hypothesis, and that only Gay-Lussac had accepted the possibility of a molecular division of elemental gases. He then suggested that such division may in fact take place when gases combine. While he apparently seemed in agreement with both Avogadro and Ampère, he discarded the term 'particule' adopted by the latter in favor of 'molécule'. He added that the more direct result of Ampère's conjecture 'consiste à considérer les molécules des gaz simples comme étant susceptible d'une division ultérieure, division qui se produit au moment de la combinaison et qui varie suivant la nature des composé'.[132]

In saying so, he departed from Avogadro whose division of the 'constituent'[133] is unrelated to the nature of their compound or 'integral' molecule. For Avogadro, moreover, and this mechanism was not suggested by Dumas, the compound molecules split into two or more parts composed of one-half, one-quarter, etc. of the number of elementary molecules. The concept then that two divisions were taking place, one of the constituent molecule at the moment of their combination, the other of the integral molecules after the combination, appeared very clear in Avogadro's mind, although less intelligible in his own words.

This concept, left undiscussed by Ampère, certainly escaped the attention of Dumas.[134] A further ('ultérieure') division of the molecules of the elemental gases which he mentioned, may be interpreted as meaning that for him such molecules are the result of a previous division. But, if this is the case, was he following Ampère and thus thinking of 'particules' formed by 'molécules'; therefore did he postulate the existence of three classes of particles (atomes, molécules, particules)? Or was he equating the terms molecule and atom?

A possible answer to these questions was given a few paragraphs later when Dumas introduced Berzelius and his 'plan general qui consiste à répresenter leur atomes comme s'ils étaient formés par des *atomes simples* réunis, toujours en nombre entières'. Now a distinction was apparently made between atoms and 'simple atoms', while Ampère's particules and Avogadro's molecules were put aside. Thus, according to this 'plan général', water was formed by two atoms of hydrogen and one of oxygen, etc. Are these 'simple atoms'? No doubt they are according to Berzelius. However, when Dumas afterwards tried to reconcile the combining ratios approach proposed by Berzelius with his 'idées énouncées sur la constitution des gaz', he reached the most alarming conclusion. Since, he argued, the formula of a compound should always represent what is entering in one volume of that substance taken at the gaseous state, then water must be formed 'par un atome d'hydrogène et un demiatome d'oxygène, l'acide hydrochlorique par un demiatome de chlore et un demiatome d'hydrogène'.[135] This short sentence would have the most serious consequences on the future of the molecular hypothesis. Due to the importance of the publication in which it appeared and the prestige of the scientific circles to which its author belonged, it did not escape the attention of many interested people.[136]

Dumas, in his attempt to combine Ampère and Avogadro, muddled the concepts of both.[137] Towards the end of his memoir he seemed to realize how unsteady was the ground on which he walked: despite the 'immense' benefits that natural philosophy had drawn from the works of Gay-Lussac, Berzelius, Dulong, Mitscherlich, Petit, as well as from the theoretical views of Ampère and Avogadro,[138] he concluded 'nous sommes bien éloignés encore de l'époque oú la chimie moléculaire pourra se diriger par des régles certaines'.[139] Two years went by before Berzelius' criticism fell on the head of Thenard's young pupil. In 1828, in the seventh volume of the *Jahresberichte*, appeared the review of the 'Mémoire sur quelques points de la théorie atomistique'.[140]

It is only fair to say that if Dumas did not succeed in promoting the cause of the molecular hypothesis,[141] and actually he was responsible for making it less convincing and less coherent than the prevailing doctrines of the time, Berzelius' contributions, at least in this case, seem scarcely productive either. His comments in the introductory part of the memoir were quite harsh: although possibly they did not attain the level of severity he used on other occasions, nevertheless his opinion of Dumas' conclusions was of explicit condemnation. 'Ehemals war es gewöhnlich, dass man eine Hypothese, sobald sie ad absurdum führte, als widerlegt betrachtete.'[142]

Undoubtedly the young French chemist had pushed the issue too far in postulating the existence of half atoms of hydrogen and chlorine. It is understandable that such incongruity could hardly be accepted by any natural philosopher of the time. As for the two main assumptions stated by Dumas, the first one – that at the same temperature and pressure, the particles of all gases, simple and compounds, are equidistant – was reproduced by Berzelius without comment. He used, however, the expression 'solid atoms' to indicate the particles that Dumas had defined as 'molecules'. Moreover, he cited as an example, nitrogen oxide, whose 'atom' is formed by 'half atoms' of its components, which is not even present in Dumas' original text – a quite peculiar way, one should say, of reviewing somebody else's work.

No less confusing is the fact that, after correctly reproducing Dumas' second generalization – that molecules of the elemental gases undergo division when they enter into combination and such division is in agreement with the nature of the compound thus formed – Berzelius employed in this case the term 'molecule' and not 'solid atoms'; this second 'Vorstellungweise', though, is immediately rejected, because its application would lead to the unacceptable composition of water suggested by Dumas. Quite clearly there was no alternative in the mind of Berzelius: since atoms cannot be cut, then the molecular division of the elements in the gaseous state is impossible. Thus, there remained only the volume theory, whose application to the determination of atomic weights, as experimentally proved by Dumas, could still be considered valid.

As for Avogadro's hypothesis, it received, through the involved presentation of Ampère, the erroneous conclusions of Dumas and the severe critical analysis of Berzelius, its most serious blow seventeen years after it had been proposed. A possible explanation can be found for this setback and the long interval in the reaffirmation through his writings of the main proposition of his molecular hypothesis: Avogadro's major chemical essays were published between 1811 and 1823; only in 1838 in the second volume of the *Fisica dei corpi ponderabili* did he return to the equal volumes–equal numbers postulate.[143] On that occasion he openly regretted that 'many chemists'[144] had not accepted from the beginning his attempt to link the laws of definite proportions and of combining volumes. There is then a definite change in his earlier feelings[145] about, for instance, the position of Berzelius towards the validity of his basic assumption.

Gaudin, Mollet, and Baudrimont

After the 1826 memoir by Dumas, one has to wait until 1845 to find Avogadro distinctly mentioned in France as the true author of the molecular hypothesis. In that particular year, M. A. Baudrimont, then professor of chemistry at Bordeaux, submitted to the *Académie* a short 'Mémoire de réclamation'[146] concerning an earlier note[147] read at the same Academy by Auguste Laurent. In it, after observing that 'the elements of the atoms by combining in this or in that proportion, yield several atoms or equivalents having different properties', Laurent expressed the hope that this concept would not raise any objection, because 'it has been a long time since Ampère had admitted the divisibility of atoms;[148] nobody, however, as far as I know, has drawn from this fact the conclusion that I have emphasized'.

Baudrimont in his 'Réclamation' strongly objected to this statement; he recalled two notes on the same subject which he had earlier presented to the Academy and, in complaining that Laurent had neglected to examine his *Traité de Chimie* where the whole question was discussed in detail, reminded Laurent that in 1811 Avogadro had anticipated Ampère in the assumption of 'divisibilité des pretèndues atomes chimiques'. Baudrimont quoted three works of Avogadro published in French periodicals in which the concept of molecular division had been supported. He concluded his note by stating that he had proved through 'preuves irrecusables' that the so-called 'chemical atoms' are divisible.[149] From the historical point of view this represented a rare case for those years: a French scientist who, in dispute with a compatriot, upheld the priority of a foreigner over a Frenchman on a very important matter.

In the interval of time between the 'Mémoire sur les masses' and Dumas' 'Mémoire sur quelques points', another French physicist had reached the same conclusions as Avogadro. In a memoir which appeared in the February 1820 issue of the *Journal de Physique*, Joseph Mollet, then Secretary of the Academy of Lyon, discussed quite extensively the physical properties of both elementary and compound gases.[150] His approach to the problem was typical of a calorist of the time. He moved from strictly physical considerations to his first conclusion :

> that the particles of every gas, under the same conditions, are all equally separated and that a set space cannot contain more of a certain gas than another; hence, the difference in specific gravity derived not from the unequal distance among particles, but from the really unequal weight of said particles in different gases.

Mollet, who was mentioned by Avogadro in the 'Nouvelles considérations', drew the conclusion from very simple calorimetric experiments that, at least in the cases of air, hydrogen, and carbon dioxide, their specific gravities – while inversely proportional to the space occupied by their particles – were, on the other hand, directly related to the amount of free space existing among those same particles.[151] By assuming that these intraparticular spaces were occupied by the caloric, then a relation could be obtained between the specific gravity of a gas and its capacity for caloric. This actually had been experimentally proven by the data reported by Delaroche and Berard in 1813 suggesting that hydrogen had a lower capacity for caloric than air, and the latter even lower than that of carbon dioxide.[152] Mollet did not quote Avogadro or Ampère, but his main assumption was strikingly close to that set forth by Avogadro nine years earlier: 'Les particules de tous les gaz, dans les mêmes circonstances, sont sans doute en égal nombre dans un espace donné.' There was no attempt, however, to apply these concepts to chemical problems as in the case of both Avogadro and Ampère. This was strictly a physical work.

It is worth noting Mollet's axiomatic view of the equal volumes–equal numbers generalization. The assertion that 'it is necessary to reject any explanatory system which might be in contrast with this principle' certainly did not please those natural philosophers who supported the opposite view that hypotheses should be deduced only from facts and figures. In any case, Mollet's essay escaped the attention of most at the time: it was unnoticed by Dumas, Ampère, Laurent, Gaudin, among others, and by the French historians in the second half of the nineteenth century. Later on, Cannizzaro did not mention him in his 'Sunto' nor Graebe in the 'Entwicklung der Avogadroschen Theorie'.

In the third decade of the century, halfway between Mollet's article and Baudrimont's note to Laurent, in the *Annales de Chimie* appeared a paper with a very ambitious title, written by a twenty-nine year-old former pupil of Ampère and Dumas.[153] A parallel between Marc Antoine Gaudin, the author of this paper, and Avogadro may appear obvious at first: they were both young and unknown when they presented their new and unorthodox views; they also shared limited personal ambitions and scant means. The similarity between the two, however, ends here. The degree of acceptance of Gaudin's ideas cannot be compared with that of Avogadro.

Gaudin was a former student of Ampère, whose influence over the academic world continued to be profound, to say the least, until the end of his life. Through Ampère with whom he probably maintained good relations, he became known to the *Acadèmie*. Thus it is understandable how his first major paper was approved,[154] although with some reservations, in 1832 by Gay-Lussac and Antoine Becquerel, the *Commissaires* appointed by the same body. In this way, the young physicist had the unique opportunity of publishing his ideas in one of the most influential scientific periodicals of the time. Gaudin's generalizations were also better accepted because they were undoubtedly more lucidly expressed than Avogadro's, and furthermore, they appeared twenty-two years later when the notion of two different classes of particles – although still debated – was less repellent to chemists than in the past.

The first half of the article was the more important from the historical point of view; in it, both atoms and molecules were clearly defined, and their difference was correctly stated.[155] Quite significant and revealing in the introductory paragraphs was the passage where Gaudin supported the view that – at least when dealing with the theories – every chemist, regardless of how 'small' he may be, is entitled to put forward his ideas even at the cost of 'combattre' his masters.[156]

For Gaudin, atoms were small, indivisible, homogenous, spheroidal bodies which might gather together to form the molecules. He then proceeded to find the 'nombre atomique-moléculaire' (or the number of atoms contained in the molecules) for several elements and compounds. The foundation of his reasoning, he indicated, rested on the law of Gay-Lussac on the combination of gases in simple ratios and on Ampère's 'principle' of the equidistance of all the molecules of gaseous bodies under equal conditions. Nothing was said of Avogadro who, by then, had already published at least ten papers dealing with his own gas hypothesis; this seems amazing if one considers that Ampère himself had recognized the priority of Avogadro in his 1814 note to Berthollet.

Gaudin then proceeded to illustrate his ideas by using the formation of hydrochloric acid from its components as his first example.[157] While this explanation is convincing, less so are those offered for the formation of water and ammonia, where again a variation of value is observed. Although his considerations in proving the diatomic nature of nitrogen, hydrogen and oxygen were of pure geometrical essence, they appear, in general, much simpler and more intelligible than those previously set forth by others on the same subject. In the case of mercury, for instance, he reasoned that from the analysis of its oxide it seemed that such a compound could not be formed unless one considered the 'particule' of oxygen as divisible by two and the particules of mercury vapor at a mutual distance equal to $\sqrt[3]{2}$.[158] Gaudin felt this was absurd, because in such a case mercury would be diatomic too and behave just like the gases and vapors of compounds. Hence, the 'particule' of mercury cannot be divided, and therefore it is an atom. Since this atom can be compared to a diatomic molecule of oxygen, then the weight of an atom of mercury is 12.6 times that of oxygen.

As for sulfur, Gaudin, on the basis of its specific gravity determined by Dumas, argued that the formula for sulfuric acid would be equal to $SO_9 + 3\ H_2O$[159] if one assumes that the 'particules' of sulfur, oxygen, and hydrogen are indivisible. However, since this formula is illogical, then the only possible conclusion to draw was that the 'particules' could be divided and were, accordingly, not atoms.

By moving from considerations similar to those used to prove the diatomic nature of chlorine, Gaudin concluded that bromine and iodine could also be diatomic. He then discussed several compounds of silicon and boron. For the latter element he reached the conclusion – in disagreement with Berzelius – that its 'particule' is divisible or, in other words, is diatomic. This leads to a weight for boron of 0.66 times that of oxygen,[160] a figure which is substantially correct, even if erroneously obtained. Berzelius' evaluations of the composition of silica and of the atomic weight of silicon were not accepted by Gaudin, who, more confident of the experimental value reported for silicon chloride by Dumas, realized through a simple calculation that if silicon chloride contains four atoms of chlorine, then the oxide should have two atoms of oxygen for each of silicon and not three as suggested by Berzelius. The weight of silicon, which he considered monatomic, Gaudin set at 1.849 times that of oxygen, a figure which is closer to the current one than any other proposed so far; Berzelius had estimated a value of 2.77.[161]

In the last paragraph of his paper, Gaudin – just before properly recognizing that most of the atomic weights discussed in his paper had been taken from Berzelius – again challenged the atomic weights of four more elements as proposed by the latter. For him the atomic weights of sodium and silver should be halved and, in this particular instance, no explanation was offered for his position.

It may be worthwhile at this point to examine how Avogadro and Berzelius reacted to Gaudin's article in view of their close involvement with the ideas proposed by the Frenchman. In his second extensive review of Gaudin,[162] which was published in the 1835 *Jahresberichte*, Berzelius was surprisingly polite, especially if one considers Gaudin's disagreement with some of the atomic weights which he had established himself. In essence, his main criticism was directed toward Gaudin's philosophy rather than his conclusions. These are all axioms, Berzelius felt, the main one being the assumption set forth by Ampère, and must be treated as such. Axioms, he continued, have an advantage over facts, for one does not need to accept or reject them. From these axioms a hypothesis evolves whose correctness or not will be determined only by time. Berzelius thought that the irreconcilability, observed when single gases combine with each other, between the number of their atoms and the number of their volumes had been overcome by Gaudin when he adopted Dumas' concept of 'divisible atoms'.[163] Needless to say, Berzelius was at fault here, because both Ampère and Avogadro had preceded Dumas in assuming the division of 'molecules'.

Further confusion was added when Berzelius in his review attributed to Ampère the notion that in all gases the distance between 'atoms' was the same, and later on to Gaudin the same idea with the difference that, now, the distance among 'molecules' was the same. In fact, Gaudin had clearly stated[164] that for Ampère the distance between molecules was the same in gaseous bodies under equal physical conditions. Berzelius, however, may have been justified in his mistake if one recalls that Ampère in his 'Note to Berthollet' had termed as 'molecules' what he should have defined as 'atoms', while he called 'particule' what later on Gaudin called 'molecule'. Dumas, on the other hand, identified the molecules with 'divisible atoms'. The jumble could not have been worse, and here even Gaudin shares part of the responsibility; a short note on his part would have been sufficient to explain that his 'molecules' were in fact Ampère's 'particules'.

When he reviewed the examples given by Gaudin, Berzelius seemed to have grasped their meaning: the molecule of water is triatomic, that of ammonia is tetratomic, etc. However, in his concluding remarks, although the idea of clustering together the atoms of elementary gases may appear as 'something attractive', he defines the whole presentation of Gaudin as 'nur ein Spiel der Phantasie'.[165] Here, *in nuce* emerges the candid and general opinion that the most influential chemist of his time had on the molecular hypothesis. There was simply no room for fantasy or imagination in the mind of the brilliant and resourceful Berzelius and, unfortunately, he was not the only one to feel that way.

In the second volume of his *Fisica dei corpi ponderabili*, which appeared in 1838, two years after Berzelius' review in the *Jahresberichte*, Avogadro devoted several paragraphs to Gaudin.[166] It is only fair to say that the tone of these remarks proves once again the mild nature of their author. Before discussing Gaudin's interpretation of mercury vapors as monoatomic, Avogadro pointed out that the French physicist 'combines in his work the fundamental basis on gas constitution with several assumptions aiming at simplifying its application, but which do not seem supported by sufficient reasons.' He added that Gaudin 'visualizes the way in which the elementary atoms may be conceived when placed in these different kinds of gaseous integrant molecules'. One may possibly detect here Avogadro's regret for not having supplemented his ideas with a graphic explanation that might have helped to make them clearer.[167] Quite revealing and somehow puzzling was the concluding sentence of the review:

> It is clear, however, that these [Gaudin's] ideas are very hypothetical, and that the general law of the constitution of gases is by itself independent from these assumptions in its application to single bodies; therefore, I will not take the time to mention the general results to which the same assumptions are leading Mr. Gaudin.[168]

This seems to indicate that for Avogadro: (a) the concepts stated by the Frenchman appeared too conjectural to be taken into consideration – a stone that certainly he could not cast at other natural philosophers, and (b) there was no correlation between Gaudin's speculative framework and the essence of the gas hypothesis which he had formulated in 1811.[169]

Cannizzaro, in examining mercury in his '*Sunto*', recalled that Gaudin had defined this element as monoatomic and supports the view that, by following Ampère and Avogadro, it would have been possible to produce the same result; this proves that Cannizzaro was not familiar with the *Fisica dei corpi*; otherwise, he would have known that for Avogadro mercury was diatomic. As for sulfur, this element was not discussed at all by Cannizzaro; its degree of 'atomicity' was still under examination at the time of his writing, and it would remain so for several years.[170]

Quite definitely, the important distinction between atoms and molecules made by Gaudin in 1833 – and that he also graphically illustrated – did not escape Cannizzaro. Yet, it is fair to say that Cannizzaro in his *Sunto* did not give Gaudin proper credit for his essential contribution which in retrospect should be considered as a key step forward for the acceptance of the molecular hypothesis.

Gerhardt and Laurent

In 1833 in the same year as Gaudin's memoir, Dumas reported the possibility of replacing hydrogen by chlorine in organic substances as far apart as turpentine oil and ethyl alcohol.[171] Further research led him to the conclusion that the substitution for an atom of hydrogen by an atom of oxygen or a halogen is definitely possible in a number of substances. These investigations were further advanced by Regnault and Malaguti between 1836 and 1845. In 1838 Dumas prepared for the first time trichloroacetic acid with properties very similar to those of acetic acid.[172] Berzelius reacted to this in the same year, denying in the *Annales de Chimie* that chlorine with its strong electronegativity could enter an organic radical: 'his idea', he argued, 'would violate the basic principles of chemistry'.

The theory of types,[173] stated again by Dumas in February 1840, followed the substitution theory almost as a necessary corollary. In disagreement with Berzelius he believed that the electrical polarity of atoms is susceptible to change. 'No view', he says, referring to the dualistic system, 'was ever more fitted to retard the progress of organic chemistry',[174] and, naturally by taking this position, he found himself in open conflict with Berzelius, who shortly afterwards ironically asked Dumas why in carbon tetrachloride, carbon too could not be replaced by chlorine. For Berzelius the active group in acetic acid was ($C_2O_3 + H_2O$) which is present in chloroacetic acid considered by him as $C_2Cl_6 + (C_2O_3 + H_2O)$. This explained why chloroacetic acid displays the same properties as acetic acid; nevertheless, the difficulty of accounting for the replacement of hydrogen by chlorine remained. Berzelius' attempts[175] in this direction failed, and resulted only in more confused and elaborate formulas. In 1839 he wrote to Liebig that the laws of combination found in inorganic chemistry were the only safe guide to the composition of organic compounds; eighteen years earlier Avogadro – it should be noted – had reached the same conclusions in his neglected 'Mémoire sur la manière de raméner'.

The dualistic theory was attacked by Liebig and Dumas in 1839, while the research conducted by Berzelius in organic chemistry appeared more and more sterile. In discussing the theory of substitution, Liebig said in 1839 that he could not share Berzelius' interpretation since it rested upon a 'mass of hypothetical assumptions of which proof of any kind is lacking',[176] a serious accusation especially when addressed to a man who had been the leading chemist of his time.

Such a severe criticism of the dualistic theory meant, of course, the beginning of the end for the concept of electrical polarization of atoms which had represented one of the main obstacles to the acceptance of polyatomic elemental molecules. The period between 1830 and 1850 is characterized in a way by the great personal animosities between the chemists of the new and the old generation. Dumas, Liebig, Thenard, and Gerhardt were all involved. A review of the literature shows that their quarrels were endless, while their science often behaved as a ship without a rudder.

During those years Laurent had proposed a theory of substitution by reporting the evolution of hydrochloric, hydrobromic, and nitrous acids when a certain number of substitution reactions take place. More important, however, was his article of February 1836[177] on the 'Théorie des composées organiques' which was severely criticized by Berzelius; in it Laurent had attempted the first rational arrangement of organic compounds by classes such as methylene, naphthalene, etc. In his doctoral dissertation (1837) Laurent offered a nucleus theory which could also be used to classify organic compounds; in essence, he suggested a geometric model, namely a regular prism whose eight vertices represented eight atoms of carbon, while at the centers of the twelve edges he placed the atoms of hydrogen. These edges could be replaced by edges of other elements such as halogens and metals; not only are actions of substitution possible, but also of addition; in the latter case, the properties of the nucleus are radically modified.

Once more the nucleus theory was harshly criticized by Berzelius, but this time, however, he had Liebig as an ally. Then Dumas, whom Berzelius had chastised as Laurent's mentor, came into the picture, denying his support of the substitution theory and giving the responsibility for it solely to Laurent.[178] A further attempt to classify organic compounds had been made in 1844 by Laurent when he proposed the adoption of an *arbre chimique*, whose branches were all related to a common nucleus; the 'chemical tree' contained five fundamental types such as $C_{20}H_{20}$ (fundamental nucleus) . . . $C_{20}H_{20}O_2$ . . . + O_4 . . . + O_6 and $C_{20}H_{20}H$ (fundamental prometallide).[179]

A few months before the 'Classification chimique' was published[180] a profound friendship began between Laurent and Gerhardt which lasted the rest of their short existences. Gerhardt, briefly a pupil of Liebig and later an assistant of Persoz and of Chevreul, had set to work in 1842 on the equivalents and a year later published[181] the first results of his studies. This paper was preceded by the now renowned essay 'Recherches sur la classification chimique des substances organiques',[182] where he suggested doubling the equivalents of several elements, because in the reaction of double decomposition of organic compounds, he had observed the formation of double molecules such as C_4O_4 (instead of C_2O_2), H_4O_2 (instead of H_2O) etc.; these formulas were exactly double those set by Berzelius, and hence were considered more reliable by Gerhardt.

Accordingly, he decided to halve the formulas of organic substances in order that they be consonant with the inorganic compounds obtained from them. As for the equivalents generally used until then for organic compounds, they would have to be replaced by atomic weights. Gerhardt's position, however, was such that he had not yet extended to elements the rule that the smallest constituents of compounds under gaseous conditions should be referred to two volumes; to some extent, he still considered the terms 'equivalent', 'atom', and 'volume' synonymous. One should not neglect to mention at this point that Gerhardt in the first volume of the *Précis de Chimie Organique* (1844–46) indicated – thirty-three years after Avogadro had reached the same conclusion – that a correlation was possible between the experimentally determined densities of chemical compounds and their compositions by volume. Unfortunately, he muddled this important point by over-extending his views, such as by assuming that the composition of all metal oxides should be related to the formula of water or be such as Me_2O. In doing so he not only modified Liebig's equivalents, but also tampered with Berzelius' atomic weights.

Almost simultaneously, Laurent published his article 'Sur les combinaisons azotées'[183] generally recognized as momentously relevant for the subsequent development of the molecular hypothesis. He was then working in Balard's laboratory at the *École Nationale*. In his essay a very clear distinction is made between equivalents, atoms, and molecules, and the divisibility of the latter is recognized. For inexplicable reasons he attributed *tout court* to Ampère the merit of being the first to conceive of the polyatomic molecules.[184]

On the same occasion he rationalized the reaction taking place between hydrochloric acid and a metal with the subsequent development of hydrogen, one of the so-called reactions in 'the nascent state' which until then had been unaccountable.

Considerable support for Avogadro's ideas came in 1846 from the experimental work of two young physicists also working in Paris; P. Favre and J. Silbermann had concluded from their research[185] on the combustion heat of carbon that the molecule of oxygen can 'split' or 'atomize' and therefore that oxygen should be considered as O_2.

This experimental evidence plus the distinction made by Laurent indicate that the year 1846 was ripe for the acceptance of the molecular hypothesis. However, the majority of chemists and physicists of the time remained faithful to the notion of equivalents, supported more than ever by their influential exponents, the only possible exception represented by Berzelius then in the twilight of his life.

Williamson, Bineau, and Cannizzaro

More evidence was on its way to prove that through chemical investigation it was possible to determine molecular formulas as accurately as through the measurement of vapor densities. Four years after Laurent's essay on the nitrogen-containing compounds, there appeared in the *Report of the British Association* a series of articles all devoted to the same subject, etherification. They were written by a twenty-six year-old, English chemist, a former student of Liebig and Gmelin.[186]

In very lucid terms, 'alcohol is water in which one half of the hydrogen is replaced by carburetted hydrogen',[187] Alexander Williamson proposed his water type theory which represented another important step for the eventual acceptance of Avogadro's hypothesis. In his definition of an alcohol and of an ether, he clarified the process of etherification which thus far had been confusing. He proved that through the use of the water type theory it would be possible to interpret the formulas of acetic acid and several other organic compounds; at the same time, he discarded the equivalents and adopted the atomic weights proposed by Gerhardt.

With a new generation of chemists – Williamson was sixteen years younger than Laurent – the degree of approval for the volume theory, still generally assigned to Ampère, became more general after 1850. However, obstacles had arisen again from the inorganic chemists: the experimental evidence indicated that for a number of salts, especially chlorides and carbonates, their vapor densities were much lower than those expected. Research conducted by Regnault, Mitscherlich, Playfair, and eventually by Troost and Deville after 1856 suggested that vapor densities of both elementary and compound substances changed according to the temperature at which they had been determined.

The first rational explanation for this was offered in 1839 by A. Bineau[188] who assumed that the substance obtained by heating ammonium carbonate comes from the dissociation of such a compound, since it remains in the state of vapor even at a lower temperature than that of dissociation. Bineau had arrived at a similar conclusion by studying the behavior of sulfuric acid, whose density decreases by raising the temperature. He explained that it probably dissociates into water and an anhydrous component. Few years later in his 'Mémoire sur les combinaisons azotées', Laurent, shortly after defining the term molecule, says: 'as for the very rare abnormal compounds which are encountered, they must be thus represented: chlorhydrate of ammonia $\frac{1}{2}ClH_4N$ indicating that the molecule is ClH_4N, but that, under the influence of heat, it is divided in two on passing to the gaseous state.'

Generally neglected by historians of the period[189] is the fact that in 1857 Cannizzaro wrote a short article[190] for the *Nuovo Cimento* in which he criticized some of Deville's earlier conclusions regarding the anomalous vapor densities of a number of compounds. To Cannizzaro it seemed likely that, under the effect of high temperatures, these compounds actually dissociate into their components as, for instance in the case of ammonium chloride, ammonia and hydrochloric acid. Under these conditions, then, the equivalent of ammonium chloride occupies for him eight volumes, four of acid and four of ammonia; in a similar way, phosphorus perchloride, when strongly heated, yields a mixture of chlorine and phosphorus protochloride which combines again when the temperature is lowered.

Cannizzaro clearly perceived – only on the basis of theoretical considerations – what had escaped Deville and his school. He argued that their reluctance to accept the coexisting presence of ammonia and hydrochloric without combining could explain the incorrect interpretation of these facts. Therefore, Gerhardt's interpretation in 1851 that molecules may occupy a variable number of volumes (from $\frac{1}{4}$ to $\frac{1}{3}$, 1, 2, 4) was faulty, and there were no exceptions to the equal volumes–equal numbers generalization.

In May 1858[191] Cannizzaro vindicated himself as the first to understand the shortcomings of Deville's work. His article appeared shortly after Kopp had, in Germany, published[192] similar considerations on the unusual behavior of the vapor densities of some compounds; on the same occasion he wondered also how his 1857 note could have escaped the German's attention.[193] 'The apparent anomalies to Avogadro's hypothesis disappear when they are subjected to rigorous examination',[194] he said, and the error of some chemists had been that of taking for vapor densities of some substances the weights of a volume of mixture of their decomposition products. Others, including Mitscherlich, Bineau, and Marignac, Cannizzaro continued, had earlier established that such decompositions take place, but their conclusions had been either misinterpreted or neglected.[195]

The only true fact remaining was that 'the gaseous densities of both simple and compound substances are constantly proportional to the weights of the molecules, and accordingly that it is convenient to express both densities and weights with the same numbers'.[196] The same concepts had been stated in a more extensive way by Cannizzaro in his 'Sunto di un corso', but this went almost unnoticed.

Afterwards, for almost two decades a spirited polemic on the concept of dissociation divided the chemists into those, like Cannizzaro and Kekulé,[197] who favored its acceptance, and those who opposed it. This debate, with the *Annalen der Chemie* on one side and the *Comptes Rendues* on the other, continued through the 1860s and early 1870s in defiance of the experimental evidence offered by L. Pebal in 1862, by Than in 1864, and eventually by Wurtz in 1873.[198]

When faced with the issue of the anomalous vapor densities, Gerhardt's stand, around 1850, appeared less forceful than that of Laurent. In the *Comptes Rendues de Laurent et Gerhardt*[199] he stated that the principle concerning the relation between equal volumes and gaseous substances should not be thought of as a 'verité moléculaire', because while it may be considered true for the organic compounds where molecules correspond to two volumes, the term 'molecule' is not necessarily synonymous with two volumes. Logically this synonymy is not essential: there are molecules with one, two, and four volumes; for instance PCl_3 is equal to two volumes, NH_3 also to two volumes, NH_3HCl to four volumes, etc. Later on, in 1853, he omitted any reference to this interpretation in his *Traité de Chimie Organique*. When Laurent discussed the topic of the anomalies of vapor densities in the *Méthode de Chimie* (1854) he reached the conclusion that to determine the weight of molecules, it is necessary to use two volumes, unless the formula shows fractional numbers; in such case, one recurs to four volumes, for example, $P_{1/2}Cl_{5/2}$.

At Laurent's death, his position and that of his friend Gerhardt were, nevertheless, closely aligned and their conclusions quite similar. Both regarded all metal oxides in terms of the general formula Me_2O, both indicated a clear approach to the concept of the divisibility of molecules thought to be biatomic in general, and both considered the gaseous volumes as the only guide for the determination of the molecular weights. By the mid-fifties, working from chemical considerations, they reached a common position which represented the first real application of Avogadro's ideas and was of pivotal importance to Cannizzaro shortly afterwards.

Cannizzaro's 'Sunto di un corso'

For a better perspective on the status of the molecular theory at this time, it may be recalled that Cannizzaro, who lived in Paris from the fall of 1849 through to the spring of 1851, not only worked in Chevreul's laboratory, but also attended the lectures of Regnault. It is likely that through them he became convinced of the validity of Dulong-Petit's Law of the specific heats as an indispensable instrument for verifying the accuracy of the atomic weights determined by chemical means. As Cannizzaro pointed out in his arguments over the correct evaluation of the atomic weight of mercury,[200] a comparison of the specific heats of this metal with those of bromine and iodine showed that mercury had a weight of 200 rather than 100 and therefore was monoatomic.[201]

The 'Sunto di un corso di filosofia chimica'[202] appeared in 1859 in the *Nuovo Cimento* when Cannizzaro, then thirty-two, had been professor of chemistry at Genoa for three years. Composed in the form of a letter addressed to his friend S. De Luca,[203] it was clearly the result of several years of meditation over the large mass of information gathered for half a century by chemists and physicists all over Europe. The analytical data of Gay-Lussac, Berzelius, Dumas, and Thenard, the speculations of Ampère, Gaudin, Prout, and Avogadro, the physical research of Mitscherlich, Dulong, Regnault, and Petit: all this was laid before Cannizzaro. Although personally he did not introduce any new piece into this vast puzzle, he nevertheless had the intellectual capability of coordinating all the ideas at his disposal, making order and light where before there had been only confusion and obscurity. As is properly stated in the preface to the English translation, 'one is impelled to the conclusion that Cannizzaro's students of 1858 must have had clearer conceptions of chemical theory than most of his scientific colleagues of a much later date.'[204]

In the 'Sunto' the development of the molecular hypothesis through the preceding fifty years was outlined. In this attempt Cannizzaro generally succeeded in his objectivity, recognizing the merits and the shortcomings of the single contributions. He recalled, however, Avogadro collectively with Ampère, even if the priority of the former in advancing the gas hypothesis was beyond any doubt; at the same time he briefly mentioned Laurent despite his considerable contributions. As for Dalton, who gave the atomic doctrine to chemistry and who came very close to anticipating Avogadro, he was almost completely neglected.

In Cannizzaro's view, Berzelius' attempt to reconcile Lavoisier's dualism with Gay-Lussac's Law of gaseous volumes led him (Berzelius) to assume that equal volumes of gaseous elements in general contain an equal number of atoms; later this was restricted to permanent gases only, but in any case his ideas had nothing to do with those of Avogadro. It would have then been sufficient for Berzelius – Cannizzaro argued – to distinguish between atoms and molecules to find an agreement with the experimental data at his disposal. He failed to observe, however, how insurmountable this distinction was within the framework of the system upheld by Berzelius.

In recalling Dumas' work on vapor densities, no mention was made of his conclusions' negative impact upon the eventual reception of Avogadro's generalization. The half-hearted acceptance by Gerhardt of such generalization which left the door open to twenty more years of debate over the anomalous vapor densities of some substances, was only mildly reproved by Cannizzaro. However, the Frenchman was criticized for considering, in general, all elements as polyatomic molecules and all metals, including mercury, as biatomic. Had he been more familiar with Avogadro's writings, Cannizzaro would have known that Avogadro shared Gerhardt's views on the polyatomic nature of all elements.[205] As will be shown, Cannizzaro later also in his 'Sunto' reached the conclusion that some metals could be diatomic. From the historical point of view, it should be observed that 'Sunto' never stressed how little praise had been given to Avogadro in the preceding forty-seven years by all those who, rightly or wrongly, had adopted his hypothesis; from Gaudin to Dumas, Laurent and Gerhardt, nobody ever cared to give him the proper credit.

In his review of a series of elements and compounds whose weights had been determined through their vapor densities by applying Avogadro's assumption, Cannizzaro compared them all to the weight of a half-molecule of hydrogen taken as reference. His were, thus, molecular rather than atomic weights. There were some anomalies among these figures especially for those dealing with allotropic elements (sulfur, ozone, etc.), but Cannizzaro already knew how they could be explained,[206] and they did not particularly worry him.

A very clear definition of atoms as distinguished from molecules was offered in the 'Sunto',[207] probably for the first time in the history of the physical sciences. To identify the atom as 'the smallest quantity of each element which enters as a whole into the molecules which contain it', one must know the weights of all or most of such molecules and their composition. Furthermore, by comparing the composition of equal volumes of gaseous substances under the same physical conditions, it might be established that 'the different amounts of the same element contained in equal volumes, either of an element or its compounds, are whole multiples of a same amount'.[208] This statement represented a most remarkable contribution to the clarification of the issues debated at the time concerning the relations between volumes, atoms, and molecules in both organic and inorganic compounds; in fact, the molecular weights were identified for every substance with the weights of equal volumes under the same physical conditions.

Cannizzaro wanted to impress two concepts especially on his students: (a) the notion of a gaseous volume which could be rendered into that of a molecule, and (b) the sharp difference between atoms and molecules. This is very important, because, while it is easy to determine the atomic weight of an element, it is not always possible to decide its molecular weight, particularly when its vapor density is not available.[209]

In dealing with the problem of the graphic expression of chemical elements and compounds, Cannizzaro discarded as too cumbersome the use, as proposed by some,[210] of molecular rather than atomic symbols; thus, for him H_2O should be preferred to $HO_{1/2}$. For many chemists this suggestion unfortunately fell on sterile ground and was not taken into proper consideration. With his system of formulas, the author of the 'Sunto' emphasized the absolute validity of Avogadro's hypothesis: formulas, in fact, which represented the weights and the composition of molecules of both elements and compounds, and also indicated the weights and the composition of equal gaseous volumes under the same physical consideration.[211]

Whenever the experimental data on the vapor densities were not obtainable, Cannizzaro made skillful use of Dulong-Petit's Law. The 'Sunto' contained a long series of tables mostly of halogen compounds – bromides, chlorides, and iodides – showing their specific and molecular heats, and how from these one could obtain the formulas for the same compounds.[212] It may be interesting to note that in these tables the expression 'calorici specifici' (rather than 'calori specifici') was adopted, and such heats were defined as 'calori impiegati dalle unita' di peso'.[213]

In the case of copper, although the chemical analyses and its atomic heat pointed to an atomic weight of 63, Cannizzaro admitted that he could not be as sure with this conclusion as he had been with mercury. From their combination weights with chlorine and their specific heats, he drew the correct conclusions about the atomic weights of potassium, lithium, sodium, and silver. Again, through the application of Dulong-Petit's Law and in spite of the fact that the vapor densities of their chlorides were unavailable, the correct atomic weights of several other metals – such as lead, manganese, platinum, calcium, etc. – appeared in the 'Sunto'.

The concept of 'capacity for saturation' of the simple metals which had been proposed a few years earlier by Frankland[214] was expressed by Cannizzaro in a more lucid way. In each group of metallic atoms existed a 'constant equivalence' – he used this expression – when they do combine either with oxygen, hydrogen, or sulfur; this property can be interpreted in terms either of 'their own nature' or of 'the state in which they are placed before combining'.[215] Mercury and copper, for instance, take different 'capacities for saturation' when they form their monochlorides and their bichlorides. According to their behavior with hydrogen and with halogen, elements can be electropositive or electronegative, monoatomic and biatomic: thus, hydrogen, potassium, sodium, lithium, silver, mercurous and cuprous radicals are all monoatomic and electropositive, while chlorine, bromine, and iodine are monoatomic and electronegative. It is sufficient to replace the terms mono- and biatomic with mono- and bivalent to see the validity of Cannizzaro's ideas, while Frankland was still struggling with Gmelin's equivalents.

Less successful was the extension of these ideas to some organic compounds: both ethylene and propylene are assumed to be biatomic radicals (in analogy with the radicals methyl and ethyl) because both combine with a whole molecule of chlorine. If the formula of the Dutch oil (ethylene dichloride) is $C_2H_4Cl_2$ – Cannizzaro argued – then the formula of mercuric chloride must be $HgCl_2$, because in each reaction one volume of ethylene (or of mercury) combines with one volume of chlorine. But, if so, then the formulas of zinc, lead, and calcium chlorides must be also of the type $MeCl_2$ or, in other words, all these metals are 'biatomic' as ethylene and propylene.

In applying his reasoning to the monoatomic electropositive radicals, Cannizzaro observed that they all behave as hydrogen; since their capacity for saturation is equal to that of hydrogen, this means that they cannot exist in the 'isolated state'; therefore the 'molecules of such radicals – as sodium, lithium, potassium and silver – must be like that of hydrogen or, in other words, they must contain two atoms. On the other hand, elements such as zinc, calcium, lead which like mercury can exist in an isolated state, are all made of one atom'.

These considerations led Cannizzaro to conclude that the metals whose molecules enter as a whole into compounds are biatomic, while those whose atom is half a molecule are monoatomic. (To avoid misunderstandings, terms such as 'monoatomic' and 'biatomic' are used here in the current meaning, not in Cannizzaro's definition.) Earlier in the 'Sunto', he had mentioned to De Luca[216] that the question of whether sodium, lithium, etc. were Na_2, Li_2, etc. or Na, Li, etc. could not be solved without the knowledge of their vapor densities. Why, then, did he attempt later to do so remains obscure; Cannizzaro was not, by nature, a man prone to groundless speculations or unnecessary inferences.

The general observation that two molecules of hydrochloric acid always react with one molecule of a metallic element is correct; if this is monovalent ('monoatomic' in Cannizzaro's terms), then two molecules of monochloride are formed; when the metal is bivalent ('biatomic' for Cannizzaro) then only one molecule of bichloride is formed.

The 'capacity for saturation' can be detected also in organic radicals such as methyl, ethyl, acetyl (all monoatomic or monovalent) and, as earlier mentioned, ethylene and propylene. In their chemical exchanges all these radicals behave as metallic elements. As for organic bromides, chlorides, and iodides, they react with metals to give either organo-metallic compounds or more complex organic compounds. Surprisingly, Cannizzaro did not mention here Frankland's extensive and renowned contribution to this field of research,[217] and asserted that nobody had yet proved the existence of compounds such as methylethylzinc obtainable from the combination of zinc with ethyl chloride and methyl chloride.[218] In dealing with both organic and organo-metallic compounds, since their vapor densities are generally known, there are more probabilities of obtaining correct formulas than with inorganic compounds and this, in general, was amply proved by Cannizzaro in the second half of his Essay. He stressed again and again throughout the 'Sunto' that his formulas corresponded to equal gaseous volumes in full agreement with Avogadro's generalization.

To those familiar with its content there is no doubt that the 'Sunto' maintained a balance between the physical and chemical considerations successively adopted by Cannizzaro in his train of thought. There is little support for the view[219] that, unlike Kekulé, he had relied on physical rather than on chemical premises in reproposing the molecular hypothesis. In this regard the argument raised in many histories of the period concerning the relative merits of chemistry and physics over the eventual general approval of Avogadro's ideas appears superfluous.

The molecular hypothesis, based upon both chemical and physical facts, was neglected for over fifty years by both chemists and physicists; then, although reproposed by a chemist, Cannizzaro, and independently confirmed by a physicist, Clausius, it passed again through several decades of mistrust and disregard. Eventually, when the intellectual circumstances for its acceptance ripened, the breakthrough – thanks mainly to the contributions of experimental physics – came in a natural way. Avogadro's generalization was then recognized as a general scientific law. But to witness this, one had to wait until the turn of the twentieth century.

NOTES

1. See F. Greenaway, *John Dalton and the Atom* (Ithaca: Cornell University Press, 1966), p. 116.
2. See G. Buchdahl, 'Sources of Scepticism in Atomic Theory', *Br. J. Phil. Sci.* **10** (1959), 120 and W. H. Brock and D. M. Knight, *Isis*, **56** (1965), 5.
3. See R. Fox, *Brit. J. Hist. Sci.* **4** (1968), 18.
4. In the first edition of his *Traité de Chimie Elementaire* (1813) Thenard did not mention the atomic theory which was introduced only in the second edition.
5. *Phil. Mag.* **37** (1811), 411.
6. In 1812 Davy in his *Elements of Chemical Philosophy* expressed his support of R. Boscovich's concept of point atoms. According to R. Siegfried (*Acts of the 10th Congress Intern. Hist. Science, 1962*) Davy's attitude toward the atomic theory was based on his reluctance to accept the large number of simple bodies upon which this theory was based. For L. P. Williams, *M. Faraday* (New York: Basic Books, 1965), p. 72. Davy placed the Daltonian atoms in the same category as imponderable fluids and as such were unacceptable. But even more important, the existence of atoms represented a violation of the law of continuity.
7. In his anniversary address, quoted by F. Greenaway, *John Dalton and the Atom*, p. 188.
8. See Brock and Knight, *loc. cit.*
9. In the 'Essay to ascertain the fixed and simple ratios in which the constituents of inorganic matter are combined' published between 1811 and 1812, reprinted in Ostwald's *Klassiker der Exacten Wissenschaften*, no. 35.
10. It is a well-known fact that Berzelius always had very little regard for Dalton's experimental skill and accuracy, whereas he considered Gay-Lussac's procedures and results very reliable.
11. Berzelius to Dalton, London, 13 October 1812. This comment was based on Wollaston's research on salts of oxalic acid.
12. *Ann. Phil.* **2** (Dec. 1813), 450.
13. Ibid. **5** (Feb. 1815), 123.
14. Ibid. **5** (Jan. 1815), 10.

15. Berzelius, it is likely, never forgave Thomson for this remark.
16. Dalton, *Ann. Phil.* 9 (Mar. 1817), 189.
17. Ibid. 193.
18. When Berzelius in 1812 wrote to Dalton that the theory of multiple proportions could not be explained with the atomic theory, see H. Guerlac, 'The background of Dalton's atomic theory', in *John Dalton and the Progress of Science*, ed. D. S. Cardwell (Manchester University Press, 1968), p. 85, he evidently did not realize that the same could be said of the law of combining gases in relation to the molecular hypothesis.
19. See I. Freund, *The Study of Chemical Composition* (Dover, 1968), p. 335.
20. The quotations here reported are from the third French edition based upon the fifth German edition of the *Lehrbuch*, published in French by Firmin, Didot Fréres, Paris, 1845, as *Traité de Chimie.*
21. The permanent, the coercible and the nonpermanent (represented by vapors).
22. *Traité de Chimie*, **1** (Paris, 1845), 62.
23. Ibid. 66.
24. The terms 'molecule' and 'molecular volume' even when referring to elements are used by Berzelius earlier in his *Traité* (p. 31) during a discussion of the forces of combination and aggregation.
25. *J. Phys. Chim.* **73** (1811), 60–1.
26. Berzelius argued that of the two forces acting on the elementary particles, the 'aggregation' is completely destroyed in the gases, leaving them subject only to the heat's repulsive force, making the existence of 'aggregate' atoms impossible. This is interesting because no mention is made of electrically homogeneous and thus repelling forces.
27. *Traité de Chimie,* **1** (Paris, 1845), 64.
28. In his 1826 'Mémoire sur quelques points', *Ann. Chim. Phys.* **33** (1826), 337.
29. This was the expression used by Berzelius in commenting on Gaudin's paper in 1835.
30. Berzelius was not familiar in 1834 with *Chemistry, Meteorology, and Function of Digestion* of W. Prout which had been published in that year.
31. *Jahresbericht*, **13** (1834), 63.
32. Ibid. **14** (1835), 84. Ampère is quoted as saying that 'in every gas the distance between the atoms is the same'. Ampère in fact is referring here to the 'particules' which were molecules and not atoms.
33. Ibid. **16** (1837), 1. In his short introduction to Ampère's essay, Berzelius did not mention the latter's death which occurred in June 1836.
34. Ibid.
35. Ampère quotes his letter to Berthollet as published in the *Annales de Chimie* in 1814.
36. 'Nuchterne', used by Berzelius, see note 33, literally means 'sober', but also 'moderate', 'calm', and 'levelheaded'.
37. Gaudin's paper will be discussed in detail later on (pp. 166–70).
38. *Jahresbericht,* **14** (1835), 84.
39. It is interesting to recall here that the same expression 'ausgetauscht' was used by Berzelius in his review of Gaudin's work and in the fifth edition of the *Traité de Chimie*, vol. 1, p. 64 (S'échangent).
40. 'Der Entwicklungsgang der Avogadroschen Theorie', *J. Prakt. Chemie*, **87** (1913), 145.
41. In the fifth German edition of the *Lehrbuch* as quoted by Graebe, 'Der Entwicklungsgang', p. 175.
42. A.Ihde in *Great Chemists*, Interscience, 1961, while speaking of Berzelius, supports

the view that he 'persistently refused to accept Avogadro's idea of diatomic molecules of elemental gases'.

43. J. H. Brooke, in his study of organic chemistry in the 1840s, attempts to minimize the conservative attitude of Berzelius towards, for instance, Laurent. He maintains that there was some rationale behind his irrationality, and that he was often wrong with good reason. See J. H. Brooke, *Studies Hist. Phil. Science,* **4** (May 1973), 90.
44. Berzelius strongly disagreed with him and with Dumas on the oxygen theory of acids. This controversy had reached its peak in 1837.
45. As reported by E. von Meyer, *Storia della Chimica*, Ital. trans. (Milano: Hoepli, 1922), p. 222.
46. Graebe in 'Der Entwicklungsgang', p. 154, mentions specifically the *Annales de Chimie*; strangely he forgot that Avogadro's first two chemical essays appeared in the *Journal de Physique*.
47. As earlier mentioned (see under heading: The neglect of the molecular hypothesis, p. 105), the *Journal de Physique* also published articles on chemistry.
48. From *'Neues Journal für Chemie und Physik'*, see *Jahresbericht*, **2**, 73.
49. Originally published in 1823 in *Memorie Reale Accademia delle Scienze di Torino* (cited as *MRAST*) **27** (1823), 43. This memoir is examined in another section of the present work.
50. *Jahresbericht,* **13** (1834), 32. Title 'Mémoire sur la force elastique de la vapeur du mercure à differens températures', originally published in *MRAST,* **36** (1833), 215.
51. The original appeared in *MRAST,* **39** (1836), 57.
52. In the same section of the *Jahresbericht* a much longer article is devoted to the confutation (not the first one) of Thomson's work on the determination of the atomic weights of hydrogen, oxygen, nitrogen, and sulfur.
53. *MRAST* (1839), 2nd ser., **1**, 179. Title: 'Mémoire sur les rapports entre le pouvoir conducteur des liquides pour les courans électriques et la décomposition chimique qu'ils en éprouvent.' The content of this essay is examined in chapter 2 of the present work.
54. In this connection it is quite ironic that Avogadro, whose fundamental contribution had been by then largely ignored for almost thirty years, be blamed by Berzelius for his omission of the German physicist's work.
55. Original 'Mémoire sur les volumes atomiques et sur leur relation avec le rang que les corps occupent dans la série électrochimique', *MRAST*, 2nd ser. **8** (1846), 129. *Jahresbericht,* **26** (1847), 42.
56. Out of a total of forty articles and memoirs published in that period.
57. As reported by Graebe, 'Der Entwicklungsgang der Avogadroschen Theorie', p. 155; 'Ebenso wenig habe Ich finden konnen, dass sich Berzelius in seinen noch nicht veroffentlichen Schriften aus jener Zeit irgendwo über Avogadro Hypothese ausgesprochen hat.' (From a letter of H. G. Söderbaum to Graebe, no date reported.)
58. As transpires from the 'Réflexions sur la Théorie électrochimique de Mr. Berzelius' 1813), one of the few Avogadro essays which were directly published in the influential *Annales de Chimie*.
59. See, for instance, the 'Mémoire sur les volumes atomiques', *Ann. Chim. Phys.* ser. 3, **14** (1845), 330.
60. The thirty-five pages of 'Nouvelles recherches sur la chaleur spécifique' which appeared in *Annales Chimie et Physique*, **57** (1834), 113, deal mainly with physicochemical considerations.
61. Domenico Morichini, born in 1779, was professor at the Sapienza (Univ. of Rome) in 1814 when he met Davy during the latter's trip to Italy.

62. Michael Donovan (1790–1876) was professor of chemistry at Dublin, where he succeeded William Higgins. He wrote *Elements of Chemistry*, London, 1832.
63. See *Elements of Chemistry*, chap. 4, part 3, p. 379.
64. Dumas had done the same six years earlier in *Ann. Chim.* **33** (1826), 337.
65. Unlike solid compounds, Donovan says, when two gases combine, even like first 'dose' of B bears a similar simple ratio to A, being either equal, double, or triple, etc., provided that volumes are considered. This does not happen when quantities are measured by weight.
66. *Elements of Chemistry*, chap. 4, part 3, p. 381.
67. This is probably the case, because the only reference given by Donovan (p. 379) is for an article published in 1827 in *Giornale di Fisica, Chimica, Storia Naturale,* **31** (1827), 1, entitled 'Sur la densitê des corps solides et liquides comparée avec la grosseur de leurs molécules et avec leurs nombres affinitaires' which appeared in vol. 31 and not vol. 8 as quoted by Donovan.
68. Thomson and Berzelius also used Gay-Lussac's Law as a basis in their determination of the atomic weights of gases. See W. H. Brock, *J. Dalton and the Progress of Science* (Manchester Univ. Press, 1968), D. S. Cardwell ed., p. 248.
69. One may agree with the opinion that Prout's paper was badly written and very confused. (See Brock, *op. cit.*, p. 245.)
70. W. Prout, 'On the relation between the specific gravities of bodies, *Ann. Phil.* **6** (1815), 323.
71. In the February 1816 issue of the *Ann. Phil.*, Prout made several corrections to his previous paper; he assumed then for ammonia one atom of azote and three of hydrogen, but in the same table he also suggested the presence of a half atom of carbon in carburetted hydrogen.
72. As mentioned by S. Gladstone, *J. Chem. Educ.* **24**, Oct. 1847.
73. Published as *Bridgewater Treatise VIII* in Britain in 1834 and in the same year in Philadelphia by Carey, Lea, and Blanchard.
74. See Philadelphia Edition, p. 57.
75. For instance, the same capacity for heat of the same volumes of all gases under the same pressure.
76. W. H. Brock saw no reason to dispute Prout's position in this instance. See *J. Chem. Educ.* **40** (Dec., 1963).
77. Published in 1832.
78. Ampère had followed the same terminology in his 1814 letter to Berthollet; however, he had clearly set forth the existence of two classes of particles.
79. Actually representing Daltonian atoms.
80. As a characteristic example, under the same conditions, a given volume of steam contains the same number of self-repulsive molecules as a similar volume of air, which for Prout was a chemical compound.
81. Second Edition, London, p. 123.
82. Graebe fails also to recall Donovan's position. After an extensive review of Gaudin's paper published in 1833, and a summary of the objections raised to his ideas by Berzelius and eventually by Avogadro, Graebe goes directly into a discussion of the *Introduction a l'étude de la chimie moléculaire* by J. Persoz which appeared six years later. For Graebe, apparently, nothing of sufficient worth had been said on the molecular hypothesis during that period. While it is possible that he overlooked the eighth volume of the *Bridgewater Treatise*, published in 1904, nine years before Graebe's essay, it seems quite correct to draw the conclusion that he also ignored Andrew Meldrum's work on Avogadro and Dalton (in which Prout is remembered for his formulation and adoption of Avogadro's hypothesis).
83. April 1816.

84. 'Nouvelles considérations', *MRAST,* **26** (1821), 40.
85. He mentions Prout, however, in his *Fisica dei corpi ponderabili*, vol. II (1838), p. 868.
86. W. H. Brock, *J. Chem. Education,* **40** (1943), 652.
87. Avogadro was and remained a calorist for most of his life, and caloric – as extensively pointed out by R. Fox in *The Caloric Theory of Gases*, (Clarendon, 1971) – appears as a characteristic manifestation of the French style of science, until the 1820's.
88. The famous essay containing Ampère's ideas on the molecular constitution, which will be discussed later on, was written in the form of a letter and addressed to Berthollet. The clanical study of Laplace and his intellectual milieu is by M. Crosland, *The Society of Arcueil* (Harvard Press, 1967).
89. See J. W. Herivel, *Brit. J. Hist. Sci.* **3** (1966), 121.
90. See R. Fox, *The Caloric Theory of Gases* (Clarendon, 1971), pp. 225 ff.
91. This is the view supported by the dissertation of J. K. Bonner, 'Amedeo Avogadro: A Reassessment of His Research and Its Place in Early Nineteenth Century Science' (Doctoral dissertation, Johns Hopkins University, 1974). Bonner defines as 'physicalist' the Laplacian-mechanical tradition and as 'non-reductionist' the opposite position followed, in his opinion, by Lavoisier, Fourcroy, Haüy, and outside France, especially by Berzelius. My understanding of the situation is that, with the exception of Fourier, the most important French scientists, contemporary with Avogadro, developed, inside the mechanico-molecular tradition, a framework which reached its zenith under the Empire and then slowly passed away with the decline of both the caloric theory of heat and the particulate theory of light. Within this school of thought, the distinction between chemistry and physics was so tenuous that any attempt to make the former independent of the latter seemed unnecessary. For Avogadro and the Laplacians, chemistry was not only a branch of physics, but it played an ancillary role. Not until the 1830s did chemistry progressively acquire its personality and a short-lived independence from physics.

    In his dissertation Bonner raises a number of substantial points. It is possible to argue, as he does, that Avogadro could not differentiate between atoms and molecules. But such a flaw was true for all his contemporaries, as an examination of the physical and chemical literature of time indicates. The concept of atom in the Daltonian meaning was also unclear at the time. Bonner argues that if it is true that polyatomic elemental molecules were not the molecules we conceive today, the Daltonian atoms, too, were not the atoms recognized by current physics and chemistry. In fact, to evaluate physical entities as identified in the past in terms of our present knowledge of the same is, to say the least, a serious historical pitfall.

    Strictly speaking, both the atomic theory and the molecular hypothesis were unsubstantiated until a more definite knowledge of the submicroscopic structure of matter became available at the end of the nineteenth century. For Bonner, Avogadro's integral molecules are a substitute for Daltonian atoms (Dissert., pp. 208–9), and they mean clusters of elementary molecules as well. My understanding is that Avogadro went through a process of modifying his terminology. In 1811 he distinguished between elementary molecules, constituent and integral molecules, each having a well-defined meaning. Berthollet, on the other hand, used molecule in a general way. Later on, Avogadro dropped the qualification 'constituent', and only in the 1830s did he occasionally use atoms to represent molecules in general. However, he differentiated the chemical atom from the physical atom, the last term referring mainly to the polyatomic molecule.

    The very strict adherence of Avogadro to Berthollet's system, as seen by Bonner, suggests a number of considerations which seem, in part, to challenge his view. For instance, the existence of compounds having their components in variable ratio – such as the series of nitrogen oxides – was not taken into consideration by Avogadro who never assumed alloys or solutions to be compounds, or that the process of chemical transformation was continuous, as Berthollet had done. Furthermore, while Dalton rejected volumism and both Berthollet and Gay-Lussac criticized atomism, Avogadro leaped over these seemingly antithetical positions to create a synthesis and unify the opposing views. There is no evidence that Gay-Lussac contemplated the formulation of the gas hypothesis that Dalton discussed and discarded as absurd.

In commenting on Avogadro's 1824 'Mémoire sur l'affinité', Bonner states that its author, in basic agreement with Berthollet's ideas, 'had succeeded in quantifying the concepts of force and mass' (p. 283). It is true that Avogadro recalled Berthollet in his argument over the table of affinity, but it is also true that he recognized (as Bonner later admits) the inability of chemistry alone to arrive at quantitative determinations of such properties as the neutralizing powers of both elements and compounds that could be achieved only through physics. This is an important difference between the two men.

Furthermore, it is beyond doubt that in the debate on the elementary nature of both chlorine and nitrogen, Avogadro followed, possibly alone with Davy, a path distinct from that of both the 'reductionists' and the 'non-reductionists' of his time, including Berzelius, and his stand on these issues was quite independent from that of Berthollet, Gay-Lussac and Thenard. (See Thenard, *Traité de Chimie*, vol. 1, 1827, p. 253.)

Bonner mentions that Haüy's notion of a geometrical form as an inherent property of the integral molecule could not be accepted by Berthollet (Dissert., p. 118). Avogadro, in my view, not only did not criticize Ampère who held a similar idea, but aligned himself with Haüy when the latter took exception to Mitscherlich's isomorphism.

The argument raised by Bonner that Avogadro missed the essential feature of the Daltonian theory (Dissert., p. 185) may be debated. In fact, most physicists of the first half of the nineteenth century and even later, considered, as far as I can see, Dalton's atoms only as a theoretical model; therefore, here Avogadro was within the prevailing tradition of the time.

The notion of composite elemental molecules was conceived neither by Lavoisier's non-reductionist school nor by Dalton and his supporters. In this sense only, can I subscribe to Bonner's impression that Avogadro worked outside rather than inside the context of the Daltonian Doctrine (Dissert., p. 205).

In discussing the views held by Avogadro on caloric, Bonner recalls (Dissert., p. 246), among other things, the subtle difference between Dalton's interpretation of the attraction of caloric by particles and Avogadro's. The latter scheme assumed that different molecules attracted different quantities of caloric, since they had different attraction for caloric. At this point Bonner adduces that for Avogadro, despite the above mentioned assumption, the molecules of two different gases, considering also their atmospheres of caloric, had the same size. However, I understand Avogadro's text in the sense that, although the *quantity* of caloric that surrounds a gas molecule varies from gas to gas, its *volume* remains the same for all gases; hence, molecular equidistances, under the same conditions of temperature and pressure.

The observation that it is not true that the molecular hypothesis was generally rejected or ignored (Dissert., p. 302) as proved by the appearance in 1814 of a similar suggestion by Ampère and also by Berzelius' acceptance of the volume hypothesis, I have already discussed at length in other sections of the present work. Here I need mention only that several circumstances prove that Avogadro's generalization on gaseous molecules was in fact either ignored or rejected by most. Probably it would be fair to qualify its appearance in 1811 as untimely; only when the cultural atmosphere changed, was it slowly recognized and accepted. The molecular hypothesis, in my opinion, was neglected even *before 1820*, thus disproving Bonner's point that Avogadro's work failed to attract the attention because its framework became obsolete *after* that time.

That scientists in the early decades of the nineteenth century used the terms atom and molecule consistently is another premise (Dissert., p. 307) I do not share. As mentioned earlier, Avogadro was not always consistent in his terminology and, late in his life, he used atoms and molecules indiscriminately. So did Berzelius, Ampère, Dumas, and Thenard at the time. What now appears confusing to us was most likely no less so for them. Gaudin, possibly, had some clear ideas on the subject, but he appears as a 'vox clamantis in deserto'. Only after the Karlsruhe Congress one may find a better understanding of the differentiation between the concept of atom and that of molecule.

From my understanding of Avogadro's treatment of the subject, it is possible to concur with Bonner's statement that Avogadro may have only hypothesized the existence of elementary molecules – whose number within the integral molecules remained undefined – but likely he did not think of them as though they really existed (Dissert., p. 309).

However, Avogadro seemed to favorably consider some structural concepts and particularly the one proposed by the scarcely known Emmet, that elementary molecules were at the apexes of the crystalline forms of chemical compounds. And, of course, Avogadro saw them as real particles and not just unsubstantial, *ad hoc* entities.

Conversely, I find myself in agreement with Bonner in his assessment of Avogadro's scientific philosophy which led him to generalizations only partially supported by experimental evidence.

The erroneous feeling given by some historians (especially of chemistry) that very important expressions and concepts dealing with atomism and molecular structure have the same meaning today as they did in the early part of the last century, is, beyond doubt, rightly appreciated by Bonner (Dissert., p. 307), and this impression should be corrected.

I would like to conclude this brief analysis by attempting to straighten once more the faulty, but widely accepted, notion that Avogadro was a chemist and worked within the French chemical tradition of his time. He was never a chemist, he never tried to become one, and his work – both bad and good – belongs to the history of physics and, when and if it is ever written, to the history of physical chemistry.

The same, in my view, applies to Dalton and atomism. As a matter of fact, chemistry did do pretty well in those years without both Avogadro and Dalton. Yet, with few exceptions, current textbooks of general chemistry deal with Avogadro and Dalton (and sometimes Cannizzaro) linking all of them together; while they generally devote little space to the real 'bench chemists' of the time, Berzelius, Davy, Chevreul, Gay-Lussac, Thenard, and later Dumas.

92. As appears from his scientific MSS, Avogadro read very carefully the *Essai* of Berthollet.

93. A detailed review of this essay appears in another section of the present work.

94. See *Chemical Statics* (London: Mawman, 1804), p. 47.
95. See *J. Phys.* **73** (1811), 74: 'que de la réunion de certain nombres précis de ces molécules dans des rapports simples il doive résulter un degré précis d'oxygénicite'.
96. Ibid., p. 75. If one keeps in mind Avogadro's obscure position in Vercelli at the time, he certainly cannot be reproached on this occasion for his modesty.
97. See *MRAST*, **26** (1821), 21.
98. Bonner quite correctly qualified Gay-Lussac's reasoning as 'circular'. (Doctoral Dissertation, p. 160.)
99. See Section III of the 'Essai d'une manière'.
100. For more detail, see the section of this work devoted to an analysis of the 'Essai d'une manière'.
101. Whose abstract he read in *Bibliothèque Britannique* in 1813, as he says himself.
102. See *Fisica dei corpi* (Turin, 1838), vol. II, first book, and my review of this treatise in another section of the present work.
103. See *Fisica dei corpi* (Turin, 1837), vol. I, pp. 159 ff.; vol. II, pp. 3, 12, 197, 806; vol. III, pp. 20, 674, 800.
104. For instance, in the August 1811 issue Davy's fundamental memoir on the elemental nature of chlorine was published; two months later Berzelius' 'Essai sur la nomenclature chimique' appeared.
105. The only exception is represented by the very short 'Idées sur l'acidité . . . ', also published in *J. Phys.* **19** (1809).
106. The rather preposterous assertion of A. Wurtz made in 1868 ('La chimie est une science française') from the *Historie des doctrines chimiques* is a good example of this attitude. As for Napoleon, there was a great deal of political expediency in his open-mindedness towards foreign scientists.
107. Quoted by M. Crosland, *The Society of Arcueil*, (Harvard, 1967), p. 270.
108. 'If equal measures of azotic and oxygenous gases were mixed, they would form nearly two measures of nitrous gas having the same weight as the two original measures; the number of ultimate particles could at most be one-half of that before the union.' *New System of Chemical Philosophy* (London, 1808), chap. 1, sec. 4, p. 71.
109. See the concluding paragraph (VIII) of the 'Essai d'une manière', *J. Phys.* **73** (1811), 76.
110. The results of his researches on electrodynamics, in fact, began to appear in 1820.
111. Published in 1738.
112. In speaking of Ampère's hypothesis, Graebe says, 'Dadurch ist er auch gelangt, fur die Verbindungen eine viel kompliziert Zusammensetzung anzunehmen.' 'Der Entwicklung. der Avog. Theorie', *J. Prakt. Chem.* **87** (1813), 157.
113. See *Opere di Ampère*, edited by M. Bertolini (Turin, UTET, 1969), p. 555. From *Correspondance*, edited by Henriette Chevreux, 1875).
114. L. DeLaunay of the Académie de Sciences, *Revue de France* (July 1921), p. 31.
115. There is no proof that Ampère ever wrote this long detailed essay. It is not clear when he began the elaboration of the ideas that he eventually exposed to Berthollet.
116. *Opere di Ampère*, M. Bertolini, ed. (UTET, 1969), p. 17.
117. This volume is preserved in the private library of the family Nomis di Pollone in Turin.
118. The 'Mémoire sur les masses relatives' published in January 1814 in the *Journal de Physique*, 78.
119. Only in the 'Nouvelles considérations' (1821) was the content of Ampère's note extensively discussed by Avogadro. The 'Memoria sul calore specifico' appeared

in the *Biblioteca Italiana,* **4** (1816), 478.

120. Both Dalton and Thomson are quoted by Gay-Lussac in the 'Memoir on the combinations'.
121. He had used the term 'molecule' on this occasion.
122. Dalton took exception to the apparently incoherent behavior of many gaseous compounds, such as nitrous oxide, whose number of compound particules was apparently one-half that of the particles of component gases. He refused to explain these facts, because for him equal volumes of gases could not contain the same number of particles.
123. On the basis of Gay-Lussac's assertion that ammonia is obtained from three parts by volume of hydrogen and one of nitrogen.
124. Dumas, 'Mémoire sur quelques points de la Théorie atomique,' *Ann. Chim.* **33** (1826), 337. A. M. Gaudin, 'Recherches sur la structure intime des corps . . . ' Prem. partie, *Ann Chim.* **52** (1833), 113, see footnote. A. Wurtz, *Leçons de Philosophie chimique* (Paris, 1863). F. Hoefer, *Histoire de la Physique et Chimie* (Paris, 1863). M. Delacre, *Histoire de la Chimie* (Paris, 1820), p. 293. L. E. Grimaux, *Théories et notations chimiques* (Paris, 1884). R. Taton, *History of Science*, vol. III (New York, 1964), p. 270.
125. A few decades later, Maxwell, in discussing the dynamic theory of gases, attributed only to Gay-Lussac the merit of discovering the 'Law of the equivaient volumes of gases' and Kay-Shuttleworth did the same in 1868; at the turn of the century, Tilden supported the view that the molecular hypothesis had been formulated by Dumas, Ampère, Prout, Gerhardt, Clausius and Krönig. Meldrum, instead, clearly indicated that Avogadro was the only proposer of such an assumption. In the United States J. Cooke also pointed this out in his Lowell lectures. For a few years, after the Karlsruhe Congress, William Odling at Oxford ignored both Avogadro and Ampère. Edward Divers remained an even more vigorous and unyielding adversary of the gas generalization; in 1902, then President of the Chemical section of the British Association, he was still supporting the belief that Avogadro had been anticipated by both Dalton and Gay-Lussac. Ampère, he disregarded altogether. Pattison-Muir, on the other hand, mentioned only Avogadro.

    In 1868 Mendeleev attributed the gas hypothesis to both Gerhardt and Avogadro, although there was a gap of thirty-two years between them. In Germany, the very influential Leopold Gmelin in his widely read and quoted *Handbuch der Chemie*, up to the 1848 reprint, recalled only Ampère; Avogadro's ideas were unnoticed by both Hermann Kopp and Gerding in the reprints of their histories until 1873, while, after the Congress of Karlsruhe, he received unmitigated approval from Lothar Meyer.
126. Whom he succeeded in 1832 as professor of chemistry at the Sorbonne.
127. *Ann. Chim.* **33** (1826), 337.
128. Graebe, 'Der Entwicklungsgang . . . ', p. 160.
129. *Ann. Chim.* **33** (1826), 338.
130. *J. Phys.* **78** (1814), 131.
131. *Ann. Chim.* 1st, **90** (1814), 46.
132. *Ann. Chim.* **33** (1826), 338.
133. For 'constituent molecule' Avogadro means molecule of an element.
134. It was reproposed quite clearly twenty-two years later by Gaudin and by Laurent again in 1846 (*Ann. Chim.* **18**, 266).
135. Ibid., p. 339.

136. It is quite possible that the minutes of the 'Mémoire sur quelques points' were discussed by Dumas with Thenard, Gay-Lussac, and even with Ampère, who was also teaching at that time at the École Polytechnique.
137. A similar view is supported by Noel G. Coley, 'Avogadro and the Molecular Hypothesis' (Dissertation, University of Leicester, 1964, p. 106).
138. This is the only time when Avogadro is quoted by Dumas who, however, does not seem familiar with his 1811 Essay.
139. Ibid., p. 391.
140. Ibid., p. 79 ff.
141. I am in complete disagreement with I. Freund, *The Study of Chemical Composition*, (Dover, 1969), repr., p. 339 when she characterizes Dumas as making a gallant attempt to rescue Avogadro's hypothesis.
142. This was not the first negative review of Dumas' work which appeared in the *Jahresbericht*. In vol. 2 (1823), p. 41, commenting on some conclusions reached by the French chemist on the atomic volumes and the procedure to determine them, Berzelius says, 'diese Resultate verdienen aber kein Zutrauen'.
143. See *Fisica dei corpi ponderabili*, vol. II (Turin: Royal Print. House, 1837), p. 853.
144. Ibid., p. 868. 'The idea I have proposed has been more or less distinctly adopted by many chemists and physicists, especially French. However, it was not taken up by Dalton or by the British physicists in general'. (With the exception of Prout, as he pointed out. As for the French, this apparently is an overstatement by Avogadro, because only Ampère and Dumas had taken it into consideration.)
145. See for comparison the introductory section of the 'Nouvelles considérations'.
146. *Comptes Rendus*, **20** (1845), 960.
147. Ibid., p. 851.
148. Here he repeats Dumas' error of 1826 which had incensed Berzelius so much.
149. One wonders if such a harsh attack by a provincial professor would have occurred against one of the members of the 'inner' circle, but it is a well-known fact that Laurent was not part of that group.
150. *J. phys.* **90** (1820), 113.
151. He never used the terms atom or molecule.
152. This is the same essay which inspired Avogadro's 1816 paper on specific heats.
153. 'Recherches sur la structure intime des corps inorganiques définis et considérations générales sur le rôle que jouent leurs dernières particules dans les principaux phenoménes de la nature, tels que la conducibilité de l'électricité et de la chaleur, le magnetisme, la refraction simple ou double et la polarisation de la lumière.' *Ann. Chim.* **52** (1833), 113.
154. An earlier article by Gaudin was published in 1831 in the *Lycée* and also reviewed by Berzelius (see note 195 below).
155. Besides the terms atom and molecule, Gaudin uses 'particule' to indicate a very small piece of matter; eventually he replaced this with the term molecule.
156. He referred here especially to Dumas whose earlier enthusiasm for the atomic and molecular theories in general had begun to fade.
157. Later on, in the same paper, he reached the conclusion that hydrogen, chlorine, and hydrochloric acid have molecules which are 'at least diatomic'.
158. He examines only mercuric oxide.
159. This formula has been interpreted by others, such as M. Giua, *Storia della Chimica* (UTET, 1946), p. 126, as $SO_3 + H_2O$; on the other hand, there are clearly nine dots above the symbol of sulfur in Gaudin's paper.
160. He calculated that if boron chloride weighs 3.57 times the 'atom' of oxygen, then from the weight of two molecules of boron chloride, minus the weight of

three 'atoms' of chlorine, he could obtain the weight of a 'particule' of boron. In current symbols: $2BCl_3 - 3Cl_2 = 2B$ or B_2.

161. It has been said by some, see J. T. Partington, *A History of Chemistry*, vol. IV, 1964, p. 220 and S. Mauskopf, *Isis*, **60** (1969), 60–1, that this was the first correct interpretation of the composition of silica; actually Avogadro came to the same conclusions in the 'Nouvelles considérations . . . ' (1821). Graebe in the 'Entwicklungsgang' suggests (p. 170) that this argument over the correct formula for silica was perhaps one of the reasons which kept Berzelius from accepting the main concept of the molecular hypothesis. On the same side with Berzelius was Mitscherlich, who said that on the basis of the specific gravity of the vapors of silicon chloride it was impossible to decide whether such a compound had three or six atoms of chlorine for each atom of silicon *Pogg. Ann.* **29** (1833), 193.
162. The first one was published in *Jahresbericht,* **12** (1833), 61.
163. Ibid., **14**, 84.
164. 'Recherches sur la structure intime des corps inorganiques'. *Ann. Chim.* **52** (1833), 116.
165. *Jahresbericht,* **14** (1835), 86. As S. Mauskopf points out, nobody took Gaudin's theory seriously, see *Isis,* **60** (1969), 70.
166. *Fisica dei corpi ponderabili*, vol. II (Torino, 1838), p. 869.
167. Gaudin used Daltonian symbols to indicate the gaseous elements rather than the chemical symbols already adopted by Berzelius and other chemists.
168. *Fisica dei corpi ponderabili*, vol. II, p. 869.
169. It should be recalled that in the same volume of *Fisica dei corpi* Avogadro expressed his disagreement with Gaudin's interpretation of the nature of the elementary particles of mercury and sulfur. He mentioned that the Frenchman had considered the 'molecule' of mercury as monoatomic and that of sulfur as hexatomic on the basis of their vapor densities, while his own conclusions indicated that the former element had a diatomic molecule and the latter had a triatomic molecule.
170. Around 1870 (see Wurtz, *The Atomic Theory*, for instance) sulfur was still considered as hexatomic at temperatures of 500° C.
171. 'Sur les camphres artificielles des essences de terebenthine . . . , *Ann. Chim.* **52** (1833), 400.
172. *Comptes Rendus,* **7** (1838), 474, 'Acide produit par l'action du chlore sur l'acide acetique'.
173. Now known as the earlier type theory, because a second type theory (water, ammonia, hydrochloric acid, hydrogen types) was proposed by Gerhardt in 1853.
174. *Comptes Rendus,* **10** (1840), 149, 'Mémoire sur la loi des substitutions'.
175. *Jahresbericht,* **19** (1840), 370.
176. *Ann. Pharme,* **31** (1839), 119, 'Uber die Subtitutionstheorie'.
177. *Ann. Chim.* **66** (1836), 175. This is the only essay of Laurent recalled by Avogadro in the *Fisica dei corpi*, vol. I.
178. The polemic between Dumas and Laurent on the type theory lasted to the end of Laurent's life.
179. *Comptes Rendus,* **19** (1844), 1089, 'Classification chimique'.
180. Ibid., p. 1089.
181. *Ann. Chim.* **7** (1843), 129, 'Considérations sur les équivalents de quelque corps'.
182. *Comptes Rendus,* **15** (1842), 498, 'Recherches sur la classification chimique'.
183. *Ann. Chim.* **18** (1846), 266, 'Recherches sur les combinasions azotées'.
184. Partington, A *History of Chemistry*, vol. IV (1964), p. 422 mentions that Laurent wrote this essay after an 1840 article by Regnault on the atomic heats and another by Avogadro in 1845 on the atomic volumes.
185. Favre was teaching at the *École de Medécine* and silberman at the *Conservatoire*

des Arts and Métiers.

186. Williamson's life has been examined by W. H. Borck and J. Harris, *Ann. Sci.* **31** (1974), 95.
187. *Phil. Mag.* **37** (1851), 350. Also *J. Chem. Soc.* **4** (1852), 229.
188. Editor in 1836 of the *Annales de Chimie* and editor of Dumas' *Leçons sur la Philosophie Chimique.*
189. M. Giua, *Storia della Chimica*, 1946, p. 149 gives a transcription of part of the content of this article.
190. *Nuovo Cimento,* **6** (1857), 428, 'Della dissociazione ossia scomposizione dei corpi'.
191. Ibid., 7, 375.
192. *Ann. Chemie Pharm.* **105** (1858), 390.
193. Three months later Cannizzaro reported in the *Nuovo Cimento* that probably Kopp had not received the issue of the 1857 Journal on time as its mailing abroad had apparently been delayed.
194. *Nuovo Cimento,* **6** (1857), 430.
195. Bineau had also detected the decompositon of ammonium chlorocyanide, but had not perceived that this was only a mixture of the two substances.
196. *Nuovo Cimento,* **7** (1858), 376.
197. *Ann. Chemie Pharm.* **105** (1858), 390.
198. *Ann. Chemie,* **123** (1862), 199. *Ann. Chemie,* **131** (1864), 129. *Comptes Rendus,* **84** (1877), 977.
199. *Pour le 1851*, 129.
200. 'Sunto', *Nuovo Cimento,* **7** (1858), 301.
201. Cannizzaro argued that, due to the biatomic nature of hydrogen, the molecular weight of volatile substances could be obtained by multiplying by two their vapor density relative to hydrogen. In the case of elements, their atomic weights are in agreement with the law of atomic heats (Dulong–Petit); this law was adopted by Cannizzaro whenever non-volatile substances were involved.
202. See also *Jahresbericht* (1858), 11; Avogadro is not mentioned.
203. S. DeLuca, 1820–80; from 1857 professor of chemistry at Pisa, moved to the University of Naples in 1860.
204. *Alembic Club Report*, no. 18, Edinburgh, 1969.
205. See *Fisica dei corpi ponderabili*, vol **I**, chap. 4, p. 307.
206. *Nuovo Cimento,* **6** (1857), 428.
207. Ibid., **7** (1858), 328.
208. Ibid., **7** (1858), 330.
209. As an example, Cannizzaro gave that of carbon whose atomic weight was known, but nobody had established whether its 'molecule' was monoatomic or not.
210. For instance, Dumas in his 'Mémoire sur quelques points . . . ', *Ann. Chim.* **33** (1826), 337.
211. *Nuovo Cimento,* **7** (1858), 335. Thus, H_2 indicated the weight of one molecule and one volume of hydrogen, while O meant the weight of half molecule and half volume of oxygen.
212. Exceptions represented by the double formulas for aluminum, iron, and chrome chlorides.
213. See *Nuovo Cimento,* **7** (1858), 339, 343, 344, 346, etc. These definitely cannot be considered as typographical errors. At the time of the 'Sunto's' appearance, the caloric theory had been set aside; the last paper supporting it had been published in Geneva by E. Ritter in June 1845, see R. Fox, *The Caloric Theory of Gases* (Clarendon, 1971), p. 305. Nevertheless, 'Calorico' was misused for 'calore' (heat) in Italian, as is the case with Cannizzaro.
214. *Phil. Trans.* **142** (1852), 417, 'On a new series of organic bodies'.

215. *Nuovo Cimento,* 7 (1858), 350.
216. Ibid., p. 345.
217. *Phil. Trans.* **145** (1855), 259, 'Researches on organo-metallic bodies'.
218. Frankland had prepared on the other hand, mercury ethyliodide, and zinc methyloxide, among others.
219. Graebe, 'Entwicklung, d. Avo. g. Theorie', *J. Prakt. Chimie,* 87 (1913), 188.

CHAPTER FIVE

THE MOLECULAR HYPOTHESIS: AFTER THE KARLSRUHE CONGRESS

For some historians of science and a great number of historians of chemistry, the appearance and diffusion of Cannizzaro's 'Sunto' at the Karlsruhe Congress in 1860 represented the moment when 'the scales fell from the eyes' of the scientific community[1] and a turning point in the development of the atomic and molecular hypotheses. However, a more critical approach to the subject and a review of what was said and written after Karlsruhe indicate that these hypotheses remained anything but generally accepted doctrine through the following five decades, on both physical and philosophical grounds.

Among chemists, especially in France, the debate over the validity of the equal volumes–equal numbers generalization lasted until the late seventies; in a characteristic way, it dealt primarily with the physical rather than chemical behavior of the substances under discussion. Eventually, the stand taken by Deville and his supporters could no longer be maintained. By 1879, A. Naumann reported his conclusive findings on the dissociation of chloral.[2] Yet Marcellin Berthelot, 'le grand patron du Collège de France', remained an equivalentist and philosophically critical of both Dalton and Avogadro until the last decade of the century.

In Germany, chemists were not unanimous either, despite the wide-spread influence of Lothar Meyer's *Die Modernen Theorien der Chemie*, a work published shortly after Karlsruhe which was largely favorable to Cannizzaro's and, therefore, to Avogadro's ideas. One may recall, for instance, the views set forth by Kekulé in 1867 about the nature of atoms, which will be examined in detail later on. It is also revealing that in the 1877 edition of his *Kurzes Lehrbuch*, Kolbe completely ignored not only Cannizzaro and Avogadro, but also Gay-Lussac.

The chemists and physicists in Britain were much more divided than their counterparts in Germany. Men such as Frankland, Williamson, Odling, Brodie, Miller, and Mills found themselves supporting very different ideas concerning the ultimate nature of matter. However, their disagreement, unlike that between Wurtz and Deville in France, had its roots mostly on the metaphysical side of the atomic-molecular doctrine.[3] In the second half of the nineteenth century and more specifically after Karlsruhe, it is possible to observe that the investigations concerning the composition of matter became more and more the domain of physicists. Until that time the line between the fields of research in chemistry and in physics had been rather indistinct: physicists were involved in chemical questions and chemists were studying physical problems.

After 1855, the caloric theory and its interpretation of the corpuscular doctrine was replaced by a new model, the dynamic theory of gases. As a result, men such as Krönig and Clausius were able to find a new and independent approach to, and a confirmation of, the molecular hypothesis. For them and their British predecessors – John Herapath[4] and John Waterston – the so far very vague and little defined notion of molecules acquired a sense of reality. Also in Britain and ten years before Cannizzaro's 'Sunto', the first quantification of a molecule – its velocity – was determined by James Joule, who incidentally had been a pupil of Dalton.[5] It seemed then, for a moment, that a half-century after its announcement, Avogadro's hypothesis had found its more convincing support in the kinetic theory of gases.

Molecules – a concept so shadowy that it could not even be identified and *a fortiori* distinguished from that of atoms – now acquired a concrete meaning. These entities were, at least mathematically, defined: the relation between their diameters, their number in a given space, and their mean length of path could now be measured. In 1863, further proof of the existence of molecules, this time experimental, was offered by L. Wiener's interpretation of the Brownian movement[6] and two years later by the researches of Loschmidt leading to the determination of molecular free paths from viscosity measurements.[7]

Theories of the Structure of Matter

While the prevailing opinion during the 1860s in the German-speaking countries, through the works of Lothar Meyer in chemistry and of Krönig and Clausius in physics, emerged as largely favorable to the molecular hypothesis, this was certainly not the case in Britain. Here, as earlier mentioned, the leading chemists of the time found themselves involved in endless debates which, in a different form, kept going almost to the end of the century. But probably more relevant for its implications was the position taken in those same years towards the atomic-molecular doctrine by the British physicists and especially by their two prominent representatives, W. Thomson and J. C. Maxwell. Involved here – and to some extent this is true also for the discussions among the British chemists – were not only questions of physical character concerning the ultimate components of matter, but also problems having profound philosophical connotations.

Doubts followed doubts, queries other queries: do atoms really exist and if they do, what are they? Could physics or chemistry do without them? And if so, how and with what could they be replaced? This wave of criticism and, even more, of skepticism towards atoms had deep-seated origins. It could be traced all the way back to the antithetical interpretation of the corpuscular theory offered in the second half of the seventeenth century by Newton and his contemporary Leibnitz. The theory of hard bodies as ultimate components of the universe, set forth by the former, stood at variance with the theory of elasticity supported by Leibnitz. For Newton, atoms were absolutely solid, indivisible and impenetrable; this was also the view that Dalton incorporated in his own atomic chemical doctrine much later.

The Newtonian approach was followed on the Continent by Christian Huygens, adopted later in the eighteenth century by D'Alembert and eventually in France by a group of its most faithful disciples, among them Dulong, Poisson, Petit, and the Carnots, both father and son. Although he never expressed himself distinctly on this topic, Avogadro, by his adoption of Poisson's considerations dealing with the mechanics of hard bodies,[8] suggested that he also belonged to the same school of thought as the French mathematician. Avogadro said in his physics textbook that the force which holds the molecules together is the force related to the cohesion of bodies and called 'molecular affinity' or 'molecular attraction'. The relation existing between this force and the intramolecular distance r is stated in agreement with Poisson's exponential term $A\ e^{-r/\alpha}$; when r is very small, then $e^{-r/\alpha}$ may be equal to 1. When r becomes sensible and a greater multiple of α, this function will not have any sensible value, because $e^{-r/\alpha}$ will be then a very small fraction.[9] For Avogadro, the real way molecular attraction decreases was unknown, but he argued that, in any case, it is of a different nature than the force which takes place among parts of matter in general, which is indirectly proportional to the square of the distance and is the cause of terrestrial gravity and universal gravitation. Molecular attraction then, unlike the Newtonian attraction, varies from body to body according to its chemical nature.[10]

Very few natural philosophers felt inclined to follow the hard atoms theory in Britain in the first half of the nineteenth century besides Dalton; among the chemists, Thomas Thomson was an early supporter of Daltonian atomism. John Herapath was also one of them and, like Dalton, incurred the wrath of those who supported the opposite vision of the universe. Herapath in his *Mathematical Physics* proposed a kinetic theory of gases based on the existence of a great number of extremely small particles moving in all directions.[11] Herapath's kinetic views, which will be discussed later in more detail, were later adopted by Joule who, under the influence of William Thomson, discarded in his system the hard particles in favor of the elastic ones.

Thus far, it has been possible to trace the atomic-molecular doctrine back to a Newtonian tradition. For both Dalton and Avogadro, atoms are solid, but their systems have something else in common: their particles are surrounded by a variable amount of caloric atmosphere which, also composed of particles, is the *deus ex machina* in accounting for all structural adjustments when solids become liquids and liquids become gases. Both Dalton and Avogadro in formulating their theories based upon two fundamental characteristics of matter – solidity and discontinuity – not only gave to chemistry a most valid foundation, but also provided the premises for the strongest criticism to their views. Although this transpired earlier with Davy and Wollaston, it especially took shape in the second part of the nineteenth century.

In the development of modern physics after 1850 and in the rise of physical chemistry with its emphasis on the kinetic theory and on thermodynamics, there was infinitely more room for a theory of elastic particles than for the billiard-balls approach adopted by Dalton and his followers. Actually there was no need in thermodynamics to assume any kind of particle, as Josiah Gibbs proved in his now famous papers of 1876, which at the time went unnoticed.

In addition to the solidity of the Newtonian chemists and physicists another completely opposite tradition, based upon a philosophy of continuity, had found, as previously stated, its proponents, since the middle eighteenth century, in Johann Bernoulli and forty years earlier in Leibniz. The 'Lex continuitatis' that Leibniz formulated in 1687 upon purely mathematical considerations ('datis ordinatis etiam quaesita sunt ordinata') indicated that the physical space is a 'plenum', thus denying the existence of a 'vacuum formarum'.[12] Accordingly, not only in motion but also in every order of things, no leaps can take place. In his wholly antimaterialistic interpretation of the atomic theory, Leibniz refused to accept the assumption that matter was composed of a finite number of very small parts. For him it was impossible to attribute a finite extension to atoms which, in essence, should be considered only as centers of force or, even, as mathematical points; there is nothing of substance or reality, he would affirm a few years later. As for 'les atomes de matiére sont contraire à la raison'[13] he added that only 'les atomes de substance' are conceivable and may also be called 'metaphysical points', indivisible, exact, real, and 'sans eux il n'y aurait rien de réel . . . . ' In 1714 Leibniz assimilated his monads to these unextended substantial atoms as also indivisible, with no shape, no parts, no extension. 'The monads', he stated 'are the true atoms of nature'[14] and there is a continuous modification from the lowest to the highest.

Ten years later, in an essay submitted to the Académie Royale of Paris, Johann Bernoulli – under the influence of Leibniz with whom he was in personal contact – presented a similar line of reasoning. Since the creation of the universe there has been an immutable and perpetual order, he argued, which may be interpreted as the law of continuity and 'all that takes place does so by infinitely small degrees'.[15] Naturally, the law of continuity excluded a discrete view of nature and thus Johann Bernoulli, too, rejected solidity and took a very critical position toward Newtonian atoms. These were merely 'imaginary corpuscles' for him, as they would be for a number of physicists and chemists in the second half of the nineteenth century.

Johann Bernoulli was still alive in 1745 when another important solution to the problem of the ultimate constitution of bodies came forth. Although Roger Boscovich indicated that his aim was an attempt to conciliate the Newtonian position with that of Leibniz,[16] in fact his mathematical points which, being immaterial, have no extension, appear closer to the monads than to solid bodies. However, there are finite distances among them, with changing forces of repulsion and attraction also. All these assumptions are scarcely in agreement with the Leibniz's 'Lex continuitatis'. Boscovich's hypothesis, later on more extensively elaborated in his *Theoria Philosophiae Naturalis* (1758), was neglected throughout the second part of the eighteenth century, especially on the Continent. Although Boscovich established a clear relation between force and distance (infinite repulsion at infinitely small distances and Newtonian forces at very great distances), he could only attempt an acceptable explanation of, for example, the gravitational attraction at finite distances, the cohesion and stability of point masses, and the attractive and repulsive forces observed in electricity and magnetism.[17]

As for a field more related to chemistry, Boscovich argued that there are 'force patterns' which intervene in the combination of atoms to form molecules and that chemical reactions take place when the constituent molecules have such force patterns as to allow their union.

The atomic theory of mathematical points neither attracted the attention of Ampère,[18] nor of Avogadro in his *Fisica dei corpi*. In France, Buffon failed in his attempts to have his countrymen accept Boscovich's hypothesis, thus leaving its revival to Davy after the turn of the century.[19]

In the points theory, Humphry Davy saw an agreeable alternative to the hard Daltonian atoms which he could never accept; moreover, he thought that the forces of attraction and repulsion among points could be explained on the basis of positive and negative electrical charges.[20] Davy adhered to the Boscovichean points throughout his life, as clearly stated in the *Consolations*.

Among Davy's contemporaries in Britain was a man of wide scientific reputation (and a good chemist also) who took a similar position in relation to the atomic theory. Wollaston, often quoted in the history of chemistry for his experimental work leading to a confirmation of the law of multiple proportions, never accepted the conclusions reached by Dalton,[21] and suggested in 1814 the substitution and adoption of the term 'chemical equivalent' for atomic weights. At this point the difference of opinion between Wollaston and Davy on one side and Dalton and his followers on the other was, however, much deeper than one of simple terminology;[22] it dealt with two opposite interpretations of the structure of matter which had roots in the previous two centuries. Wollaston came to think that atoms were just a bold assumption, while they simply did not exist for Davy, at least in the billiard-balls form proposed by Dalton.

It may be interesting to recall here that when Cannizzaro, in 1896, gathered his notes concerning the historical development of the atomic and molecular theories he wrote that:

> Wollaston had been originally a supporter of the principles of the atomic theory by invoking physical evidences to prove a limit in the actual division of matter; he had tried to convince Davy of the importance and the usefulness of Dalton's new hypothesis; one cannot say that he ever changed his opinion. . . .[23]

an account that does not seem fully coherent with the actual facts.

In Britain, through Wollaston and Davy, the antagonism to the Daltonian tradition of hard bodies passed to Faraday who found in his electrical and electrochemical researches the most important evidence against the apparent indivisibility of atoms. In his mature years, he finally revealed a position quite similar to that of Boscovich; by seeing atoms as centers of force he identified in the force the essence of matter.[24] Atoms, he said,[25] have no meaning when separated from their force; they are in fact points with an atmosphere of force around them and the properties of a body – such as hardness, specific gravity, and conduction – belonging not as much to the point (or nucleus) but to the forces themselves.[26]

In 1837 Faraday, who was then forty-six and, thanks to his works on electromagnetic induction and electrochemistry, widely recognized as one of the foremost scientists of his time, took the side of Sir William Rowan Hamilton when he publicly attacked the notion of material atoms. The occasion was a meeting of the chemical section of the Royal Society where Hamilton assailed the use of the term atom, in the Daltonian sense, in chemistry.[27] This was a very important date in the course of the history of British scientific circles and their subsequent aversion to the Daltonian doctrine. Not too many years went by before Faraday, in an 1844 article in the *Philosophical Magazine*,[28] fully disclosed his criticism of the atomo-mechanistic philosophy and clearly set forth his 'speculation' on the nature of matter.

The structure of matter assumed by the supporters of the corpuscular theory, he said, is that atoms are provided with a certain volume and are separated from each other by space; this must be quite large in some cases as proved by the relatively easy compressibility of gases. Thus, taking for granted a continuous space which surrounds and enmeshes the single particles, the main problem remains that of establishing the nature of the space, especially in view of the electrical behavior of bodies. Faraday argued that, according to the atomic theory, a typical non-conductor, such as a piece of shellac, is probably made up of atoms of shellac which must be surrounded by an insulating space; otherwise, one could not explain the properties of shellac as an insulator. On the other hand, metals – as conductors – have a structure formed by their atoms enmeshed this time in a conductive space. Now, it is obvious that space cannot be a conductor in some cases and an insulator in other cases; either it is or it is not, should one want to remain faithful to the 'ordinary' atomic theory. Finally Faraday concluded: 'any ground of reasoning which tends to such conclusions as these must in itself be false. . . .'[29]

To this reasoning based on electrical considerations follows another one which comes instead from chemistry. Here Faraday attempted to prove, by referring to the example of potassium and potassium hydrate, that the conduçtivity of metals cannot be explained by simply surmising that their atoms are packed closely together. In fact, through a simple calculation, Faraday demonstrated that, for instance, in a cubic inch, atoms of metallic potassium are much further apart, despite its excellent conductivity, than those in the same volume of potassium hydrate, a non-conductor.[30] This would support the existence of a conductive space in metals, but at the same time would contradict the electrical properties of potassium hydrate whose atoms are much more closely packed. The only solution to this quandary could be found, according to Faraday, by overcoming the dualism of the corpuscular theory and by assuming that matter is continuous with space. As mentioned earlier, he accepted the Boscovichean system of points of force as the more adequate answer to the problems left unsolved by the Daltonian system of atoms.

A few years later, in a letter entitled 'Thoughts on ray vibrations' and addressed to his friend Richard Phillips, he further elaborated his thoughts on the subject.[31] If the atoms are considered centers of force, he suggested, then they have no definite size as do the hard atoms, because what represents size in these centers of force 'may be considered as extending to any distance to which the lines of force of the particle extend: the particle indeed is supposed to exist only by these forces and where they are, it is'. On the same occasion he examined the relation between the matter of the ether and ordinary matter; for instance, copper, an electrical conductor. There are two approaches to their 'essential constitution': either they were both composed of little nuclei, considered in the abstract as matter, and of force of power associated with these nuclei, or they both consisted of mere centers of force of the Boscovichean model. 'There is no reason to assume that the nuclei are more requisite in the one case than in the other. . . .'[32] There was an obstacle, however. The force of gravitation acts on copper and consequently this is ponderable, but not on ether which has no weight. How could this be explained? For Faraday the answer lay in the fact that 'of all powers of matter, gravitation is the one in which the force extends to the greatest possible distance from the supposed nucleus, being infinite in relation to the size of the latter and reducing that nucleus to a mere centre of force'.[33]

It is only fair to say that this speculative framework left many people unconvinced and, accordingly, the reactions to Faraday's ideas in the 1840s were lukewarm at best.

In Britain in those same years, the kinetic theory of gases, which would soon prove to be the most important foundation offered by nineteenth-century physics to Avogadro's hypothesis, had found its early proponent in a man whose speculative leanings seemed closer to the hard atoms tradition. John Herapath attempted in the late 1840s to prove that the billiard-balls approach, although challenged by Faraday on electrical conduction considerations, could still find its application in the study of the gaseous state's behavior. For Herapath, gases were composed of 'very minute, perfectly hard particles which fly in all directions and collide with each other and with the walls of the container, thus maintaining a pressure against them'.[34] It is true that on this occasion he supported the view that a similar condition could be explained even by assuming that gases are formed by 'either perfectly or imperfectly elastic' particles; in other words, that hard atoms are not indispensable for his system. For Faraday's position this was not the case, however, because he excluded electrical phenomena that could be interpreted only with hard atoms.

Herapath's conclusions concerning the composition of gases had been published in 1847; in May of the same year Joule, in a lecture given at St. Ann's Church,[35] publicly offered his opinion of describing the 'particles or atoms of the bodies' as in motion. Their dimensions were so minute that their motion could not be observed, due also to the extremely high velocity of such particles. No distinction was made between atoms and molecules, as indicated by the example given for the velocity of the 'atoms' of water which he sets at one mile per second. The velocity of the particle should be considered, he pointed out, as an average one, because, quite likely, while some of them are at rest, others are moving very fast. As for the temperature of a body, it was generally proportional to the velocity of revolution of the particles. No indication was offered by Joule at this time about the intrinsic nature of his atoms. He did not decide whether they are hard or elastic, nor was he explicit in his 1848 paper read to the Manchester Literary and Philosophical Society[36] in which he discussed the constitution of elastic fluids. He remarked there that the 'elastic force or pressure must be the effect of the motion of the constituent particles in any gas'.

While Davy hypothesized a rotary motion[37] which applied to the particles of fluids and elastic fluids, Joule indicated that the assumption made by John Herapath in 1847 – the pressure of a gas being represented by the random impact of its particles against the walls of the vessel – is 'somewhat simpler'; accordingly, he used it as a basis for the calculations that lead eventually, in the course of the same paper, to an evaluation of the velocity of hydrogen particles.

It is quite evident that at this time Joule had not yet taken a clear stand on the question of the elasticity of atoms. Although in the paper just mentioned he recalled Davy, whose views on the corpuscular theory were certainly known to him, he chose to remain on neutral ground. Later on, possibly as a result of his close collaboration with William Thomson,[38] he was even more reluctant to discuss this subject.

From the unnoticed manuscript of John Waterston,[39] through Herapath, an early supporter of the hard atoms, and Joule's agnostic attitude, the kinetic theory of gases finally found the proper intellectual condition for it to develop in the German-speaking countries. In a philosophical tradition that went back to Leibniz, there was no room for hard bodies. At the same time, the Boscovichean points seemed too metaphysical to be taken into consideration.

Then a German physicist proposed the assumption that *elastic* molecules could be formed by *rigid* atoms. In the interpretation offered by Clausius in 1857 in 'Über die art der Bewegung',[40] molecules would be subject to both a translational and rotational movement, while their components, the atoms, would undergo a vibrational movement, seemingly the most apt explanation of the physical behavior of gases.

The kinetic theory in the system proposed by Clausius introduced a new differentiation between molecules and atoms, thus far undetected by his predecessors. He attributed the difference between the three states of matter both to intramolecular motion and to intermolecular distances.

A parallel can be drawn at this point between the conclusions that Cannizzaro reached in his 'Sunto' (1858) and the considerations that Clausius had offered the year before. The Italian chemist[41] realized that the physical studies conducted by Clausius[42] indicated that molecular distances for bodies in the gaseous state are related only to the temperature and pressure to which they are subjected, and are wholly independent from the nature of the molecules and their masses and the number of atoms they contain. Moreover, Cannizzaro stressed that, by differentiating between atoms and molecules, one could put an end to the confused state of chemistry in those years. Thus elementary gases and some other elements have polyatomic molecules which break up when they react chemically.

There was a similar breakthrough in physics for Clausius. Of his predecessors, Joule seemed still undecided on the characteristics of his own 'constituent particles', while Krönig described particles as elastic molecules moving in a straight line at constant velocity. The next step – and a long one indeed – was taken by Clausius through his full analysis of the kinetics involved at the corpuscular level. Besides the simple translational movement envisioned by Krönig, Clausius assumed the presence of both a rotational movement for the molecules after their mutual impact, and a vibrational movement. The latter can be described in several ways, but occurs essentially among the rigid atoms composing the molecules.[43]

Through this collection of assumptions, with its substantial mathematical basis, the kinetic theory reached its maturity in 1857. The originality of

Clausius' thought lies mainly in finding a model which would satisfy the requirements of gaseous behavior and at the same time preserve the Newtonian tradition of hard atoms.

Maxwell

After the Congress of Karlsruhe, through the efforts of Cannizzaro, chemists had at their disposal a satisfactory definition of molecule as distinguished from atom, and also a convincing justification for the polyatomic nature of several elementary molecules. Almost at the same time the notion of molecule, again as an entity differentiated from atom, became a well-established physical fact, thanks to Clausius. As William Thomson pointed out,[44] Clausius not only investigated the mutual impact of molecules, but also studied the relation between their diameters, their number in a given space, and the mean length of path from impact to impact. For physicists, at least, molecules were a reality whose behavior could be statistically analyzed, as Maxwell did in 1860.

In the 1860s, the arguments supporting the atomic-molecular doctrines, from both the chemical and the physical points of view, seemed on the verge of prevailing. This has induced several historians of science to consider the period as one of full vindication of Avogadro's hypothesis. In fact, there is slight evidence in favor of this thesis.

Although one has to wait for some time for the major critics to appear – Mach still followed then an 'orthodox' position towards atomism and Ostwald, of course, had not emerged yet – there were still widespread feelings of skepticism and agnosticism towards the ideas of both Dalton and Avogadro. As will be shown shortly, this was especially true in Britain between 1860 and 1890. The apparent success of the atomic-molecular theories in chemistry failed to convince the skeptics because atomic weights seemed to have a mathematical meaning but no relation to reality.

There are two considerations which should be made concerning the impact that, according to historians of the period, the dynamic theory of gases had on the re-evaluation of Avogadro's generalization. The first one is that neither Clausius nor Maxwell recalled Avogadro; his name is completely ignored by the latter. As for Clausius, he eventually recognized that Dumas, Gerhardt, and Laurent had preceded him in their assumption of polyatomic molecules, but neglected to remember that they, in turn, had been anticipated by Avogadro and Ampère.[45]

The second observation is the vague distinction between molecule and atom still present in Maxwell's writings. This is in contrast with Cannizzaro's clear definition of the two terms in the 'Sunto', at least from the chemical point of view.

Maxwell in 1859 used the expression 'particles' in discussing the important relation between their velocities, the absolute temperature, and the specific gravities of gases.[46] He intended to prove the laws of motion of an indefinite number of small, hard, and perfectly elastic spheres acting on each other only during impact. One may arrive at the same conclusions, he said, by assuming the equivalence of the particles with centers of force whose 'action is insensible, except at a certain small distance when it suddenly appears as a repulsive force of very great intensity'.[47]

This was, of course, a Boscovichean position that Maxwell set apart for the sake of his argument in favor of the alternate hypothesis of perfectly elastic spherical bodies. Are they atoms or molecules? Maxwell did not say; with only one exception, he refrained throughout this important essay from adopting either of the terms, despite the fact that his 'Illustrations' were inspired by Clausius' 1857 work, where the distinction was clearly made.[48]

It may be appropriate to follow here in more detail Maxwell's reasoning. In proposition XII the query was to find the pressure per unit of area of the side of the vessel caused by the impact of the particles upon it. Maxwell, after determining the momentum of the striking particles, calculated from this the whole pressure $p = 1/3MNv^2$.[49] But $MN = \rho$ and $v^2 = 3k$, and thus $p = 1/3\rho .\ 3k = \rho k$ which represents Boyle-Mariotte's Law. By supposing that elastic particles move in a straight line and that $v^2 = kT$, and, at constant temperature, $v^2\text{-}k(1/\rho)$, then at the same temperature and pressure the value NMv^2 is the same for all gases; but it has been proved (in proposition IV) that in two systems of particles moving in the same vessel the mean *vis viva* of each particle will become the same in the two system, or Mv^2 has the same value in each system of particles. Hence, the number of particles in a unit of volume is the same for all gases at the same temperature and pressure.

Maxwell had thus arrived at Avogadro's generalization; he concluded by saying that 'this result agrees with the chemical law, or law of equivalent volumes, that equal volumes of gases are chemically equivalent'.[50] There was no mention of Avogadro here, nor was his name mentioned a few years later when Maxwell examined the historical background of the atomic theory from Lucretius onwards.

Thirteen years later, in an article which appeared in *Nature*,[51] Maxwell defined molecule as a 'modern word'[52] and continued: 'According to the received doctrine, in each molecule of water there are two molecules of hydrogen and one of oxygen. Whether there are or not ultimate atoms I shall not attempt to decide. . . .' Shortly afterwards he added:

> . . . every substance, simple or compound, has its own molecule. If this molecule be divided, its parts are molecules of a different substance or substances from that of which the whole is a molecule. An atom – if there is such a thing – must be a molecule of an elementary substance. Since, therefore, every molecule is not an atom, but every atom is a molecule, I shall use the word molecule as the more general term.[53]

It is quite clear at this point that Maxwell had still serious doubts about the real existence of atoms. Furthermore, he recognized the molecule as the essential constituent of matter, not only in compounds but also in elementary substances, regardless of their physical state. He did not see any reason to adopt other terms which could provide only a subordinate meaning. In the same article he credited Gay-Lussac as author of the equivalent volumes of gases law, whose 'dynamical' explanation is that a cubic centimeter of every gas, at standard temperature and pressure, contains the same number of molecules.[54]

Later on, there is a further important qualification of molecule which, although 'indestructible' is not a hard, rigid body but is capable of internal movements.[55] As a consequence of these movements, the molecule emits rays whose wavelength is a function of its vibration time. In the case of hydrogen, for instance, this wavelength has been proven to be the same regardless of the source of hydrogen used. What Maxwell meant by 'indestructible' molecules is not clear, because in the same lecture he discusses the electrolytical 'dissociation' of the water molecule along with the formation of oxygen and hydrogen. He was probably referring to the 'elementary' molecules as 'indestructible', but this qualification is not used here.

In the 'Dynamical theory of gases', read on 31 May 1866, Maxwell indicated that he intended to consider the molecules of a gas 'not as elastic spheres of definite radius but as small bodies or groups of smaller molecules repelling each other with a force whose direction always passes very nearly through the centers of gravity of the molecules.' A series of experiments on the viscosity of air at different temperatures had led him to modify his earlier theory on gas constitution. In discussing the probability of finding a law relating the temperatures of liquid and solid bodies with their molecular energy, Maxwell mentioned that in such a case the molecules of a gas 'are those portions of it which move about as a single body', and hinted that these molecules may be pure centers of force provided with inertia, or mere points, or they may be 'systems of several such centers of force bound together with their mutual actions'.[56] In such a case the whole energy is that of translation. On the other hand, he added, one may consider that the molecules are small bodies of a determinate form; this, however, would require the introduction of a 'molecular theory of the second order', because in such a case one would have to assume a new set of forces binding the parts of these small bodies together.

In his article on 'atom' written for the ninth edition of the *Encyclopedia Britannica* (1878), Maxwell revealed many of his views on the atomic-molecular theory. He devoted the first section to a rather detailed historical review of atomism from Democritus and Lucretius, through Descartes and Boscovich, to Helmholtz and W. Thomson; remarkably absent is the name of Dalton among others. The second section of the article was devoted to the 'modern molecular science' and plunged, after defining the kinetic energy of a 'particle', into a mathematical approach to Boyle's Law, dropping the term atom altogether. 'Bodies', he said on this occasion, 'are made up of small parts which act on each other in a manner consistent with the principles of conservation of energy.'[57] These 'small parts' are in motion, as may be observed by the diffusion of gases. Maxwell indicated here that he did not know whether the small parts are all of the same size or whether they have extension or shape. Their masses, however, can be measured, and they act on each other when they come near enough to do so.

Later on, in the same article, Maxwell once again mentioned Gay-Lussac as the one who established that 'the weights of the chemical equivalents of different substances are proportional to the densities of these substances when in the form of gas'.[58] At this point, very significantly, he stressed that the definition of molecule as employed by Gay-Lussac is by no means identical with the definition of the same word as used in the kinetic theory of gases. Chemists ascertain by expreiment the ratio of the masses of the different substances in a compound;[59] from this they deduce the chemical equivalent of the different substances taking hydrogen as unity.

> The molecules of the substances are in the ratio of their chemical equivalents. [He added] this reasoning is purely chemical reasoning, it is not dynamical reasoning, it is founded on chemical experience, not on the laws of motion . . . our definition of a molecule is purely dynamical.[60]

Through the entire article, Maxwell used chemical equivalent, never the expression 'atomic weights', even when discussing the Dulong–Petit Law of atomic heats.

As for his interpretation of molecule, it ranges from a system of elastic spheres to one of a finite number of atoms kept in their places by attractive and repulsive forces. The concept prevails in his thinking that, in the atomic-molecular structure, a rationalization must be found for the phenomena detected with the spectroscope suggesting a vibrational movement within the molecules. Obviously, Lucretius' hard bodies could not fit in such a scheme and, on the other hand, the Boscovichean centers of force, indivisible and massive, cannot be imagined as capable of vibrations, unless one assumes that molecules may be composed of many such centers of force, which would mean the reduction of the function of the atoms to an action at insensible distances. Maxwell, therefore, concluded that, by keeping their characteristics in mind, the vortex atoms proposed by W. Thomson on the basis of Helmholtz's vortex rings are closer than any atom thus far conceived to satisfy the requirements that they may assume any form, vibrate, and are permanent as far as their strength and their volume is concerned.[61]

Maxwell later returned to the statement of the 'law of equivalent volumes' in a lecture given at the Chemical Society.[62] The quality of the audience and some points raised in the course of the lecture make this one of the more interesting chapters in the study of Maxwell's approach to the molecular theory. He stressed on this occasion that the law of equivalent volumes, thus far with only a chemical basis, had now, through the kinetic theory, a support on 'dynamical principles'[63], and the molecule could therefore be defined in physical terms without referring to the chemical proportions. Once the validity of this law is assumed, then it is possible to calculate accurately the relative molecular masses from the densities of gaseous substances. Needless to say, Avogadro had arrived at this corollary of the molecular hypothesis sixty-five years earlier. One of the most difficult obstacles met in the study of the molecular theory, Maxwell pointed out, is that represented by the investigation of the motion of the constituents of the molecule in relation to the center of its mass whose motion has been thoroughly examined. While the movement of the constituents of the molecule represented an important aspect of Clausius' system, Maxwell avoided this issue.

If we assume, Maxwell said, that the constituents of a molecule are atoms and that each atom is what 'is called a material point', then it is necessary to take into consideration three variables for each atom in order to determine its position and configuration inside the molecule. To pursue the mathematical investigation it is not essential, however, to assume that the molecules are made up of atoms. This is important because it reveals that the notion of the existence of polyatomic molecules in elemental gases does not represent an essential feature in Maxwell's conceptual approach to the molecular theory.

The determination of the ratio between the specific heat at constant pressure and at constant volume, which for Clausius also indicated the ratio between the whole energy of molecular motion and the energy of translation, had given a value of 1.408 for some gases.[64]

Spectroscopic observations, on the other hand, definitely suggested the very complex structure of some molecules and their vibrations and, in turn, denoted that such molecules have a number of variables even greater than six; their calculated specific heats would then further increase. In other words, the more elaborate one theorizes the molecules to be, the greater the difference between the observed ratios of the specific heats and those calculated. For Maxwell this represented 'the greatest difficulty yet encountered by the molecular theory'.[65] It also became one of the main targets of criticism of the kinetic theory, as the discussion of William Thomson's position will show.

One may be drawn to conclude here that the kinetic theory, as interpreted by Maxwell, could use elastic molecules, but could somehow discard atoms, something the chemists in the second half of the nineteenth century were not inclined to do. The hostility towards a theory of hard bodies which, as we have seen, from Davy on had lingered for decades among Dalton's countrymen – *nemo propheta in patria* – strongly survived in Maxwell's views. In stressing the contrast between his own 'dynamical' molecules and the 'chemical' molecules, he revealed only the typical attitude of his contemporaries from the same school of thought.

It may be worthwhile to recall that the Faraday Lecture of 1872 was given in London by Cannizzaro.[66] It is not known whether Maxwell, then professor at Cambridge, was present, but it is apparent that the content of this lecture had very little effect upon his opinions, as evidenced by his subsequent papers.

Clausius

As mentioned earlier, on the Continent a decisive step forward in the understanding of the relation between the ultimate particles of gaseous fluids was offered in those same years by Rudolf Clausius when he published his classic essay 'Über die Art der Bewegung welche wir Warme nennen',[67] shortly before Cannizzaro's 'Sunto'. Clausius, with the exception of an immediately preceding work by Krönig,[68] had no inkling of the ideas set forth in the previous fifty years about the chemical and physical behavior of gases. However, in his 1857 essay, he reproposed three concepts which by then had been amply discussed within the scientific community: the equal volume–equal number assumption, the polyatomicity of elemental molecules, and the combining volumes' law.

The 'Über die Art der Bewegung' remains a very important document for the historian, because it represents the first complete approach to the molecular hypothesis from the physical point of view. While it has been said that Clausius was able to derive Avogadro's hypothesis from his kinetic theory and thus make the physicists aware of it,[69] this was not actually the case, because not only was Avogadro unknown to him, but also because he had reached the conclusion that equal volumes of elemental gases contain *equal number of atoms*; which, of course, seems more the position held by Berzelius than that of Avogadro.

Clausius arrived at the equal volumes–equal number generalization in the central portion of his essay. Shortly before, he had stated that he considered of great importance the clarification of the empirical fact that 'when two gases combine with each other or a gas combines with another substance and the compound is gaseous, then the volume of the compound gas is in a simple ratio with the volumes of the individual constituents as long as they are gaseous'.[70] This statement, reviving a law by then fifty years old, was introduced abruptly without connection to the preceding portion of the essay. To Gay-Lussac, whose name was mentioned later on in discussing his other law relating pressure and temperatures in gases, no credit was given by Clausius for his 1809 generalization on the simple ratios in combining volumes.

Clausius now quoted Krönig as having proven that the pressure of a gas on the walls of a container must be proportional to the *number of molecules* present in the unit of volume and to the kinetic energy that the single molecules have through their translational movement.[71] If one applies this concept to elementary gases, he continued, and also by moving 'from other bases' ('aus anderen Gründen') – not further identified – it is possible to reach the likely assumption that in 'gleichen Volumen gleich viel Atomen enthalten . . . .'[72] Hence, he said, *atoms* of different gases must have the same kinetic energy in relation to their translational movement. It is apparent here that, while the conclusive remark is new and conceptually correct, the gas hypothesis is phrased in a faulty way. This is a bit surprising, because Clausius, as is shown from other portions of 'Über die Art . . .', seemed to have reached a very clear understanding of the differentiation between atoms and molecules, not only from the physical point of view but also in their chemical meaning. There is little doubt that he had atoms in mind when he used this term; the same is true for molecules.

Clausius then elaborated on the case of two gases combining in simple but definite ratios and stressed how–although the molecule of a gas (e.g., nitrogen oxide) is biatomic and that of the other gas (e.g., nitrous oxide) is triatomic–both these gases in equal volumes contain the *same number of molecules*, and therefore even heterogenous compound molecules have the same kinetic energy in relation to their translational movement.

However, the situation is different when two simple gases combine in equal volumes, and the experiment indicates that there is no diminution of volume while, according to the general rule, the volume should decrease in the ratio of 2 to 1. Clausius explained this on the basis of a hypothesis which 'should be further verified.' He assumed: (a) that the force which is at the origin of chemical compounds probably consists of a kind of polarity of the atom[73] which is active even in elementary substances, and (b) that also in such substances 'several atoms are bound into a molecule'.[74]

He concluded by surmising that the simpler and more likely case is that such a bond is formed by two atoms for each molecule, thus explaining the anomalies in the volume ratios observed in the experiment. He gave two examples, one to illustrate the chemical combination of equal volumes of nitrogen and oxygen with no change of volume, the other concerned the chemical combination of the same elements in the 2 to 1 ratio, leading to the formation of a compound gas whose molecule is triatomic (e.g., nitrous oxide). Clausius explained that this happens because the ratio of the number of molecules decreases from 3 to 2, during the chemical combination. Therefore, even the volume decreases in the same ratio.[75] One may say at this point that the explanation offered by Avogadro in 1811, although more elaborate, is in fact more valid and in agreement with the current interpretation of chemical reactions in the gaseous state. Clausius, who had earlier conjectured that equal volumes contain an equal number of atoms, seemed to suggest here that they also contain the same number of molecules. This is true for elementary gases, but not necessarily so for compound gases, as the example he gave indicates.

Before entering into the mathematical section of his paper, Clausius considered those elements – such as sulfur and phosphorus – which occupy a volume in the gaseous state smaller than is expected from their atomic weights. He hinted then – and here again he had been preceded by the chemical works of Gerhardt, Laurent, and Bineau, among others – that these anomalies can be explained by considering the biatomic nature of the molecules as the simplest, but not necessarily the 'only possible assumption'. Due to the rather large number of elementary and compound gases, he argued, it is unlikely to find complete agreement among them; there can be exceptions to the general rule, but their importance should not be overestimated. With the help of the hypothesis on the molecules of the elements (i.e., that they are biatomic, at least), he concluded that it was conceivable 'to link all the volume ratios of the gases to the principle that the individual molecules of gases have the same kinetic energy with regard to their translational movement'.[76]

There is no doubt that in 1857 for the first time, although starting almost from scratch, Clausius was able to put together a rather complete picture of the elaborate relationship existing between the chemical and physical behavior of molecules. His work, in this instance, should be rightly quoted as a typical example of 'independent discovery'. He was provided with that gift for conceptual synthesis which apparently was missing in both the chemists and the physicists preceding him. Some of them simply could not grasp the meaning of atoms; some could not accept molecules, and this condition continued even after 'Über die Art', as will be seen shortly. Among the chemists, the best – Berzelius and Dumas – were either confused or confusing; the physicists, on the other hand, limited themselves to the periphery of the problem, but dared not get to its center. Those few – such as Gaudin, Ampère, and Avogadro – who displayed a clearer understanding of it were either ignored or criticized.

In a long footnote, at the beginning of the 'Über die Art', Clausius recalled that in a conversation with William Siemens[77] he learned that Joule's article of 1857 on the constitution of elastic fluids reprinted in the *Philosophical Magazine* had originally appeared in *Memoirs of Manchester Literary and Philosophical Society* of 1851. Since he did not have the opportunity of seeing the original paper, he was unaware of its content and regretted that Joule had not published it in a better known journal. For Clausius, as he himself admitted, this was one more inducement to go ahead and publish his own views on the physico-chemical behavior of gases.[78] Actually, Joule had read two memoirs on the constitution of elastic fluids in August and October 1848 before the British Association and the Manchester Literary Society.[79] On the former occasion he mentioned both Davy's ideas[80] and Herapath's hypothesis, in which the particles of gas were constantly flying in every direction with great velocity, the pressure of the gas being caused by the impact of the particles against any surface presented to them. This concept, of course, had already been set forth by Bernoulli 110 years earlier, but was, at the time, unknown to either Joule or Clausius.[81]

Clausius returned to the polyatomicity of gaseous elements one year later in a memoir discussing the properties of ozone.[82] An experiment, conducted by Schönbein on the effect of ozonized air upon a strip of paper moistened with lead peroxide, indicated that eventually ozone disappeared and was converted into ordinary oxygen. In this, Clausius found confirmation of the notion earlier conceived that the properties of ozone could be explained by assuming the polyatomic nature of elementary gases. A very extensive footnote followed in which Clausius revealed that the above mentioned concept, according to a note by Verdet and Marignac[83] in the *Traité de Chimie*, had already been put forward by Dumas, Gerhardt, and Laurent. While all this he said 'mir bis dahin unbekannt war',[84] he defended his lack of awareness on the basis that he had not been able to obtain the original works of Dumas and Laurent; the same was true for the last part of Gerhardt's *Treatise of Organic Chemistry*, whose German translation was not available when he was writing 'Über die Art der Bewegung'. As a final justification, Clausius argued that Gerhardt had reached his conclusions about the nature of elemental gases by moving from 'merely chemical considerations', i.e., that free hydrogen is actually hydrogen hydride, free chlorine is chlorine chloride, and so on.

However, it is quite clear that Clausius himself had in mind the unusual behavior of volumes' ratio in some chemical combinations when he explained it by presuming the division of gaseous molecules in such reactions. Avogadro had followed the same train of thought in 1811, and his reasoning, too, was based upon the conclusions earlier reached by Gay-Lussac and drawn from chemical facts.

Altogether, the arguments raised by Clausius on this occasion seem weak; in his neglect, for instance, of the ideas of Gaudin and Ampère, published in the *Annales de Chimie*, he could not invoke the limited circulation of the publication, as he had done for Joule with the *Memoirs of the Literary and Philosophical Society* of Manchester. The truth remains that, despite the diffusion reached by scientific periodicals after the third decade of the century, the dissemination of information among men pursuing similar cultural interests was still a fortuitous event. Although on one hand there were delays in printing and distribution, the language barrier, and similar restraints, on the other hand, the subdivision and superspecialization of scientific research had not reached today's levels, and so it was relatively easy to be informed. Nevertheless, there were cases of communication gaps which defy any explanation. Rudolf Clausius had been a professor first in Berlin for five years and then in Zurich for two more years when he reported in 1857 his ideas on the dynamical theory of gases. By then, the *Jahresberichte*, which contained abstracts of works published all over Europe, had been published for thirty-five years and the *Annalen der Chemie* for a quarter of a century, just to leave aside the French journals for a moment. Apparently, Clausius was not interested in chemistry or he rushed the publication of the 'Über die Art' so much so that he he paid no attention to previous works.

William Thomson

A very serious criticism of Clausius' kinetic model came from across the Channel. William Thomson, one of the most brilliant and fertile representatives of British science in the nineteenth century, had covered almost every field of physics in his researches. It was not surprising then that in the sixties he entered the arena of debates about the intimate constitution of matter. A man of his standing and encyclopedic knowledge could not be, after all, absent from one of the critical issues facing natural philosophy at the time.

Thomson's approach to the atomic-molecular debates may be characterized by these elements: his firm denunciation of the kinetic theory of gases as developed by Clausius, Joule, and Maxwell, his repeated characterization of chemistry as 'powerless' to explain the essence of matter, and finally a substantial uncertainty which lasted almost until the end of his life as to the reality of such ultimate components of matter as atoms and molecules.

From 1867 to 1901 Thomson attempted to give a mechanical explanation of matter through his numerous models; these being just a tool he felt necessary to use in order to reach his goal.[85] Was his overall contribution advantageous for the development of the molecular hypothesis, and did he make the concept of atoms or molecules easier to understand and to accept at the time? With the benefit of hindsight it is only natural to answer these questions in the negative. The truth remains, however, that Thomson, Maxwell, and their contemporaries appeared to struggle with difficulties that they were unable to overcome conceptually and experimentally. The breakthrough, of course, came only near the turn of the century; by that time, however, Maxwell had been dead for twenty years and Thomson, in the dusk of his life, still remained bound to a nineteenth-century scientific outlook.

As for Avogadro's hypothesis, in 1900 Thomson (by then Lord Kelvin)[86] joined the then prevailing opinion, and defined it as a 'Law', but it is fair to say that, although he implicitly recognized it, he also had no use for it within his conceptual framework. This may be easily understood if one keeps in mind Thomson's unwavering criticism of the gas kinetic theory which, after all, had emerged since 1860 as the strongest support offered by physicists for the validity of Avogadro's generalization. Even more remote from Thomson's thought were the systems of spherical atoms mutually attracting and repelling at a distance and which had been the foundation of the atomic and molecular theories. While Dalton and Avogadro had also reached their conclusions through chemical considerations – the laws of definite proportions and of combining volumes – the view that chemistry could do without reference to a rigid atomic and molecular structure appeared again and again in Thomson's writings.[87]

In 1867 he published his outline of a configuration of motion in a continuous fluid and disclosed his system of vortex rings,[88] nine years after it had been suggested by Helmholtz. Thomson chose this occasion to harshly attack those 'pieces of matter' infinitely hard and rigid whose existence the 'leading' modern chemists had taken for granted. Such structure, as conceived by Lucretius and adopted by Newton, could be replaced by vortex atoms moving through a perfect homogeneous medium. So, what is called matter are only portions of such fluid provided with a vortex motion; these portions are in fact atoms which cannot be formed nor can they change or disappear. At the time, the experimental demonstration supplied by Tait with his ammonium chloride smoke rings convinced Thomson even more of the reality and validity of his theory.[89] Here he had at last a model provided with kinetic elasticity, while the endless variety of structure of the vortex atoms seemed more than adequate to explain the variety, the mutual affinity, and the allotropic forms of the elements. Moreover, radiation phenomena could be more easily accounted for through the vibration observed in a vortex-like model than through a rigid particles' system. Alternatively, the only way to shed light upon the cause of vibrations would be to consider, for instance, sodium not as a single atom but as a molecule formed by a group of atoms separated by empty spaces; however, such a molecule, Thomson remarked, would be unstable and therefore self-contradictory. On the other hand, a vibrational movement might be explained by vortex atoms, since they moved and changed their form considerably without disconnecting their constituent parts.[90]

In 1867 Thomson could not explain the change in temperature of his vortex atoms; the mathematical problems involved had not yet been solved, and a theory of the thermal expansion of the vortex was required. There were still more serious obstacles to be hurdled such as the inertia or gravitation, and the stability of such rings. Furthermore, no explanation had been attempted of their behavior in chemical combinations: for instance, since the vortexes are indivisible, how from two molecules of hydrogen and one of oxygen only two molecules of water could be obtained? What about the extreme complexity required to explain the large organic chemical structures? Thomson felt, however, that he need not account for these facts: chemistry would take care of itself. It took almost thirty years for him to reach the conclusion that the vortex atoms system should be abandoned; this is evident from his correspondence.[91]

In developing an earlier approach to the same subject in 1883 Thomson discussed extensively the problem of the 'size of atoms'.[92] Four experimental approaches were available: capillary attraction, contact electricity, the wave theory of light, the kinetic theory of gases, and all led to a set of measurements in rather close agreement. Atoms *and* molecules of matter must have a diameter of 1×10^{-7} and 1×10^{-8} cm. respectively. He apologized this time to the chemists for his misuse of appropriate terms, but at the same time he restated his view that 'chemists don't know what an atom should be'; they didn't know, for instance, whether hydrogen is made of two fragments of matter combined in making one molecule or whether it is made of simple molecules indivisible or somehow nondivisible by chemical action.[93]

The importance of the 1883 lecture lies in the fact that Thomson then clearly recognized that the 'most decisive' method of determining the atom's size had been offered, through the evaluation of the average length of molecular free path,[94] by the kinetic theory of gases. Significantly, there was no mention in the entire paper of vortex models.

The following year Thomson returned to a configuration of motion in a continuous all-pervading fluid, in his Montreal speech before the British Association. This was centered around an elaborate mechanical model to explain the elasticity of solids.[95] He then added that when the vortex theory of gases is made clearer it will not fail as the kinetic theory of gases did with its elastic solid molecules.

Among the possible alternative ways offered to Thomson to represent a molecule were: (a) as an elastic solid, (b) as a configuration of movement, and (c) as a mathematical point. By 1889 he had rejected the first and became increasingly dissatisfied with the second. So he turned his interest more and more to the possible use and contribution that the Boscovichean theory of points could offer towards an acceptable explanation of the constitution of matter. This new attitude probably did not take place without some painful readjustment, because thirty years earlier, before his introduction of the vortex atoms, he had harshly criticized Boscovich for introducing the most 'phantastic of the paradoxes', namely that contact does not exist.[96] On the same occasion he had praised Faraday for opening a new era in science and thus ending the physics of the eighteenth century. Now, in 1889,[97] after a short introduction in which he mentioned Auguste Bravais' crystallographic studies,[98] qualifying them as 'the grammar of molecular constitution', Thomson focused his attention on Boscovich.

It should be stressed at this point that he did not fully endorse the points doctrine of atoms proposed by Boscovich. In fact, this was not yet accepted; in Thomson's view, however, it could offer a kinetic and static basis upon which one would be able to attempt an understanding of the molecular structure of matter. Throughout the whole lecture, together with an extensive discussion of intra-atomic dynamics, Bravais' molecular lattices were discussed and Boscovich's points used to study a new model of elastic solids. There was no more mention of vortex atoms at this time.

In the following years, after Crookes' and Perrin's works on the cathode rays had amply proven their material nature and after Joseph Thomson's determination of the mass and the charge of the electron, William Thomson attempted to update his model of elementary particle.[99] In part this occured in the course of a well-known discourse he offered at the Royal Institution in April 1900. His topic[100] was the state of some key physical problems still without solution at the turn of the century, such as: (a) the movement of the earth through an elastic fluid or a motionless ether, while Michelson's experiment indicated the movement of the latter together with the earth; and (b) the principle of equidistribution of energy still supported by Boltzmann and Maxwell. For Thomson these two issues remained 'thick clouds' on the horizon of modern physics, despite the attempts made by Lorentz and Boltzmann respectively to disperse them.

In 1900 the model conceived by Thomson for the atom was a series of concentrical spherical surfaces having the same density. He seemed to lean now more and more on Boscovich's theory. The molecule was assumed to be a group of Boscovichean atoms whose mutual forces must be such that the group remains in a state of stable equilibrium whenever it is at rest. A gas is then made up of a large gathering of such molecules, each group freely moving through space except when it enters into collision with another group of atoms; each molecule maintains its own constituents and only when there is a collision, does an exchange of equal atoms take place. At the same time, while discussing the specific heats of gases, Thomson recognized that both hydrogen and nitrogen were, according to the 'modern chemical theory', biatomic gases consisting of two Boscovichean atoms.[101] Apparently, for him physics had not yet been capable of proving their constitution. He recalled here 'the molecular dynamics' of Avogadro's Law. Mention of this law appeared also in a footnote[102] to a lecture offered at the Royal Society of Edinburgh on 16 July 1900, three months after the 'Clouds' lecture in London.[103] The subject of this lecture was the state of motion caused in an infinitely elastic solid by the movement of a body through the space occupied by the same elastic solid. Oddly enough, in discussing the value *N* which, as Thomson reminded his audience, was the same for all gases, he qualified it as the 'number of atoms' per cubic centimeter.

The final chapter of Thomson's forty-year-long inquiry into the ultimate composition of matter came in 1904, when he wrote that 'in this twentieth century we are compelled to accept notions on the atomic origin of all things which clearly remind us of those proposed by Democritus, Aepicurus and Lucretius'.[104]

Since the 1867 definition of atoms as the 'monstrous assumption of Lucretius', Thomson had come a long way in his Diogenesian search for truth. His had been an unparalleled intellectual experience, stretching from Faraday to Rutherford, from point atoms to structural atoms. In 1907, when his life ended, the new physics was just born.

The Chemists' View

A review of what has been said and written in Britain after the 1850s indicates that the molecular hypothesis – besides the lack of interest and the hostility expressed by some prominent physicists – met also with the criticism and skepticism of the exponents of the new school of chemistry in that country. In September 1860, among those who participated in the Karlsruhe Congress was a thirty-one year old British chemist[105] who had already attained a reputation for his work in organic chemistry.[106] William Odling, who later became Fullerian Professor and eventually one of the most influential[107] men of science in Britain, returned from Karlsruhe quite unconvinced by the ideas Cannizzaro proposed there. He reported that, while Kekulé had clearly distinguished between physical and chemical molecules, Cannizzaro assumed only the existence of physical molecules that cannot be determined by chemical means.[108] As for the Avogadro–Ampère hypothesis, its importance was related only to chemical molecules or gaseous molecules. Had Odling been more familiar with the content of the 'Sunto', he would have known that Cannizzaro drew a very unequivocal distinction between atoms and molecules that nobody else had done before, with the exception of Gaudin, and that he also had very clear ideas on the chemical and physical methods to be used in order to arrive at a correct evaluation of both atomic and molecular weights.

In 1861, Odling published *A Manual of Chemistry Descriptive and Theoretical* in which, while adopting Gerhardt's weights, still failed to recognize the difference between the concept of molecular weight and atomic weight. In the same book Odling renewed his support for the ideas of Gerhardt and Laurent, recalled Ampère, but omitted to mention Avogadro's gas generalization.[109] His staunch advocacy of Gerhardt is again evident in 'Reports on chemistry', the Presidential Address presented by Odling to the British Association in 1864.[110] Here, after stressing that Gerhardt had advanced further than Berzelius by reconciling the combining proportions with the 'physical' atom, he defined him as 'the great founder of that modern chemical philosophy in the general spread'.[111] On this occasion Odling named just Wurtz and Cannizzaro and expressed his dissent with the atomic weights then supported by Dumas and Frankland.

Besides Gerhardt, he recognized the great value of Hermann Kopp, then at Giessen, for his contribution to the knowledge and determination of molecular heats. In reviewing the state of the chemical science, Odling stated, 'the comparative unanimity which prevailed before the time of Gerhardt was the unanimity of submission to authority but the greater unanimity which now prevails is the unanimity of conviction. . . .'[112] Later on, he concluded his report by saying 'scepticism still prevails, not however the sterile scepticism of resignation, but the fertile scepticism which aspires to greater and greater certainty of knowledge'.[113]

Not long after expressing these feelings, Odling became personally involved in one of the most lively public exchanges of opinion that had occurred in Britain on the subject of the atomic and molecular theories in the second half of the nineteenth century.

The stage for these debates was the Chemical Society in London, the year 1867, and the participants were some of the more prominent British physicists and chemists.[114] At the meeting of June 6 the main speaker had been Benjamin Brodie then at Oxford, who took this opportunity to criticize the atomic theory for its inadequacy and the misconceived representations to which it had led; his purpose, however, was to propose a new method of determining chemical combinations through calculus. While there is no need to discuss here the details of Brodie's proposals,[115] it may be worthwhile to recall that he chose as unit of ponderable matter the portion of such matter which occupies a volume of 1000 cubic centimeters at 0° and 760 mm of mercury. With regard to chemical substances, he said, they should all be brought to a gaseous state for the simplicity of the laws to which gaseous combinations are subject. To deal with solids and liquids, in fact, would be much more difficult. Avogadro had arrived fifty years earlier at similar conclusions, as proven by his constant reference to chemical combinations taking place by volumes rather than by weight, but this was not mentioned by Brodie.

Odling, a close friend of Brodie, had encouraged him three years earlier in 1864[116] to publish his views in order to improve the symbolic notations used in chemistry. Now, at the end of Brodie's lecture, he waited for the comment of four others present at the meeting before he stood up. In his remarks he praised Brodie because his symbols were in direct relation with chemical facts, and did not require the intervention of the atomic theory as the present method of chemical notation did. Later on, after stating his amazement at Frankland's previous questioning of the positive existence of atoms, he pointed out that he himself did not believe in atoms and intended to keep them in the background as much as possible. However, he recognized that, without the atomic theory, there would be no support for the chemical language presently used.[117]

Odling had been preceded in the course of the debate by Frankland and two physicists and was followed in turn by Alexander Williamson and by Brodie's final remarks.

Edward Frankland, then forty-two and professor at the Royal School of Mines, objected very strenuously to Brodie's intimation that together with Crum Brown he had advocated the use of graphic and glyptic formulae[118] to represent chemical compounds. Not only did he 'repudiate' the notion that symbols or models could be adopted to denote portions of matter called atoms or their position in the compound, but also he stated very plainly that he neither believed in the existence of atoms themselves nor in the existence of centers of force.[119] With this off his mind, Frankland then devoted the rest of his intervention to some minor criticism of Brodie's system which, however, had the great advantage of requiring very little hypothesis for its formulation.

As mentioned earlier, Frankland's words caused a great shock. William Odling ironically observed that, by questioning the existence of atoms, Frankland had 'grossly deceived the chemical public'.[120] There is no record of any further response by Frankland to Odling's comments.

As for the physicists present at the June meeting of the Chemical Society,

James Maxwell and George Stokes,[121] their views were generally in agreement with those expressed by Brodie. For Maxwell, the concept of unit of ponderable matter set forth in Brodie's *Chemical Calculus*[122] was not new, but had already been proposed by the chemists (though without naming him, he very likely referred here to Cannizzaro's ideas in his 'Sunto') from chemical considerations. As for the kernel of the discussion, the validity of the atomic theory, it could be decided only by relying on a dynamical approach. The qualification of 'physical dynamics' was stressed by Maxwell at this point in contrast to 'chemical dynamics'. In fact, through the assumptions made by Clausius and 'others'[123] in formulating the kinetic theory of gases, it was possible to arrive at the conclusion that at 0° and 760 mm, the number of 'atoms or molecules' in the gaseous state which occupy the volume of a liter must be the same. To accept this statement, Maxwell said, necessitates a belief in atoms; the whole theory, however, he added, is now 'under probation' among scientists. Maxwell was followed by Stokes, who praised Brodie briefly for having established his system of well-known facts and by not referring to 'superfluous hypotheses', a view quite similar to that earlier stated by Frankland.

Before the conclusion of the debate, the Chairman, Alexander Williamson, then teaching in London at University College, raised and elaborated his opinion on the subject.[124] Williamson was a student of Liebig, a close friend in his youth of both Laurent and Gerhardt, an early supporter of the types theory, and after a short period of very fruitful research, an extremely successful and influential teacher. His ideas undoubtedly stimulated Kekulé in his pursuit of views which eventually led to the doctrine of valence.[125] In the course of the debates at the Chemical Society, he among the British chemists of his time, emerged as the most unequivocal supporter of the atomic theory.

At the June 1867 meeting, Williamson limited himself to a vague approval of Brodie's system, although he confessed that he did not feel capable of accepting some of its points. On this occasion, quite peculiarly, he omitted any reference to the atomic theory and raised no objection to the skepticism earlier displayed by both Frankland and Odling. Probably he felt that the time and the circumstances were not yet ripe for a full disclosure of his own interpretation of the intimate constitution of matter.

But the occasion for this would not be far away. Two years later, in June 1869, he presented his position very clearly in a long lecture entitled 'On the Atomic Theory' which he read at the Chemical Society in London.[126] Williamson's purpose was to support the intrinsic value of the atomic theory.

That this was necessary at all ten years after Cannizzaro's 'Sunto' is only proof once more of the weak stand of those historians of chemistry who maintained that the Karlsruhe Congress represented the turning point for both Dalton's and Avogadro's theories.

The main trouble with critics of the atomic theory, Williamson said, was that they just condemn it, without either coming forward with corrections for its shortcomings, or having the courage to reject it *in toto*. The result was a state of uncertainty about the whole system. In his reasoning, Williamson considered the law of the multiple proportions a cornerstone for the foundation of the atomic theory; actually, if removed from such theory, the law in question would be meaningless.[127] However, some of those who had adopted it did not seem to realize that, by doing so, they had implicitly accepted the notion of atoms.

From the atomic theory, the concept of molecules defined by Williamson as clusters of unlike atoms combined together in various numbers, was also derived.[128] The weight of these clusters represented their molecular weight. The truth of the atomic constitution of molecules had been confirmed by the density of their vapors, and conversely the molecular weight had been found proportional to the vapor density of every compound which evaporated without decomposition.[129] Although Williamson did not specifically refer to Avogadro, the soundness of the molecular hypothesis seemed fully supported by him in his 1869 speech, as evidenced by his comment that vapors of like molecules occupy the same volume under like conditions, whatever the molecule may be.[130]

The objections eventually raised to the equal volumes–equal numbers generalization by the anomalous behavior of the vapors of some compounds were then briefly discussed and explained. As for the boiling points, velocities of diffusion, and their relation to molecular weights, they also confirmed the truth of the atomic view of molecular constitution, as did the close relationship observed between the melting points of homologous organic compounds and their molecular weights.

All the facts, Williamson concluded, lead only to the conclusion that:

> molecules have no *locus standi* in the absence of the atomic theory. They are, to use the words of Dumas, physical atoms; their existence is a necessary consequence of the atomic theory and all chemical reactions agree in proving their existence. They are also discovered by examination of the mechanical properties of gases.

And now it was up to 'the opponents of the atomic theory to explain, in some other way, the facts which point so distinctly to the existence of molecules'.[131] It should be emphasized that these statements are important historically, because they represent the clearest expression relating the concept of molecules to the atomic theory of any British chemist of the time. Later on, in discussing atomic heats, Williamson said that 'the law of atomic heats has contributed to the development and systematization of that theory . . .' and he praised the validity of Cannizzaro's suggestion of doubling the atomic weights of a number of elements such as mercury, iron, tin, lead, etc. which had shown half of the atomic heats' values observed in other metals.[132]

After an extensive review of the examples offered by the variety in the atoms' arrangements, especially among carbon compounds, there was a discussion of the changes in composition and, accordingly, in 'atomic values'[133] observed in some compounds of great stability which are dissociated only at very high temperatures. In these cases the lower atomic value corresponded to the higher temperature and vice versa.

A few of the philosophical considerations in the concluding portion are worth mentioning, since they may reveal the hold that August Comte's positivism still had on Williamson. It was suggested that, when dealing with the atomic theory, one should adhere to facts and avoid discussing as much as possible 'all that is not in evidence', such as the divisible or indivisible nature of elementary atoms. Williamson recognized that he did not know what the intimate physical properties of such atoms were and what their shape was and whether they were vortex-like or not, a position somehow more reticent than that earlier taken.[134] However, given this, it was still helpful to study the reactions occurring among atoms and the physical properties of their products. The long address came to a close with a renewed note of high praise for the atomic theory which, supported by so many different facts, was defined as the 'very life of chemistry'.[135]

Shortly afterwards, on 4 November 1869, at a meeting of the Chemical Society, a debate was held concerning the topics discussed by Williamson in his June lecture at the Society.[136] Since several prominent British chemists and physicists were present, it is valuable to examine in detail the opinions of some participants in the discussion.

Benjamin Brodie, the Chairman, asked Williamson some precise questions to begin with: whether there were atoms at all, secondly what they were, and finally whether they were provided with the properties assigned to them regarding their 'replacing power' and their value (valence). Brodie also wanted to know what form of the atomic theory – whether the original of

Dalton or that of Berzelius or of Gerhardt and Laurent–Williamson was ready to support and defend. Williamson replied that the atomic theory had existed from time immemorial without substantial criticism and that whatever changes had intervened since Dalton's reproposal of it, they had always been consistent with the original concepts. Furthermore, he stated that the purpose of his lecture was to gather all possible chemical evidence in favor of the validity of the atomic theory, and certainly not to define whether atoms are still divisible or to describe their geometrical shape.

On both these topics he could not elaborate.[137] However, he stressed that, after Berzelius, atoms were generally accepted by chemists whose researches could only confirm that compounds' molecular weights at least correspond to the smallest atomic proportions representing the actual analytical figures.[138] There was also strong physical evidence from independent sources, such as the equal volumes–equal numbers generalization,[139] the boiling points of homologous liquids and the laws of gaseous diffusion which supported the existence of atoms.

According to the minutes of the meeting, six people participated in the debate and extensively commented on Williamson's remarks. The questions discussed were essentially few, but these significantly revealed the intellectual positions of those involved. Is the atom a universal truth or is it just the head placed upon the body of chemical observations to rationalize them? Or may it be considered simply an instrument used by the chemists to enlarge the knowledge of their science?

The stand of those who spoke ranged from a reasoned support of the atomic theory to a mitigated criticism, to open agnosticism, to a clearly expressed denial of the necessity of the existence of atoms. To some extent, J. Tyndall, G. Foster, and W. Miller[140] sided with Williamson, while Brodie, E. J. Mills,[141] and Odling took a position quite similar to that of Frankland, who stated that he did not wish to be considered a 'blind believer' in the existence of atoms. As can be judged from the minutes, Frankland also raised a very strong argument by denying that the possibility exists of establishing the truth of the atomic theory. He argued this by indicating that with the atomic theory alone the action of attractive and repulsive forces upon matter could not be understood; and that, while chemistry may still find it convenient to adopt the atomic theory, this should be considered more an obstacle than an aid when attempting to explain physical facts. For instance, even Faraday had found it very difficult to account for the conducting and non-conducting properties of bodies on the basis of a system formed by solid particles surrounded by a void space.[142]

To the reasoning offered by Frankland in favor of a continuous state of the matter, Odling then added that to the chemists the keystone of their way of thinking was not the atomic theory, but the law of definite proportions which had been compared by Davy to the Keplerian laws of planetary motions. As for the molecules, the experimental evidence, as in the case of ammonia and marsh gas, suggested that, against the opinion of Williamson, their existence could be postulated even without the atomic theory.[143]

Miller intervened at this point and challenged the position of those who, while denying the validity of the atomic theory, were unable to provide a substitute which could assist in the interpretation of observed facts. Without a hypothesis, he added, it is impossible to draw any logical conclusion from facts; to explain chemistry without the atomic theory would be like attempting to find a meaning in optics without assuming the undulatory theory.

Miller was followed by Foster,[144] who stood in opposition to Odling by stating that the usefulness of a theory should be separated from its truth; under discussion was not whether the atomic theory should be considered useful, but whether it was true or not. For Foster the atomic theory – even if the existence of atoms cannot be necessarily observed – relied upon the basis that matter undergoes a change and that a relation, both qualitative and quantitative, exists between disappearing and appearing bodies.

Tyndall next argued that the atomic theory was introduced, as were the gravitational and undulatory theories, to explain well observed facts and, therefore, should be considered valid as long as it relied upon an acceptable experimental basis.[145] To this E. J. Mills replied that it was wrong to establish a parallel between the atomic and undulatory theories, because, while the notion of waves in light is a fact, there is no evidence of a limit to the subdivision of matter, as supported by the atomists. In fact, while it is legitimate to assume that matter is divisible in an infinite way, he stated, then all scientific theories, including the chemical one, should be established upon motion which is the only supreme 'generalization of modern science.'

Williamson came back with a short comment indicating that his purpose had been that of gathering all possible evidence – thus far he remained unchallenged by his opponents – on the present understanding of the atomic theory, and that he did not intend to speculate about different possible interpretations of the same facts which might arise in the future.

After summarizing his own view of the atomic theory, Brodie, as Chairman of the meeting, concluded that the concept of physical indivisibility of matter ought to be kept apart from chemical facts. Theory and observations, in contrast with Williamson's position, cannot be intertwined. He regretted and also disagreed with, the view of those, such as Miller and Frankland, who accepted the atomic theory as solely an instrument to explain the experimental evidence, but at the same time refused to consider it an absolute truth. On the other hand, he concurred with Odling that chemistry neither required nor proved the atomic theory and personally admitted that, for him, until that time came, the ultimate constitution of the material universe had not been established.[146] On this note of renewed agnosticism on the atomic doctrine, the meeting of November 4 of the Chemical Society was adjourned.

Before leaving for a while this inquiry on the position of the British chemists and physicists toward the atomic-molecular theories in the late sixties, it may be appropriate to discuss at some length an event that apparently had very little immediate effect on that position. In May 1872, three years after the atomic debates were held at the Chemical Society, the Faraday Lecture was given by a well known Italian chemist, who at first seemed to appear on the British scientific stage as the right man at the right time.

Stanislao Cannizzaro was then forty-six, close to the peak of his career; the year before he had been appointed professor of chemistry and Director of the Chemical Institute of the University of Rome which had become the capital of the new Kingdom of Italy in 1871.

Cannizzaro's lecture in London is very indicative of his attitude concerning chemical theories and, more specifically, the atomic-molecular doctrine in the years following the Karlsruhe Congress. He was especially interested in convincing chemical educators of the time that – as he had earlier stated in the 'Sunto' – a completely new approach should be followed in teaching chemistry, essentially based upon both Dalton's and Avogadro's theories.

Cannizzaro gave his speech in English, his wife's mother tongue, in the impassioned and stirring style that characterized him. Unfortunately, the reactions of the British chemists to this event were not recorded. No comments appeared in the *Chemical News* or in the *Journal* of the Society which reported the text.[147] According to Pattison Muir, Cannizzaro had advanced far beyond the point reached by Williamson in 1869; definitely a fair statement.[148] Unlike Frankland, Odling, and others among their countrymen, he did not accept the thesis of using the atomic theory as a disposable rhetorical instrument, the so-called 'scaffold' approach that they maintained during the atomic debates.

Historically, the main importance of the 1872 Faraday Lecture is that it represented the strongest and clearest display of support for the atomic-molecular doctrines offered up until that time in the British Isles. It is fair to say that for a number of British chemists, who had not attended the Karlsruhe congress (eighteen were there, among them Crum Brown, Roscoe, and Odling), it was most likely their first opportunity to hear an intelligible and precise explanation of Avogadro's hypothesis and its applications.[149]

A review of the 1872 Faraday Lecture is very revealing, above all, of Cannizzaro's philosophy as a scientist and as a teacher. In fact, to use the term philosophy here may sound ironical in view of his strong criticism of metaphysics. At a period when many physicists and some chemists extensively debated the constitution of matter, to the question of the existence of atoms, Cannizzaro replied with an unqualified affirmation. However, when faced with the problem of the static rather than dynamic character of the atomic theory, and with the issues of divisibility, size, or form of atoms, he replied that these questions were beyond the experimental findings of the time, that nobody could prove or disprove their validity, and that in teaching chemistry they should be definitely avoided. At the same time, it would be inaccurate to say that Cannizzaro was an opponent of hypotheses per se which would put him in the same position as Berthelot and the positivists, but he certainly reiterated that lecture rooms were not the place to discuss concepts unsubstantiated by facts, for they would only confuse the young minds of the students at a time when they, above all, needed clearness.

In teaching science, he advised against the inductive process of reasoning from particular facts to general conclusions; for instance, to use such empirical laws as those of definite and multiple proportions may be disadvantageous for the students, because these laws may appear to them as isolated facts not easy to correlate.

There is another peculiar facet of Cannizzaro's heuristics which deserves attention and is very clearly expressed in a statement he made in his 1872 lecture: 'It often happens that the mind of a person who is learning a new science has to pass through all phases which the science itself has exhibited in its historical evolution.'[150] He stressed that, for instance, while in teaching astronomy it is wise to follow the road of induction, because Kepler's Laws had been discovered before they were explained and bound in a single idea by the gravitational theory; in teaching chemistry it must be shown that it was the atomic theory that preceded and led to the discovery of a great number of empirical laws and that created the language by which those same laws could be expressed. Thus, in chemistry, theories came first, then the fact-finders, as was clearly the case with Thomas Thomson and Berzelius who, only after becoming acquainted with Dalton's doctrine, were able to understand the experimental findings at their disposal. From that moment it was impossible, in Cannizzaro's view, to disassociate the atomic theory from the empirical laws upon which chemical equations, formulas, and symbols were based. As the atomic debates reveal, this opinion was not generally shared by the chemists of the time.

Accordingly, it was regrettable to observe some chemistry teachers who not only leave terms such as atom and molecule undefined, but still support the equivalents; a clear criticism of Berthelot and those schools such as Leipzig and Göttingen, where the influence of Leopold Gmelin was still felt in the seventies.

More sympathetic was the assessment of Wollaston's position on the atomic theory; for Cannizzaro, Wollaston, despite his equivalentism, always remained appreciative of Dalton's doctrine. However, this seems in disagreement with what Wollaston himself wrote in 1814; in considering atomic weights a hypothetical conception, he stated that he had no desire of 'warping his numbers according to an atomic theory'.[151] At any rate, equivalents definitely have nothing to do with atomic weights and, to use Cannizzaro's words, they are a pure anachronism. He recalled that during the atomic debates of May 1866 in London, the erroneous view was supported by Brodie that there were several atomic theories. This is not so; there is only a theory set forth by Dalton. The distinction made later on between atoms and molecules and the criteria adopted to calculate the molecular weights have only helped to make such theory even more firmly established.

After these preliminary clarifications, Cannizzaro entered into the core of his lecture by restating very clearly the theory of Avogadro, Ampère, Clausius, and Krönig on the constitution of perfect gases. These four men, significantly all physicists, were associated with the generalization that equal volumes of gases contain equal numbers of molecules; a generalization representing for Cannizzaro the starting point of the theoretical teaching of chemistry.

Then, for perusal of the students, a summary of the history of the development of Avogadro's hypothesis follows, beginning with Gay-Lussac in 1809. It is only fair to say that Cannizzaro's historical analysis was anything but faultless. Dumas is recalled, for instance, but neither his erroneous interpretation of Avogadro's hypothesis nor Berzelius' equally damaging assumption – equal volumes–equal number of atoms – was mentioned.

The exceptions successively found to Avogadro's hypothesis were gradually removed, in Cannizzaro's view, by Gerhardt, Wurtz, Williamson, and Frankland through thirty years of work in organic chemistry. As for the anomalies presented by vapor densities of ammonium salts, phosphorus chlorides, and sulphuric acid, the experimental findings showed the dissociation of these substances at high temperatures. Cannizzaro, however, failed to relate the lengthy arguments that took place on this subject between St. Claire Deville, his school and their opponents.

After this historical outline there was an assertion of pivotal importance for understanding Cannizzaro's line of reasoning. In teaching theoretical chemistry, he said, one should move from physical observations to a proof of the validity of the equal volumes–equal numbers generalization rather than from chemical considerations to a determination of the molecular combinations. Thus, first, the densities in vapor state must be found before one may proceed to calculate the molecular weights. The student will recognize then that there is agreement between the molecular weights determined by physical observations and those determined by chemical means. It is surprising that Cannizzaro neglected to mention that Avogadro in 1811 and 1814 had moved exclusively from physical data in determining both the atomic and molecular weights of several substances.

The equivalentists had suggested the simple weight ratio of compounds reacting with each other, but this method did not prove the existence of molecules nor their relative weights. In view of the fact that there was no adequate number of substances whose molecular weight could be established through transformation, it was impossible to teach that the molecular weights can be understood through the law of compound equivalents. There had been instances where one had recourse to molecular weights to demonstrate the existence of atoms and then deduced from them the laws of transformation; but, obviously, this is a tautology.

The study of the constitution of perfect gases, on the other hand, offered a much more logical starting point for arriving at the notion of molecule. Cannizzaro felt that there was no need to resort any further, as many natural philosophers had done, to a distinction between a physical and a chemical molecule. In this connection, the studies of Maxwell and Tyndall were mentioned for the 'great credit' they gave to the concept of molecule. Yet Cannizzaro failed to note that the differentiation between physical molecules – which Maxwell defined as dynamical – and chemical molecules, expressed as ratios of their equivalents, remained as valid as ever in the middle seventies.

Cannizzaro then disclosed the details of his teaching program. In harmony with the idea he had expressed earlier, the course outline followed an historical pattern beginning with the end of the eighteenth century. In fact, the first key concept that must be explained to the students was that of the invariability of mass observed by Lavoisier whenever chemical transformations occur. These transformations were actually union and separation of ponderable things. At this point it was not necessary for the teacher to state whether matter was continuous or discontinuous, in rest or in motion; rather the attention of the learner must be focused upon the weights of the reactants and of the products of reaction. After these concepts had been instilled, the law of definite proportions was discussed and the experimental procedures used to obtain constant ratios were described. Subsequently, the student should be trained to compare the composition of bodies from different points of view in order to avoid Proust's pitfall when he failed to detect the law of multiple proportions. The latter, of course, did not escape the attention of Dalton who went beyond the experimental figures.

Both laws of multiple and reciprocal proportions must then be related in the students' minds to the fact that all compounds contain the elements in the ratios of the atomic weights and their multiples. This had been observed by Berzelius after he became aware of Dalton's theory; however, by proceeding only through the ponderable constitution of bodies, he obtained laboratory data which provided him with arbitrary conclusions in choosing the atomic weights and the number of atoms present in compounds. Obviously, at such a point in the development of chemistry, a way had to be found to verify the accuracy of the molecular weights and to remove the arbitrariness of Berzelius' method. For Cannizzaro, the law of combining volumes of gases reported by Gay-Lussac in 1809 offered the means to determine the molecular weights, and thanks to the simple laboratory apparatus devised by Hofmann, the mercury column, the significance of such a law can be easily understood by the students.

Again, through experimental observation, the student must learn the regular behavior of different substances, under changing temperatures and pressures, whenever they reach the state of a perfect gas. Yet, other physical properties, such as densities and limits of compressibility, remain distinctive of each gaseous substance.

Typical of Cannizzaro's pragmatic style of teaching was his suggestion that, when discussing a scientific law in any introductory course, exceptions to such law should not be discussed. In explaining the physical properties of gases, for instance, it is superfluous to mention to the students the small deviations in gas behavior. A further recommendation to teachers was, that in explaining the ratios of weights and volumes of substances, to avoid the use of actual weights and volumes of gases at a given temperature and pressure, both weights and volumes must be compared directly, thus encouraging the use of relative rather than absolute values.

The next step in the course was to compare different volumes of gases or different weights of equal volumes which, according to Avogadro, meant to compare different molecular weights. Here again it should be stressed to the student that the ratios between weights of equal numbers of molecules (or equal volumes) are only relative numbers. When the molecular constitution of gases is used to explain ratios, as for instance, in the formation of hydrochloric acid, the student must understand that in hydrochloric acid there is present only a semi-molecule of hydrogen and a semi-molecule of chlorine; furthermore, he should be able to appreciate at this point that the molecules of elemental gases cannot be identified with Dalton's atoms. To avoid further confusion on this issue, Cannizzaro suggested giving special emphasis to molecular weights deduced from gaseous volumes and postponing any discussion of the atomic theory. This does not imply, of course, the rejection of the notion of atoms, but only the oversimplified explanation of them offered by Dalton.

By assuming as unit of reference the weight of a half molecule of hydrogen – or the amount of hydrogen present in a molecule of hydrochloric acid – a table of the molecular weights of a number of elements and compounds may be prepared and explained to the students. Thus, they will learn how to calculate the weights of the elements present in compound molecules. Next, they are asked to compare the composition of different molecules containing the same element and will see that the weight of each element entering into the molecules is always an entire multiple of the same weight. This minimum quantity was defined by Cannizzaro as an *atom*. That being the case, Dalton's atoms express the weights of combining elements, as well as the composition of equal gaseous volumes. With regard to the molecules of some elements, they always contain a whole number of atoms.

Cannizzaro felt that through the logical process he had just outlined, the existence of atoms was proved beyond any doubt as a true law and this law was founded on the molecular hypothesis which assumed that there was an equal number of molecules in equal volumes of perfect gases. From a didactic point of view, this hypothesis could be set aside once it has been used and the term molecules may be replaced by 'equal volumes'. In this case, the atom represented the constant quantity of an element entering by entire multiples into equal volumes of gaseous compounds. Conversely, equal volumes, under the same physical conditions, were compounds of whole multiples of quantities constant for each element – the atoms.

At this point it was again stressed that in the study of chemical phenomena there was no need to be concerned with such properties of atoms and molecules as size, shape, distance, continuity or discontinuity; the only fundamental property of both atoms and molecules with which chemistry is involved is the weight of their masses. Physical theories may deal with atomic and molecular properties other than their weight, when they involve physical phenomena, but these theories are not 'sufficiently exact' to link them with chemical concepts. Quite a statement and a nice way to say that, while physicists can play around with their reveries, chemists should keep their feet on the ground!

Cannizzaro used the opportunity to recall Faraday – after all, this was a Faraday Lecture – and to criticize him for his opposition to the molecular and atomic theories in their 'physical form'. Needless to say, Faraday's criticism of the Daltonian constitution of matter was wider and more substantial than that. For Cannizzaro, it was helpful to remove the nonessential from the molecular and atomic theories when teaching them, and to adhere only to facts upon which these theories may evolve, and 'we may some day arrive at a conception of matter totally different from that which is now entertained and find in that conception the explanation of what we call atom and molecule . . .'.[152] Altogether, this is not an inaccurate anticipation, if one keeps in mind the turn of events that would follow seven years later when William Crookes submitted to the Royal Institution his experimental findings and his speculations on the fourth state of matter.

There were other considerations in subsequent portions of Cannizzaro's lecture that further helped in understanding his heuristics. The debated question, certainly valid in times of the accelerated unfolding of scientific ideas, concerned the boundaries within which a teacher should stay when discussing new theories with his students. To teach them as irrefutable when they are only probable – and Cannizzaro specifically referred to the atomic and molecular theories – would be misleading. On the other hand, not to mention them might be as harmful to the pupils because their imagination either could go astray and work alone, independent of reason, or could become paralyzed. This is a very revealing view of pedagogy at this time. The sensible solution then, in discussing theories, is to definitely mention their degree of probability. With regards to the notions of atoms and molecules, the only property to be stressed is their ponderability; everything else is superfluous. Again great care must be exercised in describing the wide limits existing on experimental accuracy such as the determination of some vapor densities, while it would be proper to point out that very little is known about the large majority of inorganic compounds. Thus, the rather narrow experimental basis of the atomic-molecular theories should not be concealed from the students. The importance of other experimental methods, isomorphism and specific heats, must be taught, especially in the case of elements not forming compounds whose molecular weights can be determined by vapor density.

There was a final consideration that for its significance should be recalled *verbatim*:

> We are teaching a science which is eminently progressive and, moreover, at the time of its most rapid evolution. I have always kept in view the idea that students should leave our schools not only with a certain amount of fixed and well defined knowledge, but likewise with aptitudes and preparation sufficient to enable them to follow the unceasing progress of the science and its transformations. . . .[153]

Few will disagree that the meaning of these words remain as true today as when they were articulated more than a century ago.

While the prevailing attitude of British science concerning the atomic-molecular theories did not change for some time, it is interesting to examine the view of Frederick Kekulé, one of the most influential chemists of the time and who happened to be a close friend of Williamson, Frankland, and Odling.[154] In an article published in the 27 July 1867 issue of *Laboratory*, shortly after he had set forth his hypothesis on benzene structure which would make him famous, he took a stand that, within the historical framework of the time, should be considered objective and also very close to that taken by Williamson in 1869. For a chemist, Kekulé pointed out, it is not only advisable, but 'absolutely necessary' to assume the existence of atoms. If we assign to this term the meaning of particles of matter not further divisible in chemical transformations, then chemical atoms do indeed exist. However, from a philosophical point of view, atoms in the 'literary signification of indivisible particles' do not exist, a distinction certainly not made by Cannizzaro.[155] Probably, he continued, the entities now called atoms will be explained purely on a mathematico-mechanical basis which will also help to unravel the meaning of such concepts as atomicity (valence) and atomic weights.[156]

In Kekulé's view, which he expressed in his *Lehrbuch der Organischen Chimie* of 1867, chemistry, after the laws of multiple proportions and of combining volumes had been enunciated, had been groping its way in almost total darkness. This was proven by its inability to find an answer to such fundamental questions as the cause of the difference of elements or the nature of the force which binds the chemical compounds together. Therefore, all the so-called theoretical considerations made in chemistry are actually only 'expediencies' and 'probabilities'.[157] The chemist then is left no choice but to look for an approach ('Anschauungweise') based upon the largest possible number of facts which can explain the chemical processes in the simplest and most comprehensive way.

As for the question of the ultimate constitution of matter, Kekulé argued that the 'atomic hypothesis' gives the best account of the physical and chemical changes of the matter,[158] can explain the law of multiple and definite proportions and can account for the difference between metamer and polymer substances. Both chemical and physical considerations lead to the assumption of very small particles which had been called atoms, molecules, mass particles or mass elements; it would thus appear that gases and crystalline substances are formed by gas molecules and crystalline molecules respectively. However, nothing is known about their relative size, a determination of which cannot be conducted through the study of their physical characteristics.[159] Chemistry has nothing to do with the sterile question of whether the smallest chemical particles are incapable of further division or whether they are ideally indivisible (*metaphysical atoms*).[160] Yet the determination of the relative size of the mass particles is of 'exceptional importance' for chemistry since chemical transformations not only indicate that such particles are indivisible, but also lead to the assumption that there are two different, very small quantities characterized as atoms and molecules.

For Kekulé, the definition of molecule was almost identical to the one offered by Cannizzaro in his 'Sunto'.[161] As for the molecule of simple bodies, it can be distinguished from those of compounds because it contains an agglomerate of like atoms, while in compound bodies different atoms are united into a molecule. Affinity, the inherent attractive power or chemical relationship, may explain, 'for the present', the reunion of atoms into molecules; in any case, at least two atoms are present in a free molecule. On the other hand, atoms never exist in a free state and, through the chemical action, they are neither variable nor divisible. Kekulé emphasized then that it is impossible to determine the absolute size and absolute weight of both atoms and molecules. Therefore, it is necessary to refer to a chemical unity of reference (such as hydrogen) on which to base the weight of the said particles.

Throughout this portion of the *Lehrbuch* one may appreciate Kekulé's lucid thought and an almost flawless exposition of concepts which, as far as the constitution of matter, had been too often muddled in those years.

Unlike Cannizzaro, he recognized the limitations of chemical science; more sensible than the Italian chemist about the contemporary positivist criticism, he stressed correctly that chemistry could not deal with metaphysical problems concerning the actual divisibility of atoms. Looking back to the *Lehrbuch* more than a century after its publication, it is fair to say that altogether its contribution to the re-evaluation of the molecular theory would have been highly positive were it not for the limitations which will be discussed later on, and which hampered the historical import of Kekulé's major chemical treatise.

In discussing the 'chemical' molecular weights and their determinations, found in a section of the *Lehrbuch* devoted to the relations between physical and chemical properties, Kekulé stated that for almost all compounds and especially for those containing carbon, such chemical molecular weights are identical to those obtained through the specific weights of the same compounds when in vapor form; in other words, the 'chemical molecules' are identical to the 'physical' ones.[162] After stressing that both the specific weights of vapors and the molecular weights are proportional numbers and not absolute values, Kekulé added that the 'empirical law' just set forth could also be expressed in the following terms: equal volumes of gaseous substances contain an equal number of chemical molecules. Furthermore, an equally large number of chemical molecules always fills an equal volume in gaseous substances or the specific volume is always the same in all gaseous substances.[163]

It is possible to arrive at this 'empirical law' by studying the relationships between the physical and chemical properties of both organic and inorganic compounds; more specifically, the relationships between the specific weight and the chemical composition of gaseous and vapor forming substances. From a number of physical considerations based upon a particulate structure of matter in gaseous state and from the assumptions made by Clausius on the space occupied by gaseous particles in continuous movement, their number, and mutual distance, one may conclude that the weight ratios of equal volumes of different gases represent also the weight ratios of individual mass particles.[164]

The development of the equal volumes–equal numbers generalization from the kinetic theory of gases was clearly outlined by Kekulé in his *Lehrbuch*. However, in a treatise in which a rather extensive historical outline is offered, beginning with Lavoisier and Dalton and moving through Berzelius, Dumas, Laurent, and Gerhardt to Liebig, Frankland, and Williamson, one is surprised to find no mention at all of Gaudin, Ampère, Cannizzaro, and Avogadro.[165]

Moreover, Kekulé neglected to recall Krönig, Joule, and Maxwell for their contribution to the kinetic theory. While the results of the work of these men are extensively described, their names are altogether ignored. This fact may appear even more peculiar if one is reminded that Kekulé was one of the promoters of and a participant in the Karlsruhe Congress, where Cannizzaro's 'Sunto' was read and distributed. It may be objected that the first edition of the *Lehrbuch* appeared a year before the Congress, but still it is difficult to explain Kekulé's oversight in the reprints which followed.

Kekulé deserves recognition for his acceptance of both the atomic and the molecular theories; he endorsed them and exposed them pragmatically, as 'thatsachen'. But certainly he ignored the individuals behind them, their imaginative efforts, their intuition, their boldness and, from the historical point of view, this cannot be forgiven.

Between the first and second editions of Kekulé's *Lehrbuch*, an event again took place in Germany which represented a major turning point for the acceptance of Avogadro's hypothesis. In 1864, in Breslau, Julius Lothar Meyer's *Die Modernen Theorien der Chemie* appeared, a book which was to be a great success all over Europe.[166]

Meyer had studied under Bunsen at Heidelberg and had established close relationship with Kekulé, Baeyer, Pebal, and Roscoe who were there at the same time.[167] He was thirty years old when, together with Kekulé, Strecker, and several other German chemists, he attended the Karlsruhe Congress in 1860. Fully convinced by the arguments raised by Cannizzaro at that meeting, Meyer became a vigorous supporter of the equal volumes–equal numbers generalization whose original concept he unequivocally attributed to Avogadro. Through *Die Modernen Theorien* the more progressive German schools of chemistry, especially Tübingen and Würtzburg, were eventually able to recognize the full implications of the molecular theory and to make it one of the cornerstones of their teaching systems.

Die Modernen Theorien is more than a simple textbook. It contains considerations quite revealing about some aspects of the philosophy of science at the time. Meyer seems to rejoice that, as a result of the atomic theory's development, chemistry will never be a 'dependency' of physics again.[168] He recognizes, nevertheless, the strict interrelationship existing between the two sciences and more specifically the influence that physical conditions have upon chemical transformations, and conversely the effect of the latter upon physical properties.

Against a great accumulation of facts, there have been only a limited number of generalizations drawn from chemists. And the degree of acceptance of those generalizations appears to be very low, because, as Meyer pointed out, chemists have exercised a wise prudence in their decision over which hypotheses should be accepted and which rejected.[169] Unfortunately – and this he did not say – some theories were neglected for no reason other than that they were either misunderstood or overlooked due to the modest position of their proposers: the molecular hypothesis is among them.

Meyer recognized both Berthollet and Berzelius as responsible for expressing incorrect ideas which, due to their personal positions, were held as valid for many decades.[170] In Meyer's opinion, the development of general chemical theories during the first half of the nineteenth century was also delayed by the attitude of physicists towards the intimate constitution of matter, whose discontinuity could not be accepted. As mentioned earlier, most physicists rejected the mere existence of particles; thus chemistry and physics stood apart for a long time. Paradoxically, Meyer did not point out that the men who gave the atomic and the molecular theories to chemistry were both physicists who, with a very limited knowledge of chemistry, had moved from physical and chemical facts in order to formulate their doctrines.

In his historical analysis of the kinetic theory, Lothar Meyer correctly recognized the priority of Daniel Bernoulli in establishing an empirical relation between a gas pressure, its volume, and its density; from there he moved to a definition of the dynamic energy of gases, how this is affected by an increase in temperature, and how the gaseous particles' velocities are calculated and related to the pressure of a gas. Clausius, Maxwell, Boltzmann, and O. Meyer's contributions are all duly recognized.[171]

In *Die Modernen Theorien* a full section is devoted to the very clear and convincing explanation of the origin, development, importance, and limitations of the molecular hypothesis and also, for the first time in the nineteenth century, in a textbook of general chemistry, full credit is given to Avogadro for his ideas. Meyer stressed the thesis that scientific speculations which precede experimental evidence are seldom successful. Needless to say, this point is debatable as is his view that in the first half of the nineteenth century there was no need to accept Avogadro's hypothesis, and that chemists could either accept or disregard it without consequence. The state of confusion into which both Berzelius and Dumas, after 1826, threw chemistry, by not sufficiently appreciating the distinction between atoms and molecules, can hardly justify Meyer's opinion. For a while these two terms were so discredited that it seemed as if no chemist wanted to use them anymore. Conversely, one may agree with Meyer that Avogadro's premature and bold attempt to extend his generalization to elements and compounds, which at the time could not be changed into the gaseous state, was detrimental to its acceptance.[172]

A great deal of space was devoted in *Die Modernen Theorien* to the practical application of Avogadro's generalization and to the determination of the molecular weights of a number of substances from their densities. It was stressed also that only through the chemically detected combining weights, may one determine the correct molecular weights that must be, in the case of an element, either equal to the atomic weight or to a simple multiple of it. However, since it had been observed that different gases do not always follow the same variations of volume for similar changes in temperature and pressure, then the equal volumes–equal numbers generalization may present some exceptions; this is why, in the determination of molecular weights, the values calculated from the chemical analyses are sometimes more correct than those drawn from the vapor densities.

The apparent deviations of Avogadro's hypothesis which were observed in a number of substances, both organic and inorganic, are explained by Meyer in a very convincing way through the interpretation offered by Playfair, Wanklyn, and Horstmann for the anomalies in the vapor densities of acetic acid. As for the unusual behavior of the ammonium salts and of the compounds of phosphorus with halogens, the long debates involving Mitscherlich, Gladstone, Deville, Troost, and, earlier, Cannizzaro, Kopp, and Kekulé are reported up to the decisive observations of Pebal and Than.[173]

Meyer offered in general a lucid, objective review of this period of extensive debates that involved both chemists and physicists all over Europe for more than a decade. In the conclusive portion of *Die Modernen Theorien* Bacon is quoted as supporting the view[174] that chemists have often set forth 'fantastic' theories and generalizations unsupported by sufficient experimental evidence.[175] This was also true for nineteenth-century chemistry and, unfortunately, Meyer said, periods of wide acceptance of often invalid theories have followed times in which theories of great soundness were not accepted.[176] Such was the case, for instance, of Avogadro's hypothesis and Dulong–Petit's law of atomic heats.

Against Meyer one may argue that, historically speaking, it is inaccurate to place these two generalizations on the same level. In fact, despite its shortcomings, the law of the atomic heats was received with favor and was neither ignored nor misunderstood. Moreover, it had a wide experimental basis, while this was certainly not so with Avogadro's hypothesis.

The differentiation of atoms from molecules could not in fact be established as long as chemists and physicists did not know more about their real existence. This being the case, the foundations on which both Dalton and Avogadro had built their speculations were still too whimsical, even in the 1860s, to justify their wholehearted acceptance.

On the whole, Meyer's book represents a very important and comprehensive work, which inauspiciously had very little impact in France,[177] and apparently no influence on the atomic debates in Britain.[178] As for Germany, contrary to received opinion, the fact remains that Avogadro's and Cannizzaro's ideas, after the *Modernen Theorien*, were not generally accepted even in that country.

At the University of Leipzig a man slightly older than Meyer was called to teach chemistry in 1865. After studying under Wöhler, Hermann Kolbe later became assistant to Bunsen at Marburg and worked for some time in London with Playfair and Frankland. Through his thirty-year-long teaching career and unusual experimental skill he inspired a generation of chemists in Germany and abroad; he was the author, among other books, of a voluminous *Lehrbuch der Chemie* published seventeen years after the Karlsruhe Congress. In this work, which is mostly descriptive, less than forty pages are devoted to an explanation of the stoichiometric rules and the law of multiple proportions; in a very few paragraphs the student is led from the atomic doctrine to the theory of valence, while nothing is said about the development of chemistry in the intervening period which, of course, embraces almost sixty years. The contributions of Gay-Lussac, Dulong, Avogadro, Ampère, Cannizzaro, and of those closely related to the molecular hypothesis, are completely ignored. This appeared an unusual approach, to say the least, for a manual of large circulation written by an eminent representative of the German chemical school of the time.

One may find an explanation for Kolbe's attitude by tracing his educational background.[179] He had spent four years with Wöhler at Göttingen, a crucial period for his scientific training. Wöhler, in turn, had had very strong personal and cultural ties with Leopold Gmelin, the founder of the equivalentist school which remained very influential throughout the late 1860s not only in Germany. In the second edition of the *Die Modernen Theorien* (1872), Meyer had stated that 'die Gmelinschen Äquivalente sind aus der Literatur wurschwunden'. He probably referred only to the use of the expression, because several German chemists remained conceptually equivalentists even in the early eighties, Kolbe among them.

At almost the same time as Meyer had come forward so strongly in favor of Avogadro's generalization, a very important textbook was published in the United States. The author, Josiah Cooke, Erwing professor of chemistry at

Harvard since 1850, took a very clear and resolute position in support of the views expressed by Avogadro in 1811.[180] In the first lecture of his *New Chemistry* which went through eight editions after 1872,[181] he defined Avogadro's Law as a 'great generalization' which holds the 'same place in chemistry that the law of gravitation does in astronomy'. Unfortunately, he added, it was not accepted earlier because of the immaturity of science at the time.[182]

Cooke explained the equal volumes–equal numbers hypothesis later on, and, after qualifying it as more general than both Mariotte's and Charles' Laws, proceeded to define the molecule in terms similar to those adopted by Cannizzaro in his 'Sunto'. In another work, which he wrote several years later, there appeared a very lucid interpretation of the relation existing between molecular structure and the physical properties of matter.[183] Avogadro's hypothesis is introduced as corollary of the peculiar behavior of the gaseous state when variations in temperature and pressure occur. With regard to atoms, characterized as the smallest mass of an element that exists in any molecule, for Cooke they could exist in a free state since he considered, in general, every molecule as consisting of at least two atoms, with 'few exceptions' such as the metallic elements.[184]

Guareschi pointed out that Cooke was presumedly familiar with the first edition of Meyer's *Modernen Theorien* – there is no mention of Avogadro in his earlier work, *Elements of Chemical Physics* (1860) – and had the merit of introducing the molecular theory to a new generation of American chemists.[185] Josiah Cooke, like Wurtz, Meyer, and Roscoe, remains a representative of that group of men who gave their impassionate support to the ideas which, set forth by Cannizzaro in 1858, went back to Prout, Dalton, and Gay-Lussac, and had been recently upheld by the physical investigations of Joule and Clausius. Yet, besides the position taken by the equivalentists, there was anything but a consensus of opinion at the time regarding the validity of a corpuscular theory.

Ostwald and Divers

While for some chemists – as we have seen in the atomic debates in Britain – both atoms and molecules were an unalterable reality with a volume and a weight, for others they were an entity of mere convention and, as such, an indispensable and useful device. More significant, perhaps, seems the position at the time of those scientists and philosophers of science who challenged the foundations of the whole atomic-molecular doctrine.

For these people, the question was not one of the weight or shape of the elementary particles, but of their very existence. An eminent representative of this attitude is characteristically the man who may be considered the founder of modern physical chemistry, Wilhelm Ostwald, who in 1887 became professor of physics at Leipzig, shortly after Kolbe's death. A very keen mind, a profound observer with a boundless intellectual curiosity, Ostwald had a Renaissance personality quite unique among fellow scientists of his time. From his writings is evident an overall vision of both the physical and chemical reality generally missing in the works of his contemporaries. At the same time, for instance, in discussing the atomic-molecular doctrine, he emerges as an objective and well informed observer of its development, from Dalton all the way to Mendeleev.

In the first issue of the *Zeitschrift für Physikalische Chemie* (founded by Ostwald in 1887) appeared the renowned essay of van't Hoff correlating the osmotic pressure with the gas pressure and also Arrhenius' findings on the dissociation of solutes. Two years later, Ostwald, in his *Grundriss der Allgemeine Chemie* reviewed Avogadro's hypothesis and argued that its confirmation could be found not only in the kinetic theory of gases, but also in the conclusions reached by Arrhenius and by van't Hoff. Earlier in the same book, after recalling the path followed since the law of definite proportions, he praised Avogadro for his clear distinction between atoms and molecules, and recognized Cannizzaro among those more responsible for finding an acceptable and unequivocal criterion by which to determine the correct atomic weights.[186]

Following Cannizzaro, with the assistance offered by isomorphism, vapor densities, and atomic heats, Mendeleev, with his periodic table, established a further means to correct and predict atomic weights. Nevertheless, for Ostwald (and he clearly indicated so) the molecular hypothesis remained the most useful tool used by chemists throughout those years; he concluded that as such it had been universally accepted with the exception of the 'ultraconservative' school in France.[187]

However, Ostwald's support of the atomic-molecular doctrines remained for almost thirty years only a framework or a 'scaffold' useful to support a few general laws in chemistry, and in turn was upheld by a number of physical observations, but little else. As a simple 'scaffold', it could be removed or modified according to the requirements of science. Shortly after praising the almost general acceptance of Avogadro's hypothesis, he wrote that the problem of the mutual relations of atoms within molecules is purely 'hypothetical', because the existence of atoms themselves is only an hypothesis, and since chemical molecules are complexes formed by the union of elementary atoms, then molecules, too, are hypothetical. But – he wanted to assure the reader that something is still valid – apart from 'metaphysical scruples'[188] about the atomic theory, there is a guarantee that further advancements will lead to results as useful as those observed in chemistry and physics. This was Ostwald's position in the late 1880s. In the following years, we will see a further development of his *Anschauung* of nature.

Probably impressed by the great progress reached by the researches on thermodynamics, he appeared more and more convinced that energy was the only reality.[189] In September 1895, Ostwald delivered a lecture at the congress of the German Scientific Society in Lübeck against materialism, in which he defined the existence of matter as an illusion, and supported the view that reality is only an interplay of energy in different forms.[190]

Such an attack on the entire atomo-mechanistic models of the nineteenth century by a scientist of quite considerable reputation caused an understandable reaction everywhere. The storm raised among the communities of physicists and chemists, who joined in their opposition to Ostwald's stand, reached such proportions as to even affect him psychologically.[191] From the philosophical point of view, as Ernst Haeckel well understood,[192] this represented the return of a romantic German to a Leibnitzian position: Ostwald's *Energie* being in fact his fundamental monad, a real essence which dominated both the living and non-living world and the only 'quantity which is common to all domains'.

Moving from this position, he stated that matter is nothing but a complex of energy factors which possess the quality of being mutually proportional.[193] In fact, they prove the fundamental characteristics of matter to be factors, or if one likes, the manners of expression of the energy. Accordingly, the mass is the capacity for the kinetic energy; the gravity is the capacity for volume-energy, and so on.

Afterwards, when Ostwald disclosed his natural philosophy, we will find him more and more critical of the atomo-mechanistic theory. In science how is it possible – he asked – to distinguish transient conclusions from those which are lasting and invariable in time?[194] That is easy: nature's laws are invariable; hypotheses are transient. In many fields of chemistry and physics one may expect all kinds of breakthroughs, but no actual revolution of ideas.

In the same lecture devoted to heat, the calculation of the molecular velocities conducted by Clausius (qualified as a very important scientific discovery) is defined as 'einer unbewiesenen Voraussetzung' which is not supported by actual facts.[196] After setting aside Clausius' views on molecular velocities, Ostwald asserted that the foundation of a heat theory can be found in Julius Mayer's first essay.[197]

In the spring of 1904 Ostwald went to London to give his Faraday Lecture which may be considered the epitome of all his scientific philosophy. In the home country of the founder of the atomic doctrine he renewed his total rejection of any form of matter that is defined again as a complex of energies found in the same place. Space is full of energy which may be either homogenous or in 'grained form'; possibly, the latter interpretation is closer to the one offered by the atomic doctrine.[198]

On the same occasion, after discussing the importance of chemical dynamics from which all principal chemical laws can be deduced without recurring to the assumption of atoms, Ostwald separated himself from Faraday's stand on the constitution of matter. For Faraday, atoms are mathematical points, and if one removes forces from them, there is nothing left at all; in this sense, Ostwald said, Faraday is still bound to the atomic hypothesis although within a Boscovichean framework. The reaction to this Faraday Lecture can easily be imagined by recalling how Ostwald's fellow scientists had taken exception to his Lübeck speech in 1895.[199]

In Britain, however, after the turn of the century, there were chemists of some renown who still rejected Cannizzaro's interpretation of the atomic-molecular doctrine, or to say the least had a very confused idea of the subject. Among them emerges Edward Divers who, as President of the chemical section,[200] read a rather singular paper before the Belfast Meeting of the British Association in 1902.[201] Divers should be recalled also for his steadfast refusal to recognize – forty years after the Karlsruhe Congress – Avogadro as the author of the equal volumes–equal numbers generalization that he credited to Dalton and Gay-Lussac.[202]

It may be interesting to examine at some length the content of Divers' lecture in order to somehow assess the standing of the corpuscular theory in those years.[203] From the very beginning Divers tried to make a distinction between the atomic hypothesis and the atomic theory, the latter thus far built upon the 'metaphysical basis' that matter is made of discrete and extremely small particles; as such, it had been regarded with skepticism. He pointed out that he did not intend to discuss the merits of the atomic hypothesis which nowadays should no longer be treated as metaphysical speculation. He limited the term 'atomic hypothesis' to the old view of a discontinuous matter and he adopted the term 'atomic theory' for his discussion of the present form of the Daltonian doctrine.

If substances are made of a great number of indivisible small particles having weight, as the atomic theory assumes, then the laws of combining weights can be deduced from this assumption. Yet these same laws do not prove the validity of the atomic hypothesis which is not explained by them. Why substances interact according to certain definite proportions is a fact which cannot be made clear for Divers, who reproached the atomic hypothesis for concealing science's lack of knowledge on the subject.

Then, if one does not consider the atomic hypothesis as necessary to the atomic theory, it is 'most desirable' that theoretical chemistry, which in essence is the Daltonian atomic theory, should be interpreted without taking into account the physical constitution of matter. Or, if the atomic hypothesis is rejected, so should the atomic theory in view of their very close relation. For Divers, many 'eminent chemists' support the view that a theory of chemistry can be developed even disregarding the atomic hypothesis. Other scientists, instead, use the atomic hypothesis as an indispensable teaching tool without necessarily accepting it.[204]

In the atomic theory, in the form supported by Divers, associated with the chemical interaction of simple or compound substances, are two concepts: indivisibility and chemical equality, which cannot be set apart, for the former is essentially subordinate to the latter. As an example of chemical equality, Divers gave the molecule of ammonia which is 'equal' to that of aldehyde, because it combines with it and disappears with it.[205]

After offering his definition of mass and matter, he stated that there are only two ways of measuring 'quantitative equivalency' between substances: the dynamical (which is related to masses) and the chemical or molecular. As for the 'molecule', Divers defined it as a 'term of relation' which stands for a chemically equal quantity.

Gaseous volumes, osmotic pressures, and other properties of substances must then be considered as correlated equalities and, as a means of molecular measurement, independent of chemical change. In general, the molecule is determined first by physical methods; this does not mean that the balance is useless in chemical operations, quite the contrary. The accurate tables of atomic weights, from which molecular weights are calculated, are also important. The balance thus is a necessary tool, but it records facts in units which are not those of chemistry.[206]

At this point, Divers asserted that neither Richter's Law nor Avogadro's Hypothesis, but the Daltonian atomic theory and Gay-Lussac's Law of simply related gaseous volumes in chemical combinations are responsible for the theory of chemical molecules and their relation to the equalities between weights and gaseous volumes of different substances. The equal volumes–equal numbers proposition was then restated by Divers. He said that if it was stripped of all hypotheses – a distinction in other words should be made between what Avogadro 'really put forward' as new and what he took for granted – then this proposition may be regarded as a 'law' and recognized as 'an advance upon Gay-Lussac's Law similar to that of the chemical molecular theory upon the atomic theory of Dalton'. However, the equal volumes–equal numbers generalization is not without faults. Besides the exceptions to it observed in a number of elements, there is no reason why such generalization should not be extended to all elements in order to make it more consistent with the molecular theory, and accordingly why should the chemical molecule of mercury or sodium not also be polyatomic since other elements are?[207] This raises the question of how there can be a chemical union between two parts perfectly alike, an objection, needless to say, much earlier brought forward by Berzelius. Therefore, the notion that a molecule may be chemically composed of two substances simpler than itself cannot be defended.[208]

The essence of this extensive lecture, through all its meanderings, was an attempt to prove that chemistry can be explained with its symbols and nomenclature without reference to any hypothesis concerning the chemical and mechanical structure of matter. One may have serious doubts about the success of such an undertaking. As is plainly evident from a review of this lecture, Divers' ideas were, to say the least, outdated; besides, he seemed amply provided with the gift of very poor intuition as when he passes judgment on Gomberg's theory of radicals[209] and on Kekulé's ring theory, which, for him, 'has no spatial significance'. As for his criticism of Avogadro's theory – that elementary molecules cannot be biatomic – this was based upon objections which might have been valid in the 1820s, but not so at the turn of the century. In general, he seemed to favor the notion of molecule over that of atom, although whether this applies also to elements, is not clear.

Divers' lecture of 1902 is seldom mentioned by historians of chemistry. The fact remains however, that while in those years the atomic-molecular hypothesis appeared to have the general support of both chemists and physicists, especially from those on the Continent, excepting Ostwald and his school on one side, and Ernst Mach along with the positivist group on the other, the president of the chemical section of the British Association was still extremely critical of it. The only possible justification for Divers' obsolete views is that, after spending twenty-six years of his professional life in Japan, he was out of touch with the ideas of his time when he returned finally to Britain in 1899.

Mach and Stallo

It is time to mention, for the significance of its content and wide cultural repercussions after the 1870s and throughout the first decade of the twentieth century, the attitude towards atomism adopted by the positivist school of physics and especially by its founder, Ernst Mach. In his younger years, while still a lecturer at the School of Medicine in Vienna, Mach made extensive use of atomic theory which he generally considered – as did others – a useful teaching instrument. He hypothesized that intramolecular forces could be related to their mutual distances and in terms of such forces he explained the density and viscosity of liquids. At the same time, from the analysis of the spectral behavior, he thought it possible to obtain information about molecular structure and especially about intra-atomic action. As for the atoms, Mach assumed the existence of two kinds – atoms of ether and atoms of matter – respectively repelling and attracting each other, the atoms of matter being the result of an agglomeration of atoms of ether. From the way they gathered together one could then explain the chemical differences between the individual atoms.

His writings of the 1860s seem to indicate that through the atomic theory it was possible to explain not only the undulatory theory – since heat is due to the vibration of ether atoms – but also the electrical and magnetic phenomena. Nevertheless, in Mach's mind, the most serious obstacle for his unconditional approval of the atomic theory remained the absence of conclusive experimental evidence to support it, which was the crucial test. And when at twenty-six, he moved to Gratz to teach physics at the university, apparently he already had considerable doubts about atomism.

His philosophical views in the following years led him progressively to the conviction that there was no ground for assumptions which cannot be proved by our senses. Accordingly, the validity of the whole atomic-molecular theory became more and more dubious after the 1870s; at that time Mach wrote that we were not justified in assigning spatial properties to entities (such as atoms and molecules) that we could not see or touch. In a lecture in 1871 he stated that 'the so-called molecular theories have been threatened by the progress of the theory of senses and of space and we may say that their days are numbered'.[210] Ten years later the concept that the atom must remain a tool for representing phenomena was reiterated.[211] In essence, this is again the utilitarian (empirical criticism) approach to the atomic theory which had been adopted by other chemists and physicists of the time, as clearly emerges from the atomic debates in Britain.

A possible explanation for Mach's gradual divergence from atomism after 1862 may be found in his increasing interest in acoustics, the physiology of the ear, etc., which had little or no use for a corpuscular theory.[212] Certainly chemistry, which relied mainly on physics, did not occupy a great place in his mind; there is no mention in his writings of either Dalton or Avogadro, Gay-Lussac and so on, although he seemed impressed by the conclusions of Mendeleev and Lothar Meyer in their studies on the classification of chemical elements.

If no analysis of the historical development of the atomic-molecular theory can ignore the impact of Mach's thought on the extent of its acceptance in the later part of the nineteenth century, not less important, although on quite a different level, are the considerations set forth at that same time by an American scholar of German origin. It may be convenient at this point to introduce Johann Stallo, particularly since Mach subsequently accepted his ideas.

Although not in sympathy with the Comtian current of positivism – he found Comte superseded – Stallo held a position in the philosophy of science quite similar to that advocated by Spencer and Mach. With Mach, though apparently not familar with his writings, Stallo shared a very strong antiatomism and with Spencer he agreed about the fallacy of the concept of potential energy. There is only dynamic energy, he said, and potential energy is a disguised form of kinetic energy. This energetic view, together with his denial of the validity of the atomic-molecular theory and with the overall relativistic approach to knowledge, represents the pillars of Stallo's thought.

Born in 1823, Stallo emigrated at an early age from Germany to the United States where he studied and practiced law for many years while essentially educating himself in science and philosophy. Later on he became involved in the political affairs of his adopted country and spent the last portion of his life in Italy where he had been appointed ambassador by President Cleveland. By the age of twenty-five Stallo had contributed articles to the *Popular Science Monthly* as well as writing a book on German philosophy. In 1881 his main work and fullest exposition of his thoughts, *The Concepts and Theories of Modern Physics*, first appeared. Despite the criticisms and quizzical reactions, the book went through several reprints and translations.

The Concepts is a very important document in any investigation of the development of the atomic-molecular theory in the second half of the nineteenth century, because it is the clearest statement of the position of the anti-mechanistic philosophy of the time. Out of the 300 pages of *The Concepts*, more than one-third are devoted to a logical, coherent demolition of the concepts of atom and consequently of the atomic constitution of matter.

Stallo considered the elementary units of mass as characterized by their simplicity; atoms are simple entities and, as such, are equal in all aspects and cannot have moving parts. Accordingly, they cannot contract or expand and should be considered as inelastic and hard. These elementary units appear to be mutually inert; since a mass which is inert regardless of its position cannot generate motion, then all energy must be kinetic. Moving from these premises, Stallo then proceeded to prove how the prevailing theories of the time, both in physics and in chemistry, appeared to be in full contradiction with the basic tenets of the atomic constitution of matter.

Aside from the growing evidence of the fallacy of the unitarian views of matter supported by Prout, Dumas, and others, there were spectroscopic researches which pointed out that, at the astronomical level, stars and planets in their evolution are shifting to more heterogenous structures. Not less important, Stallo said, is the argument against the simple nature of elementary units of mass provided by the 'cardinal principle' which is in direct conflict with the absolute equality of atoms. Such a principle, the Avogadro-Ampère Law, suggests beyond any doubt that the ultimate atoms of different chemical elements are of different weight.[213]

Stallo, to be sure, did not attempt to demolish Avogadro's Law, which he considered the only one capable of explaining the variations of gaseous volumes in both physical and chemical operations. Rather he qualified it as a simple hypothesis, but argued that such law is not reconciliable with Thomas Graham's assumption that interprets differences in densities of primordial atoms in terms of unequal volumes, these resulting from different velocities of movements with which the several kinds of atoms are provided.[214]

For Avogadro, on the other hand, there are inequalities of densities (and therefore of masses) in equal volumes. To Stallo this fundamental principle of modern chemistry is antithethical to one of the canons of the atomic-mechanistic theory, the equality of all atoms.[215] But a much more subtle criticism of Avogadro's hypothesis was next raised on physical rather than on chemical grounds. Avogadro's assumption, while it sets the minimum numbers of atoms, does not determine the maximum number of atoms present in each molecule.[216] The result is that molecules of greater weight can be extremely complex in their nature. In the case of gaseous substances, according to thermodynamics, heat is a form of energy which may be attributed in part to the translational movement of their molecules and in part is dependent upon the oscillatory or rotatory movements of their component atoms.[217]

Only in the case of monoatomic elements (for instance mercury, whose molecular weight coincides with its atomic weight) will their C_p/C_v determined through experiment coincide with that calculated, thus proving the simplicity of their molecules.

Stallo's argument consisted in stressing the paradox that theoretical chemistry's fundamental canons – he alluded here to Avogadro's hypothesis – tear to pieces the critical notion concerning the equality and simplicity of the elementary units of matter. Thus, he concluded, one either sets aside the equal volumes–equal numbers generalization or discards an essential tenet of the atomic theory. Needless to say, Stallo seriously misapplied Avogadro's hypothesis here; there are not seventy-one atoms in the chlorine molecule, as his example suggests, but its weight is seventy-one times that of a hydrogen atom. Structurally, then, the molecules of all elementary gases are equally simple, regardless of their molecular weights, a concept not always lucid in Avogadro's writings, and yet in general well appreciated by chemists after Cannizzaro. As for the differences between the calculated ratio of the specific heats and that found through experiment, they could not be explained in terms of Avogadro's hypothesis and would have to wait some time before their correct interpretation.[218]

Beside being equal and simple, the elementary units of mass, in harmony with the mechanistic theory, must be hard and inelastic. Atoms, it was said in *The Concepts*, cannot be elastic; were they capable of variations of their volumes, this would mean that they have parts. But in such a case one could not define them as atoms anymore. On the other hand, by taking for granted the existence of perfectly solid atoms only, how can the kinetic theory of gases be interpreted since a perpetual motion of these hard particles would not be possible?[219] Here appears the variance between the kinetic theory of gases and the assumed hardness of the elementary particles. No less difficult to explain is the principle of conservation of energy, because in case of mutual impact of solid bodies there would be a loss of motion. This could be explained by the partial transformation of translational energy into the internal agitation of 'minute parts' forming the bodies in collision. However, there are no parts in *atoms* or *molecules*, and consequently no transformation of the kind mentioned above may occur. A new dilemma: either one has to assume the validity of the kinetic theory and, accordingly, the elasticity of gas particles or one must accept the mechanical theory and its solid and inelastic atoms.

After mentioning the vortex-ring atoms' hypothesis advanced in those years by Thomson, as a 'valiant attempt' to bridge the gulf between the dynamic theory of gases and the atomic-mechanistic doctrine, Stallo continued by stating that such a system is 'illusory'.[220] He argued this by stressing that the fluid proposed by Thomson was an impossible vehicle of real motion, since his vortex atoms have no inertia, an essential property of matter. He also questioned the possibility of obtaining from such vortex-rings a solidified mass displaying the properties of matter. Thus, the solution offered by Thomson was in fact sterile and the essential dualism between mass and motion remained unsolved.

Stallo returned in the seventh chapter of *The Concepts* to the atomic theory. This hypothesis – for him, as for Mach, it was nothing but a hypothesis – was founded on three fundamental premises: (a) atoms are indivisible, unchangeable, indestructible, and simple; (b) matter is essentially discontinuous and accordingly atoms are separated from each other. These intervals increase or decrease when a body respectively expands or contracts; (c) weights are assigned to atoms and they correspond to the equivalent weights of combination.[221]

How does the atomic hypothesis explain these three propositions and how is it reconcilable with them? The first one deals with the indestructibility and impenetrability of matter. That matter is indestructible is proved experimentally through the use of the balance,[222] in fact, the quantity of matter is measured by its weight which appears to be unchangeable. Yet the constancy of weight is nothing but the constancy of a relation between two bodies mutually attracting. While the mass of a body does not alter, its weight can be reduced by simply moving its position. Thus, what is usually described as the indestructibility of matter delineates only the persistence of the effect of the action of forces in which masses are truly measured. Now, Stallo argued, this persistence finds no explanation in the atomic hypothesis. Atomic weights may then be considered as only relative to the physical position (or location) where they are determined. They are not absolute values and the indestructibility of matter is not proved by the atomic constitution of matter.

As for its impenetrability, this would be proved if the experimental science could indicate an absolute limit to the compressibility of bodies. Apparently this is not the case for some liquids and solids, but there are also examples of fluids[223] whose volume is reduced by merely mixing them.[224]

In the case of gases, an experimental limit to their compression is observed only when the pressure applied to them causes liquefaction or solidification; the fact that two or more gases chemically inert and placed in a given container may diffuse through each other as if they were alone and without change in their volume also confirms the fact that the impenetrability of matter is only a 'generalization' deduced from superficial observations.[225]

On the other hand, for those who interpreted the atomic theory in Boscovichean terms which assumed the atoms as points of force, it is simpler to explain either that matter is not penetrable since attractive and repulsive forces among such points are a function of their mutual distances or that matter is penetrable by simply modifying the laws which rule over those same forces.[226]

In both the arguments offered to disclaim that atoms are hard and indivisible, Stallo was walking a narrow metaphysical path. That atomic weights are not absolute weights had been expressed, although in a different way, by both Dalton and Avogadro, and that there is no measurable change in mass during chemical reactions had been shown by Lavoisier. It is true that he had not conducted experiments at, let's say, the summit of Mont Blanc or at the bottom of a mine pit, but clearly there was no reason to do that. The difficult problem – and here Stallo had a point – was the ability to convincingly relate Dalton's, Prout's, and Lavoisier's Laws to a theory of discrete constitution of matter. But even in the absence of positive proof, the arguments in favor of such constitution seem more cogent than those offered by Stallo. The view that the impenetrability of matter, when seen against the experimental evidence, is just a generalization does not hold. The researches conducted by Thomas Andrews and earlier by Faraday and Davy on the liquefaction of gases were probably familiar to Stallo. He quoted Cauchy among the physicists who thought that the atomic theory could be reconciled with the penetrability of matter, but the excerpt of the French mathematician's writings is anything but convincing, especially when he invokes 'the Author of nature' as a possible first cause for the modification of interatomic forces.[227]

In support of the thesis, that matter is discontinuous and, therefore, that a void interspace existed between atoms, come such optical phenomena as the dispersion and the polarization of light, which cannot be understood unless it is assumed that the ultimate particles of matter are set apart.

From a thermodynamic point of view the light-bearing ether, according to Maxwell, may be considered a very thin gas filling the intervals in the intermolecular spaces which would probably possess an enormous specific heat, much greater than that of any other gas under the same physical conditions. Now, the specific heats observed in some diatomic gases – as mentioned earlier – are already less than those calculated for the same gases if formed by elastic, non-perfectly spherical molecules. However, by adding the specific heat due to the intermolecular ether to the calculated specific heats of these gases, the difference between such figures and those determined through the experiment would naturally be even greater. One may then raise serious doubts about the assumption of the molecular nature of the light-bearing ether.

Thus far, the atomic theory had been challenged by Stallo on merely physical grounds: the indestructibility and the discontinuity of the ultimate components of matter. Now he proved that even the chemical aspect of that theory rests on flimsy foundations. In chemistry, he said, we can observe three main groups of facts: the first represented by the persistence of weight and combination in definite proportions, the second by changes of volume and involution or evolution of energy, and the third by the appearance of completely new chemical properties.[228]

The second group of facts cannot be explained in terms of the atomic theory, because this is incapable of accounting for changes of temperature, latent energy, and volume. As for chemical reactions, they lead to fundamental changes in the chemical properties and to a complete destruction of the 'integrity' of masses which, on the other hand, according to the atomic theory, should remain unchanged. Only through the adoption of subordinate hypotheses, has chemistry been able to overcome this dichotomy, but by so doing, the atomic doctrine loses its simplicity and consequently its distinction as a theory.

The persistence of weight and the combination in definite proportions remain then the only proof of atomism. But these occurrences are interpreted by reiterating the experimental facts in the form of a hypothesis, a procedure which is not logically acceptable. In support of this criticism, Stallo recalled that Thomson had commented that 'the assumption of atoms can explain no property of a body which has not previously been attributed to the atoms themselves', a very significant statement which describes more than anything else the feelings of a prominent critic of the atomo-mechanistic theory.[229] The line of reasoning followed by Stallo at this point was outlined in vague terms in a short footnote. In atoms, differences of weight or density are not possible, because a difference of density represents a difference of distance between particles contained in a given space. Since there is neither a multiplicity of particles nor empty spaces in atoms, then no difference of density may occur.

The logic seems faultless but Stallo was apparently unable to rationalize, on the basis of his analysis, the difference in weight between, for example, the same volume of lead and copper. Earlier he had concluded that only relative atomic weights exist, but now even these cannot be reconciled with the atomic doctrine when strictly interpreted. For him the notion of distinct atomic weights for different bodies is in itself a *contradictio in terminis*.

In the concluding paragraphs of this section of *The Concepts*, Stallo defined the atomic theory as a working hypothesis for chemists; it still seems a matter of debate whether it is essential to them. For some scientists – and Antoine Cournot is mentioned as favoring this view[230] – the atomic theory appears more a hindrance than a help, being inadequate and erroneous in its delineation of chemical processes ('la foi dans les atoms est plûtot un embarras qu'un secours', said Cournot). This position is not too unlike that taken by his countryman J. B. Dumas thirty years earlier.

Along with Cournot, Benjamin Brodie is mentioned and quoted as writing that the atomic theory 'has proved itself inadequate to deal with the complicated system of chemical facts which has been brought to light by the efforts of modern chemists'.[231]

Although Stallo's ideas were criticized and not always understood in 1881, the importance of his book lies in giving us a clear picture of the position of an important current in the philosophy of science at the time.[232] In his discussion and merciless demolition of the atomic-molecular theory, there are weak and strong points, as one might expect.[233] He is not as convincing in dealing with the chemical approach to the atomic theory as he is in using the knotty problem of the variances in the specific heats as an argument against the assumed complexity of the molecular structure.

A detailed and correct analysis of such variances had to wait until subsequent theories were able to explain the 'deterioration' of the specific heats of vibration and rotation in the low temperature region. Needless to say, Stallo was completely off target when he took exception to Avogadro's molecular hypothesis, as leading to an extremely elaborate organization of the elementary molecules in direct function of their atomic weights. While this topic may not have been clarified in some of Avogadro's essays, there was certainly plenty of evidence in Cannizzaro's 'Sunto', and in the writings of subsequent backers of the biatomic nature of elementary gases. One is at a loss here in trying to understand Stallo's misconception of the subject.

The Concepts and Theories of Modern Physics was published in America at approximately the time when the strong antiatomistic ideas of both Ostwald and Mach were taking shape in the German-speaking countries. These two thinkers shared with Stallo the vision of atomism as simply a utilitarian system, which could be useful for practical purposes, for instance the teaching of chemical laws, but had nothing to do with reality. For them, both atoms and molecules remained only a convenient fiction with strong metaphysical overtones.

Stallo's scientific background was that of a self-taught man; he had no place in the scientific arena of his time on either side of the Atlantic, and this in part may explain the very limited impact of his work when it appeared. Yet his ideas were sound, as proven by the impression they left on Mach when he came in contact with them at the turn of the century.

In the 1880s most chemists and a good number of physicists, who believed in the sound foundation offered to their scientific systems by the doctrines set forth by both Dalton and Avogadro, were annoyed by any opposition to the atomic-molecular theory. In fact, they did not see any valid reason in these attempts to shake an almost universally accepted conception, which had taken so many years to formulate and whose experimental basis seemed to become more reliable and stable as time went by.

Nevertheless, there are two elements in both Stallo's and Ostwald's thought that cannot be neglected for their significance: the relativistic vision of the scientific knowledge supported by the former, and the theory of energy with its full refusal of matter visualized by the latter. In a convergence of these two positions, one may foresee the premises for the eventual establishment of the twentieth-century's new physics.

The Experimental Evidence for the Molecular Reality

The sequel of events which eventually brought the experimental proof of the existence of atoms and of their particulate structure had begun in 1859, the year before the Karlsruhe Congress, when at the University of Bonn, Julius Plücker, a well-known mathematician turned experimental physicist, observed the effects of an electric current passing through a gas at low pressure. Among other phenomena, he noticed the phosphorescence on the walls of the container that Faraday had earlier detected.[234] His investigations led to further and more extensive studies of the subject by one of his assistants, Johann Hittorf, and in the seventies by E. Goldstein, who reported that the rays emanating from the negative electrode peculiarly followed a path at right angles with the source; he named them cathode rays and assumed that they behaved as waves in the ether. Shortly before that, Cromwell Varley in England had also studied the same facts and attempted to explain the phosphorence as caused by the impact of the rays upon the wall of the vacuum tube.

In those same years another Englishman, William Crookes, turned his interest to this same field of investigation and by 1879 had already reached extremely interesting conclusions on the nature of cathode rays. Crookes, a former chemistry student of Hofmann in London, had only an embryonic knowledge of mathematics, and he learned physics as he needed it for his work. In his personal laboratory Crookes discovered thallium in 1861 and invented the 'light mill' or radiometer in 1875.[235] Later on, experiments with this instrument induced Crookes to investigate phenomena in gases at very low pressure. It should be said that the facilities to obtain vacuum were still rather crude in the seventies; mercury pumps, such as the Toepler and Sprengel models, required hours of manual operation to give inadequate results. However, through continuous improvements of the Sprengel pumps,[236] Crookes could obtain 10^{-3} mm. of vacuum, sufficient to allow a reasonably free path to gas molecules inside the tubes. By 1879 he concluded that the cathode rays displayed a mechanical action and he confirmed their propagation in a straight line and their deviation by a magnet.

As for their nature, he advanced the theory of radiant matter, a state in which, due to the greater intermolecular distances resulting from the vacuum, the properties of a gas would be quite different from those of the same gas at ordinary pressure. And cathode rays – or molecular rays, as he termed them – represented gas molecules charged by their contact with the negative electrode.

Within the frame of the molecular hypothesis, Crookes, in his experiments, seemed not only to confirm some mechanical properties of the molecules but also to disclose new properties such as their behavior in a magnetic field. The material nature of the cathode rays, as assumed by Crookes and earlier by Varley, met the objection of those German physicists (Goldstein, Wiedmann, and Hittorf) who, after studying the same phenomena, explained the rays in terms of wave-like radiations. Shortly before his death in 1892, Hertz put his high reputation as an experimenter behind the undulatory theory, by proving that cathode rays may pass through thin metal leaves and that their path is not deflected by an electric field.

Soon afterwards, a twenty-five year old French physicist, working on the behavior of low pressure gases subject to electrical discharges, published a remarkable paper. In 1895, Jean Perrin, then assistant to Gabriel Lippmann in Paris, proved that by placing an electroscope inside a vacuum tube, the cathode rays there formed carried a negative charge of electricity. Thus the experimental evidence seemed to favor definitely the material nature of the rays; yet it fell short of convincing those supporting the opposite view.

More conclusive proof seemed in order and this came through the intuition, the ingenuity and the persistence of J. J. Thomson, a former student of Owens College who, after going through his Tripos at Trinity, had remained at Cambridge to teach and to conduct research. He joined the Cavendish Laboratory in 1880, the year after Maxwell's death and at the time when Lord Rayleigh became its director. A few years later, at twenty-eight, Thomson succeeded Rayleigh in the chair of experimental physics and shortly afterwards he began, with Richard Threlfall, his momentuous studies on the passage of electricity through gases. In the early stage of his research, he assumed the electrical forces to be responsible for splitting some of the gaseous molecules into two atoms, one positively, the other negatively charged. However, a turning point came in 1895 when, after the announcement by Roentgen of his discovery of the X-rays, Thomson had a replica of the apparatus used by the German physicist made at Cavendish Laboratory, and decided to direct his attention to the electric properties of gases when exposed to Roentgen rays. The immediate observation seemed to indicate that the rays changed a gas into a gaseous electrolyte.[237]

Further tests strongly suggested that the conducting gas (made such upon exposure to X-rays) lost its conductivity after being subjected to electric forces. This fact induced Thomson to think for the first time that particles charged with electricity could be responsible for the gas conductivity and that these particles were produced by the Roentgen rays. As he said, successive attempts to explain the difference between the behavior of cathode rays under magnetic and electric forces prompted the discovery of the electron.

While the rays were deflected when submitted to a magnetic force, their path remained steady when exposed to an electric force. From these experiments Hertz had concluded that the cathode rays behave as a flexible electric current flowing through the ether. Thomson, on the other hand, after confirming Hertz's findings, assumed that the gas between the charged metal plates could be ionized by the cathode rays and thus generate particles both positively and negatively charged. These particles would then neutralize the effect of positive and negative electrification of the two plates, and consequently no electric force could be produced in the space between them. From this reasoning, Thomson properly deduced that the presence of gas or gases was responsible for the absence of deviation of cathode rays and that a new series of experiments should be conducted at a much lower pressure.

Hence, it became a necessity to greatly improve the vacuum pump systems then available; in the 1890s a technological breakthrough, such as the discovery and the success of the incandescent lamp, began to require the large size production of very efficient vacuum-making equipment. The availability of such pumps gave Thomson the means to prove the validity of his assumption on the nature of the cathode rays. In tubes under much higher vacuum, the ray's beams deviated, when subjected to electrical forces, in such a way as to show beyond doubt their negative charge. Subsequently, the experimental techniques elaborated by Thomson enabled him to measure the velocity of these negative particles and the ratio between their mass and their electric charge. A further series of experiments indicated that the ratio e/m remained quite constant for different gases filling the tube, and its average value came close to 10^7 or about 1000 times the ratio for the hydrogen atom in the electrolysis.

On 30 April 1897, at a discourse of the Royal Institution, Thomson reported for the first time the remarkable news that atoms are not indivisible, because much smaller particles can be detached from them.[238] These particles have all the same mass and are present in all atoms, so they denote a fundamental constituent of matter.[239] These experimental findings seemed so surprising that Thomson himself decided to postpone their publication until he became fully convinced that the data and the techniques at his disposal offered no other acceptable alternative.[240]

Thomson's discovery of a 'primordial substance' symbolized a major conceptual turning point in the development of the physical sciences. As happens in similar events, both the participants and the man himself responsible for the discovery did not fully realize at the time the impact of what they had brought to light. Thomson at first resurrected William Prout's hypothesis of a 'protyle', by replacing in fact Prout's hydrogen atom with a corpuscle one-thousandth times lighter. Less known is his assumption in 1897 that in the proximity of the cathode, gaseous molecules, when subjected to a very intense electric field, split up not into ordinary 'chemical atoms' – as he said – but into 'primordial atoms' or corpuscles; then, at least for the gases, a direct breakdown would occur from the molecular to the corpuscular level. A few years later, Thomson, in mentioning again Prout's ideas,[241] remarked that the combining weights[242] of all elements should be integers and this had not been proven. Yet, he argued, the discovery of the periodic law leads to the conclusion that atoms of different elements have something in common.

In the years between 1897 and 1903, Thomson, while continuing his laboratory work, directed his attention to the elaboration of an atomic model which would satisfy the new experimental data. He realized at once the importance of a very stable equilibrium in an atomic configuration; in such a system, how would the constituents of the atom mutually act? Through a Boscovichean system of forces or are the particles held together by a central force? Needless to say, the more he attempted to come to a rational interpretation, the vaster the nature of the physical and mathematical problems involved seemed. Things appeared rather easy as long as one dealt only with a few corpuscles at rest. But even the lightest atom apparently contained a very large number of corpuscles, and the question of finding their distribution became too complex for a mathematical solution, as Thomson himself recognized.[243] At the time he recurred, as a working model, to Alfred Mayer's floating magnets, but these again could not provide a satisfactory solution when the number of corpuscles involved exceeded fifty or sixty.[244]

The Mayer model, however, remains conceptually interesting in view of Thomson's attempt to relate the number of corpuscles, in their arrangement, with the combining weight of an element and to interpret in this way the periodic law. With the simple atomic structure at his disposal, Thomson endeavored to explain also a property of chemical elements, the valence, which by then had been intriguing chemists for a generation. Quite intuitively, he assumed that 'the valency of an atom depends upon the ease with which corpuscles can escape from or be received by the atom'.[245] As for the tendency of an atom to enter into chemical combination, this may be explained by its power of acquiring a charge of electricity. An uncharged atom is also unstable and more prone to combine chemically. Some atoms, however, though stable when uncharged, become unstable when they receive an additional corpuscle and will not be able to enter into chemical combination.[246] Argon and helium were given as examples of such atoms.

There is no doubt that the approach, despite its limitations, was correct. Attempts in those same years were also made by Thomson to explain, in terms of his atomic model, simple and double bonds in organic compounds; in view of the complexity involved, they may appear rather naive to today's chemist and they were certainly premature, but they reveal the amazing ingenuity of their proposer. The problem of the diatomic molecules of elementary gases, which had haunted Avogadro's hypothesis since its proposal, also attracted Thomson's attention. He visualized that a fast movement and the collision of equal atoms in the gaseous state could cause a number of them, those moving more rapidly and colliding more frequently, to lose some of their corpuscles; as a result, they would become positively electrified, while their released corpuscles would show a tendency to attach themselves to the more slowly moving atoms which, in turn, would acquire a negative charge. Eventually these atoms with opposite charges would form, in Thomson's view, diatomic molecules.

The discovery of the electron, at the end of the nineteenth century, had the most profound effect upon the whole philosophy of physical sciences, as it had been understood up to that time. The finding of a new particle represented the convergence of physical and chemical speculations about the structure of matter which had their roots in Dalton, Davy, and Berzelius, while vindicating the vision of Faraday, Ampère, and Avogadro of a harmony between electric, magnetic, and chemical phenomena. Furthermore, it definitely proved the materiality of the atoms and confirmed their structural relation to molecules in the gaseous state. For those supporting the atomic view of matter, there was now 'something' experimentally verified, and the metaphysical interpretation of the atom had been challenged. Nobody, of course, could see or touch the atom, but one knew that a corpuscle with a mass, a volume, a kinetic energy was part of it and could be detached from the atom. Paradoxically, the atom had proved its existence, while contradicting its own qualification as an indivisible entity.

In 1900, both Mach and Ostwald were able to witness the emergence of further experimental evidence in favor of the validity of the atomic-molecular theories. In November 1908, in an exchange of ideas he had with van't Hoff, Ostwald admitted that he had renounced his long held attitude against atoms and molecules as a physical reality, on the basis of the unquestionable results obtained shortly before by Perrin and Svedberg. How long Ostwald had been considering his conversion is not clear. Certainly, in 1904 when he gave his Faraday Lecture, he reiterated his adherence to chemical dynamics as a full replacement for the atomic doctrine. This he maintained, despite the evidence accumulated in the meantime by Thomson at Cambridge. Thus, until 1904, the leading spokesman of the antiatomistic current held his line.[247] As for Ernest Mach, in 1909, a few years before his death, he was still irreducibly critical of any 'atomistic belief of the present day'.[248] Almost at the same time, in May 1908, and September 1909, two essays by Jean Perrin, then teaching at the Faculté des Sciences of Paris, appeared in the *Comptes Rendus.*[249] Their significance places them among the most important of the time. Perrin's work had its roots in the researches conducted more than fifty years before by Wiener and related to observations made even earlier by a British botanist, Robert Brown, who happened to be a contemporary of Avogadro.[250] It was Wiener who first thought that the irregular movements of certain particles in suspension could be interpreted in terms of the molecular movements.[251] Later on, as Perrin himself related, more speculations were published in relation to the causes of the Brownian movements; until in 1888–89 a very clear and thorough investigation of the phenomenon by L. G. Gouy revealed that the action observed was independent from the nature of the particles, the effect of light, and convection currents.[252] Moreover, it could be maintained indefinitely without external cause. Gouy's merit consists in having especially understood that the incessant movement of the fluid's molecules was solely responsible for the peculiar behavior of the particles.

It was then up to Perrin to pursue these ideas in order to prove their correctness beyond any doubt. He moved from Gouy's assumption that the Brownian movement can be explained if one considers every fluid as formed by elastic molecules animated by a perpetual motion. From the works of Joule, Clausius, and Maxwell, it had been established that a relation existed between the mean kinetic energy of translation of molecules and Avogadro's constant or the number of molecules present in a gram-molecule. At the time, the researches of van't Hoff and Pfeffer on the osmotic pressure of dilute solutions indicated that at the same temperature all molecules of all fluids had the same kinetic energy which was proportional to the absolute temperature.

From here, the next step was to prove experimentally that the mean energy of molecular translation is equal to that possessed by the granules of an emulsion, in order to see if the Avogadro number obtained from the granular energy would have an order of magnitude comparable to that expected from the equations of van der Waals, Clausius, and Maxwell.

The experiments led to a series of results indicating that the Avogadro numbers ascertained from the Brownian movements of suspensions of gamboge and mastic were indeed within the limits previously calculated from completely different phenomena.[253] The average value assigned by Perrin to N from the granular energy was 70.5×10^{22}.[254]

The conclusions reached by Perrin from the experimental data had proven the validity of his assumption and are so important that one feels compelled to report them verbatim:

I think that it will henceforth be difficult to defend by rational arguments an attitude hostile to molecular hypotheses which, one after another, carry conviction and to which at least as much confidence will be accorded as to the principles of energetics . . . there is no need to set these two great principles one against the other and the union of atomistics and energetics will perpetuate their dual triumph. . . .[255]

This allusion to the 'hostile attitude' to the molecular hypotheses and to energetics indicates how aware Perrin was of the position of Ostwald and his followers. He wrote these words in the summer of 1909, and it is rather likely that he was then unaware of the change of mind taking place in the German physicist at the time. The denial of or, to say the least, the mistrust of the atomic theory is still detectable in Ostwald's *Vorlesungen über Naturalphilosophie* of 1905.[256] While it is beyond any doubt that the years from 1906 through 1908 were the critical ones for the conversion of Ostwald to atomism, the influence that the works of Perrin and Thomson had over the evolution of his thought is not less questionable. This is very clearly stated in the fourth edition of the *Grundriss der Allgemeine Chemie* which appeared in 1909, where the researches on the electron and on the Brownian movements are quoted as 'experimental proof of the atomic nature of matter'.

However, as earlier mentioned, Ostwald was more convinced by Perrin's conclusions than by Thomson's. To the Frenchman then goes the distinction of having surmounted the last remaining obstacle to the universal acceptance of the equal volumes–equal numbers generalization proposed in 1811 in Vercelli by an obscure Amedeo Avogadro.

NOTES

1. According to the often quoted comment by Lothar Meyer after his reading of the 'Sunto'.
2. A. Naumann, *Berichte,* **12** (1879), 738, 'Über chloralhydrat'.
3. As D. Knight says, those who, among chemists and physicists, refused to support an atomic theory, could not accept the molecular theory either, *Atoms and Elements* (London, 1967), p. 94.
4. His work, scarcely recognized, could not be published in the 1818 *Proceedings of the Royal Society*, because the content seemed too speculative. Such, at least, was the opinion of Humphry Davy, then president of the Society.
5. *B.A.A.S. Report* 2 (1848), 21. Joule, born in 1818 was sent as a young boy by his father to Dalton, who for many years tutored at the Manchester Literary and Philosophical Society.
6. *Ann. Phys.* **118** (1863), 79.

7. *Wien. Berichte*, no. 2, **52** (1865), 395.
8. Avogadro refers here to vol. 8 of Poisson's *Mémoires* published in October 1829 by the Académie.
9. See *Fisica dei corpi ponderabili* (Turin, 1837), p. 1, libro I, capit. 1, sez. 2. A = intensity of the force related to an infinitely small distance, r = distance between two points mutually attracting, α = finite but insensible line, e = Neperian log base.
10. Avogadro's views on this topic and others discussed in the *Fisica dei corpi ponderabili* will be examined in a separate section of this work.
11. John Herapath, *Mathematical Physics* (London, 1847), p. 237.
12. G. W. Leibniz, 'Extrait d'une lettre à Monsieur Bayle sur un principle général utile à l'explication des loix de la nature, 1687', *Opera Philosophica* (Berlin: Erdmann, ed., 1840), p. 105.
13. G. W. Leibniz, 'Systéme nouveau de la nature', *Journal des Savans* (1695), p. 294.
14. G. W. Leibniz, *La Monadologie* (Hanover: Erdmann, ed., 1840).
15. Johann Bernoulli, *Opera Omnia*, vol. III (Brosquet: Lausanne and Geneve, 1742), p. 9.
16. Roger Boscovich, *De Viribus Vivis*, Diss. (Rome: Monaldini, 1745), pp. 38 ff.
17. See M. Hesse, *Forces and Fields* (Littlefeld, 1965), p. 164.
18. See L. P. Williams, *Faraday* (New York: Basic Books, 1965), p. 186, n. 32.
19. The interpretation offered by F. Greenaway, that Boscovich's atomism had very little to offer to practical investigators, seems in fact contradicted by the support that two such great experimentalists as Davy and Faraday gave to Boscovich. See F. Greenaway, *John Dalton and the Atom* (Ithaca: Cornell University Press, 1966), p. 29.
20. In his *Elements of Chemical Philosophy* (London, 1812), Davy revealed these ideas.
21. Although he had attempted, between 1808 and 1813, to correlate geometrical arrangements of atoms and chemical atomism.
22. As for Davy, he used the term 'proportional number' instead of atomic weight.
23. S. Cannizzaro, *Scritti intorno alla teoria molecolare ed atomica* (Palermo, 1896), p. 138.
24. The liquefaction of chlorine (1823) found more justification in the theory of point atoms as Williams points out in his work.
25. See Tyndall, *Faraday: A Discoverer* (New York, 1868), p. 123.
26. Faraday, *Experimental Research*, Series II (1844), p. 290.
27. See L. P. Williams, op. cit., p. 356. W. Hamilton, the well-known Irish mathematician, had published shortly before his essay, 'On a general method in Dynamics'.
28. Faraday, *Phil. Mag.*, **24** (February 1844), 136.
29. Faraday, *Experimental Researches*, Series II (1844), p. 284.
30. Although the molecular weight given for potassium hydroxide is correct, it is in disagreement with the formula assumed, KO_2H containing four elementary atoms; Faraday was still bound to Davy's and Dalton's formulas and atomic weights.
31. Faraday, 'Thoughts on ray vibrations', *Phil. Mag.* 28 (1846), 345, reprinted in *Experimental Researches in Chemistry and Physics* (London: Taylor and Francis, 1859), p. 366.
32. Ibid., 368.
33. Ibid.
34. Herapath, p. 237.
35. And later published by the *Manchester Courier*.
36. This paper of Joule's was read in 1848, but published in 1851; in 1857 Clausius said it was unknown to him, while he was preparing his 'Über die Art'.
37. Quoted from Humphry Davy, *The Elements of Chemical Philosophy* (London: Johnson, 1812), p. 95.

38. An explanation offered by W. Scott, *The Conflict Between Atomism and Conservation Theory* (London: McDonald, 1970), p. 256.
39. This manuscript was rejected by the Royal Society in 1846 and only abstracts appeared in 1846 and 1851.
40. *Ann. Phys. Chem.* Band C, **3**, (1857), 353.
41. See S. Cannizzaro, *Nuovo Cimento,* 7 (1858), 323.
42. He is the only one recalled by Cannizzaro in his 'Sunto' together with Gay-Lussac among all those who had been involved in this field of research.
43. The content of Clausius' 1857 Essay will be discussed in detail in the next section of this chapter.
44. In his August 1871 speech before the British Association Meeting.
45. Clausius, *Ann. Phys. Chem.* **103** (1858), 644, 'Uber die Natur des Ozon'.
46. Maxwell, 'Illustrations of the Dynamical Theory of Gases', lecture read at Aberdeen, Sept. 1859, quoted from W. D. Niven (ed.), *The Scientific Papers*, vol. I (New York: Dover, 1965), 377.
47. Ibid., 378.
48. Maxwell recalled Clausius' 'Über die art' in his 'On the Dynamical Theory of Gases' (May 1866). See *The Scientific Papers*, vol. II, pp. 28–29.
49. Clausius had arrived at a similar equation in 'Über die Art'; $p = nmu^2/3v$ (see *Ann. Phys. Chem.* Band C, **3** (1857), 375.
50. Maxwell, 'Illustrations of the Dynamical Theory', p. 390.
51. Maxwell, *Nature,* 8 (25 Sept. 1873), 437, reprinted in *The Scientific Papers* vol. II, p. 361.
52. Ibid., p. 363. Actually, molecule in the meaning of small mass (moles) was originally used by Gassendi in 1658 and later on by Baumè in 1763.
53. Ibid.
54. Ibid., p. 438.
55. Ibid., p. 440.
56. Maxwell, 'On the Dynamical Theory', pp. 29 and 33.
57. Maxwell, 'Atom', *Encyclopaedia Britannica*, 9th ed., vol. III, 1878, 38.
58. Ibid., p. 40.
59. Ibid.
60. Ibid. In the 16th volume of the same edition of the *Britannica* the entry for 'Molecule' was written by A. Crum Brown. In a footnote he mentioned the importance of the agreement between gas densities and molecular weights, first indicated by Gay-Lussac, and he added that in 1811 Avogadro 'brought forward his famous hypothesis'.
61. Ibid., p. 45.
62. Maxwell, *Nature,* **11** (1875), see following note.
63. Maxwell, *The Scientific Papers*, vol. II, p. 430.
64. Since the formula to calculate the specific heat of a gas at constant pressure is $(1/2J) \cdot (P_0 V_0/273°) \cdot (n + 2 + e)$ and the formula for the specific heat at constant volume is $(1/2J) \cdot (P_0 V_0/273°) \cdot (n + e)$, then their ratio is $(n + 2 + e)/(n + e)$ (I) (where n equals the number of variables and equals 6 for a rigid body).

    Two assumptions can be made at this time: either the molecules are atoms, simple material points with no rotational and internal motion and in this case, $n = 3$, $e = 0$ and the ratio (I) is = 1.66, too great for a real gas; or if the molecules are capable of both rotational and vibrational motion, then their c_p/c_v is = 1.33, too small for gases such as H, O, N, CO, HCl.
65. Ibid., p. 433.
66. This will be extensively examined later on.
67. Rudolf Clausius, 'Über die Art', *Ann. Phys. Chem.* Band C (1857), 353.

68. Karl Krönig, 'Grundzuge einer Theorie der Gase', *Ann. Phys.* **99** (1856), 315.
69. For instance, see J. R. Partington, *A History of Chemistry*, vol. IV (London: Macmillan, 1962), 492.
70. Clausius, *Ann. Phys. Chem*, Bend C (1857), 367, sec. 11.
71. This indicates that Clausius gave to Krönig the merit of having developed this concept which, in fact, has been stated by Daniel Bernoulli in 1738.
72. Clausius, loc. cit.
73. The concept of particles in polar relation had been advanced also by H. Davy in 1814. See MMS Common Place Book, as quoted by L. P. Williams, *Faraday*, p. 79.
74. Clausius, *Ann. Phys. Chem.* Band C (1857), 369. The sentence in quotation marks is in italics in the original article.
75. Ibid.
76. Ibid., p. 370.
77. The brother of Ernst Siemens, founder of the well-known German firm manufacturing electrical machinery. He moved to Britain after 1844 to establish in that country the local branch of the company.
78. Clausius, ibid., p. 354.
79. See James P. Joule, *The Scientific Papers* (London, 1833; reprinted London: Dawsons, 1963), p. 293. Original read 3 October 1848.
80. See *Elements of Chemical Philosophy*, p. 95. For Davy, particles in solids have vibrational movements, while in fluids and elastic bodies they have both vibrational and rotational movements.
81. According to W. Scott, *The Conflict Between Atomism and Conservation Theory* (London: McDonald, 1970), p. 269, it was Dubois-Raymond who later on called Clausius' attention to Bernoulli's *Hydrodynamica*.
82. Clausius, *Ann. Phys. Chem.*, **103** (1858), 644.
83. Marcel Verdet (1824–66) was professor at the École Polytechnique. This note appeared in *Ann. Chim.*, **50** (1857), 505.
84. Clausius, op. cit., p. 645.
85. P. Duhem, *Aim and Structure of Physical Theory*, 2nd ed. (Princeton: Princeton University Press, 1954), p. 103.
86. Thomson was raised to the peerage in 1892 as Lord Kelvin. All further mention of him will be as Thomson. Lecture given by Thomson at the Royal Institution, April 1900, entitled 'Clouds of the Nineteenth Century'.
87. See, for instance, his renewed praise for St. Claire Deville in the 1884 Montreal speech and again in the April 1900 lecture at the Royal Institution, for proposing his 'stupendous' theory of dissociation without recurring to the kinetic theory.
88. See *Proc. Roy. Soc.* **6** (Edinburgh, 1867), 94.
89. As reported by Thomson himself in his vortex atoms article of 1867.
90. See the detailed study by R. H. Silliman, *Isis*, **54** (1963), 472.
91. See above note.
92. W. Thomson, 'Report to the Royal Institution'. *Proc. Roy. Soc.* **10** (3 February 1883), 185 and *Nature* (31 March 1870), p. 551.
93. Needless to say this is a very incorrect statement, because the biatomicity of elementary gases had been established by then.
94. As determined from the viscosity and the gas diffusion studies of Clausius and Maxwell.
95. Thomson, speech printed in the *British Association Report*, (1884), p. 613.
96. Lecture given by Thomson at the Royal Institution, May 1860.
97. Thomson, *Proceedings of the Royal Society* (July 1889).
98. They had been published in 1866.

99. Crookes' researches went back to the late 1870s and those of Perrin to the middle 1890s.
100. The title was 'The Clouds of the Nineteenth Century', see note 86.
101. Thomson, *The Mathematical and Physical Papers,* vol. V (Clay, 1890), 351.
102. Thomson, *Baltimore Lectures*, Appendix A, p. 468.
103. I was able to find only two references to Avogadro in Thomson's *Mathematical and Physical Papers*, rev. ed., J. Larmor (Cambridge, 1911). The first reference is to 'Avogadro's law' and is mentioned in a footnote to the printed version of Thomson's lecture of July 1900, delivered before the Royal Society of Edinburgh. See *Baltimore Lectures*, Appendix A, p. 468. On 27 April 1900, Thomson mentioned Avogadro a second time in his lecture 'The Clouds of the Nineteenth Century'. Here Avogadro is cited for his 'law of molecular dynamics'. See *Baltimore Lectures*, Appendix B, p. 486.
104. See Thomson, *Baltimore Lectures*, Appendix D, p. 540.
105. Roscoe, Williamson, and Frankland had also been invited from Britain, but the last mentioned was unable to attend the congress.
106. His papers on the constitution of acids and salts and organic radicals were published in 1855 and 1858.
107. He studied under Gerhardt in Paris in the early 1850s, succeeded Faraday in 1868 at the Royal Institution, and four years later took Brodie's chair at Oxford.
108. William Odling, *Chem. News,* **2** (1860), 226.
109. See also I. Guareschi, *Discorso Storico Critico* (Turin: UTET, 1911), p. 64.
110. W. Odling, 'Reports on chemistry', *Brit. Assoc. Report,* **33** (1864), 21.
111. Ibid., p. 22.
112. He clearly refers here to Berzelius' unyielding position on dualism and other issues.
113. Odling, 'Reports on chemistry', ibid., p. 24.
114. W. Thomson was not present, maybe by pure coincidence; the debate took place in the same year in which he introduced his concept of vortex atoms. With the exception only of Williamson, he ignored throughout his writings all the other chemists who participated in the atomic debates.
115. Among other assumptions, he considered hydrogen present in different amounts in all elements.
116. In the course of the Presidential Address previously mentioned.
117. As reported by D. Knight, Odling thought chemists were influenced only by chemical arguments and not by physical evidence until Cannizzaro's 'Sunto' (1860). See *Isis,* **56** (1965), 12, n. 49.
118. Glyptic was the term assigned to structural models of balls and rods as those shown by Hofmann in 1861 at the Royal Institution.
119. Frankland, *Chem. News* (14 June 1867), p. 302.
120. Ibid., p. 303.
121. Stokes, then forty-eight, was from 1854 to 1885 secretary of the Royal Society; his hydrodynamics research influenced to a great extent the subsequent studies of Thomson.
122. *Phil. Trans.* **166** (1866), 781, and **167** (1867), 35.
123. Not better identified.
124. He had built his reputation fifteen years earlier with his well-known theory of etherification. His biography forms the subject of a recent inquiry by W. H. Brock and J. Harris, *Ann. Sci.* **31** (1974), 95.
125. A subject extensively discussed by C. A. Russell, *The History of Valency* (Humanities Press, 1971), pp. 111 ff. and p. 126.
126. Williamson, 'On the Atomic Theory', *J. Chem. Soc.* **22** (1869), 328.

127. Ibid., p. 339. A very similar view had been held by Berzelius in 1812.
128. He defined molecules in his *Chemistry for Students* (Oxford, 1868) as the smallest cluster of atoms of any substance. On this occasion he referred only to molecules of compounds, and there is no mention of the polyatomic molecules of some elements.
129. Williamson, op. cit., p. 346.
130. Ibid., p. 349.
131. Ibid., p. 350.
132. Ibid., p. 358.
133. The expression used by Williamson to indicate valence.
134. In 1850, Williamson wrote that he discarded the 'unwarrantable' hypothesis that the atoms are in a state of rest and that he based his reasoning on a form of 'atomic motion'. *Phil. Mag.* **37** (1850), 356, 'Theory of aetherification'.
135. Williamson, op. cit., p. 365.
136. As reported in the 12 November 1869 issue of *Chem. News*, p. 235.
137. Needless to say, nobody at that time was in a position to discuss this particular aspect of the atomic theory.
138. As Cannizzaro had first done in his 'Sunto', but he is not mentioned here by Williamson.
139. Avogadro is neither recalled here nor in previous occasions.
140. William Miller, professor of chemistry at King's College. He was born in 1817, had a medical education, and was especially interested in spectroscopy.
141. Edmund Mills (1840–1921) was both an organic and a physical chemist, taught at Glasgow and at West of Scotland Technical College.
142. This has been discussed on page 200 of the present work.
143. Hydrogen in marsh gas, for Odling, could be divisible into four parts and hydrogen in ammonia only into three parts.
144. George Foster, professor of physics at Anderson College, Glasgow, and later a colleague of Williamson.
145. Tyndall was then professor of natural philosophy at the Royal Institution.
146. *Chem. News.* (1869), p. 237.
147. *J. Chem. Soc.* **25** (1872), 941. A thirteen line review appeared in the *Jahresbericht* (1873), p. 3 and in the *Amer. Chem.* **3** (1873), 245, 343, 381.
148. M. M. Pattison Muir, *A History of Chemical Theories and Laws* (London: Wiley, 1907), p. 142.
149. The English translation of Lothar Meyer's *Modernen Theorien* fully supporting Cannizzaro's views as set forth in Karlsruhe did not appear until 1888.
150. Cannizzaro, *J. Chem. Soc.* **25** (1872), 942.
151. W. H. Wollaston, *Phil. Trans.* **104** (1815), 1. 'A synoptic scale of chemical equivalents'.
152. Cannizzaro, op. cit., p. 960.
153. Ibid., p. 966.
154. Kekulé in 1867 was still professor at Ghent; he moved to Bonn the same year. A pupil of Liebig, a close friend of Gerhardt, he spent two years at St. Bartholomew's Hospital as assistant to Stenhouse. At the time he became closely associated with Odling and Williamson.
155. Frederick Kekulé, *The Laboratory*, **1** (27 July 1867), 304.
156. The rest of Kekulé's paper was devoted to a severe criticism of the hypothesis advanced by Brodie about the 'compound' nature of such elements as chlorine, bromine, and nitrogen.
157. See F. Kekulé's *Lehrbuch der Organischen Chemie*, 1st ed. (Erlangen, 1867), p. 95.
158. Ibid.
159. Joseph Loschmidt's publication on the diameter of atoms appeared in October 1865.

160. Kekulé, *Lehrbuch der Organischen Chemie,* vol. I, p. 96.
161. Ibid., p. 97.
162. Ibid., p. 233.
163. Ibid.
164. Ibid., p. 232.
165. More than 300 pages of the *Lehrbuch* are devoted to the introduction of general concepts in both physical and organic chemistry.
166. Five German editions between 1864 and 1884. It was translated into English and French. See *J. Chem. Soc.* **69** (1896), 1403.
167. Meyer was born in 1830, Kekulé in 1829, Roscoe in 1833, and Baeyer in 1835.
168. Lothar Meyer, 'Introduction', *Modern Theories of Chemistry*, trans. by Bedson (London: Longmans, 1888), p. 26.
169. Ibid., p. 27.
170. Ibid., p. 577.
171. This appears in the English translation of 1888 which includes works by Maxwell in 1868, by Boltzmann in 1868–72, and by Oskar Meyer in 1877.
172. As I said, however, in another section of this work, Avogadro's approach, through well reasoned analogies, led in several instances to correct results.
173. Lothar Meyer, *Modern Theories of Chemistry*, pp. 51, 53, 55 ff.
174. In the 54 Aphorism of the *Novum Organum*, 1620.
175. Meyer, op. cit., p. 575.
176. Ibid., p. 577.
177. Where Wurtz's *Leçons sur quelque points* had also been published in 1864. In this work the equal volumes-equal numbers of atoms generalization is maintained and credited to Berzelius and Ampère.
178. It should be emphasized that the English translation by Williams and Bedson appeared in 1888, when the atomic debates had subsided. There is no mention of L. Meyer's book in the minutes of the meetings held at the Chemical Society in the 1860s.
179. Kolbe, a very controversial chemist, remained all his life a critic of Williamson, Wurtz, and later of Kekulé. His hostility to the later and to the valence theory is discussed extensively by C. A. Russell, *The History of Valency* (New York: Humanities Press, 1971), pp. 127 ff. As for structuralism, the vitriolic attack against the young van't Hoff in 1887 in the *J. Prakt. Chem.* is reported in every history of chemistry. A parallel could be made between Berzelius' conservative stand in the forties and Kolbe's in the late sixties.
180. He had studied also under Regnault in Paris.
181. The volume contains the thirteen lectures given before the Lowell Institute in Boston.
182. Josiah Parsons Cooke, *The New Chemistry* (New York: Appleton, 1874), p. 13.
183. Josiah P. Cooke, *The Principles of Chemical Philosophy* (Boston: Allyn & Bacon, 1891).
184. Ibid., p. 38.
185. Guareschi, *Discorso*, p. 25.
186. In the chapter devoted to the 'Choice of Atomic Weights'.
187. See p. 188 of the English translation by Walker (New York: Macmillan, 1890).
188. Ibid., p. 192.
189. The experimental works of T. Thomson, Berthelot, and Le Chatelier appeared mostly after the 1880s.
190. Lecture by Ostwald entitled 'Die Überwindung des Wissenschaftlichen Materialismus'.
191. See Walden, *Berichte*, Band A (1932), p. 101.
192. Ernst Haeckel (1834–1919) the first German biologist to endorse Darwinism was,

as was Ostwald, a man of wide cultural interests. He convinced Ostwald to become president of the German League of Monists.

193. See Ostwald, *Studien zur Energetik* (Sachs. Ges. Wills., 1891), p. 275.
194. Ostwald, *Verlesungen über Naturalphilosophie*, 10th ed. (Leipzig: Van Leitz, 1905), pp. 211 ff.
195. Ibid.
196. He makes here the analogy of a man who has an annual income of 20,000 Marks, but we don't know what his real capital is, nor the ways in which he obtains his income.
197. He is probably referring to the *Bemerkungen über die Krafte*, published in 1842.
198. Ostwald, *J. Chem. Soc.* **85** (1904), 508.
199. See F. G. Donnan, *J. Chem. Soc.* **136** (1933), 316.
200. E. Divers, born in 1837, studied under Hofmann, assistant to Frankland, was a good friend of Kekulé, and moved to Japan in 1873. The history of the team of British scientists who went to Japan to help the planning and the development of Japanese education is related in detail by J. Harris and W. H. Brock [*Ann. Sci.* **31** (1974), 125 ff.].
201. Edward Divers. 'The Atomic Theory with Hypothesis', *Report of British Association for the Advancement of Science* (September 1902), p. 557.
202. Divers is recalled by Guareschi in *Discorso Storico Critico*, p. 27, and by Meldrum, *Avogadro and Dalton: The Standing in Chemistry of their Hypotheses* (Edinburgh, 1906), pp. 11–13. Meldrum says of Divers: 'there are the high priests of chemistry giving decisions which, to use plain language, are in fact in contradiction to one another'.
203. His ways of expression and his train of thought are, unfortunately, anything but lucid. The result for the reader is serious difficulty in grasping the overall meaning of the sentences.
204. He is probably referring here not only to Ostwald and his school (although he does not mention any names) but also to Brodie.
205. Which aldehyde is not specified; the combination of ammonia and formaldehyde gives hexethylentetramine.
206. A rather similar concept had been expressed twenty years earlier by J. Stallo, as will be discussed later in this chapter.
207. Had Divers been more familiar with Avogadro's work, he would have known that Avogadro generally envisaged a polyatomic structure of all elements, metals and non-metals.
208. Divers forgets here that Gerhardt sixty years earlier had defined chlorine as chloride of chlorine and hydrogen as hydride of hydrogen, due to their biatomic nature.
209. Gomberg's research on triphenylmethyl radical had been published in 1900 and led to results in full contradiction to Diver's views.
210. Lecture by Ernst Mach, 'On symmetry', Prague, 1871, pub. in T. McCormack, trans., *Popular Scientific Lectures* (Chicago: Open Court, 1898), p. 89.
211. Ibid., p. 186. Restated at the anniversary meeting of the Imperial Academy of Sciences in Vienna, 1882.
212. See E. N. Hiebert, *Boston Studies in the Philosophy of Science*, vol. IV, edited by R. S. Cohen and R. J. Seeger, (Boston: Reidel, 1970), 79 ff.
213. This comes from the fact that, according to Avogadro's Law, the weights of the molecules are proportional to the specific gravities of the gases under like conditions of pressure and temperature.
214. See Thomas Graham, *Philosophical Magazine*, 4th ser. (1864), p. 81.
215. He quotes J. Cooke's definition of 1872 that Avogadro's Law holds the same place in chemistry that the law of gravitation does in astronomy.

216. It is beyond doubt, from the examples given by Stallo, that both chlorine and hydrogen are, at least, biatomic molecules.
217. Stallo uses the terms oscillatory and rotary as equivalent, but they are not. The experiment indicates that for a gas, C_p (specific heat at constant pressure) is always greater than C_v (specific heat at constant volume) and at the same time the experimental value of the ratio C_p/C_v is always less than the corresponding value calculated according to the thermodynamic theory. An explanation for this is offered, according to Stallo, by the fact that part of the heat supplied to a gaseous body is transformed into intramolecular stirring. On the other hand, if one assumes the molecular constitution as highly complex, then the energy of intramolecular origin (within the molecule) would become enormous, and consequently the calculated specific heat of a gaseous body would be much larger than that observed by experiment. For instance, he says, such would be the case with the chlorine molecule whose weight is 35.5 times that of hydrogen – and since this is diatomic – each chlorine molecule would contain not less than seventy-one atoms. This, in turn, would lead to a specific heat (calculated) far exceeding that determined through the experiment. One may then conclude that it is not possible to assume an indefinite multiplication of atoms within the molecule.
218. Deviations from the classical theory of specific heats found an explanation twenty years later in quantum-mechanics terms by Einstein and later on by Debye and Blackman.
219. Stallo, *The Concepts and Theories*, edited P. W. Bridgman (1881; reprinted Cambridge: Harvard University Press, 1960), p. 72.
220. Ibid., p. 74. The vortex-rings theory is discussed earlier in this section.
221. In a footnote Stallo explains that he intentionally neglected the distinction here between molecules and atoms; the former being defined as 'the ultimate products of the physical division of matter', the latter as 'the ultimate products of its chemical decomposition'.
222. Stallo, *The Concepts*, p. 111.
223. Stallo mentions a mixture of water and sulfuric acid as an example.
224. Stallo, *The Concepts*, p. 116.
225. Dalton, however, in making this observation, did not have in mind proving the impenetrability of matter.
226. A. Cauchy in 1868, as quoted by Stallo, supported the view that in such a way one could foresee the hardest bodies mutually penetrating.
227. Stallo, *The Concepts*, p. 118.
228. Ibid., p. 125.
229. There is no reference by Stallo to this quotation.
230. Antoine Cournot (1801–77), French mathematician, professor at Dijon and Lyons.
231. Benjamin Brodie, *Chem. News,* **15** (1867), 295.
232. P. W. Bridgman, 'Introduction', to Stallo, *The Concepts*, p. 21.
233. Stallo's ideas had been qualified as 'most arrant nonsense' by Peter Tait, then at Edinburgh, in his review in *Nature*.
234. The adoption of the evacuated tubes with fused electrodes developed by Geissler and, after 1857, the availability of the Rühmkorff induction coil as a source of high voltage had encouraged Faraday to resume his investigations in this field of research that went back almost twenty years and are described in the XIIIth Series of the *Experimental Researches*. Faraday observed the formation of bands of light and dark inside the tubes at low pressure, already reported by W. R. Grove in 1852. These experiments were repeated in John Gassiot's house in 1858, and Faraday noticed also the effect of a magnet on the position of the bands. The deflecting

influence of the magnet, however, could not be explained by Faraday. He remained faithful to his fundamental theory of electric action where conduction is preceded by induction, while in the Geissler tubes the two seemed to be simultaneous.

He never published any conclusions that he may have had on the nature of the phenomena observed. The subject is extensively discussed by L. P. Williams in the final section of *Michael Faraday* (Basic Books, 1965), pp. 475 ff. from which the preceding information has been obtained.

235. The two findings are conceptually related, because in his attempt to determine the atomic weight of thallium, Crookes observed that small deviations in weight were caused by light falling on the balance; from this he decided to study the effect of light on delicately balanced metal pieces.
236. Giminham deserves credit for this work.
237. Roentgen in his 'Second Communication' (1895) reported the discharge of electrified bodies, placed either in air or in hydrogen through which X-rays have passed; this communication probably came to the attention of J. J. Thomson, although he did not mention it in his *Memoirs*.
238. They were published in the *Philosophical Magazine* six months later and in abstract in the *Electrician*, May 1897.
239. In Germany, Wiechert and Kaufmann at the same time obtained similar values of m/e, but they measured only the magnetic deviation which gives mv/e, but not v.
240. All information concerning J. J. Thomson, his life and research is taken from his *Recollections and Reflexions* (New York: Macmillan, 1937). In this work which is extremely detailed in some sections, there are unexplainable omissions: for instance, the author does not devote a single word to the contributions of Jean Perrin and C. Varley to the study of the cathode rays.
241. J. J. Thomson, *Electricity and Matter* (New York: Scribners, 1904), pp. 71 ff.
242. The term J. J. Thomson used instead of atomic weights.
243. See *Electricity and Matter*, p. 113. This volume contains the lectures given by Thomson at Yale in the spring of 1903.
244. Mayer models' limitations, for instance, that they are not spatial as the actual movement of corpuscles, were admitted by Thomson.
245. *Electricity and Matter*, p. 131.
246. Ibid., p. 132.
247. Possibly just a coincidence, but the Nobel Prize in 1909 for chemistry was awarded to Ostwald only five years after his 'conversion' to the atomic-molecular doctrine.
248. This quote appears in the introduction to *Die Prinzipen der Physikalischen Optik* (Leipzig, 1821).
249. Reference is made here to *Compt. Rend.* **146** (1908), 967; **147** (1908), 530; **149** (1909), 477. On Jean Perrin's scientific work, its antecedents, developments, and conclusion, see M. J. Nye, *Molecular Reality* (New York: Neal Watson, 1972).
250. Brown's report which had appeared in *Phil. Mag.* 1829 had eleven years later been commented upon by one of Avogadro's associates at the University of Turin. G. B. Botto published it in 1840, *Mem. R. Acc. Sci., Torino*, Ser. 2, **2** (1840), 457, 'Observations microscopiques sur les mouvements . . . .'
251. Wiener, *Pogg. Ann.* **118** (1863), 79.
252. L. G. Gouy, *J. Phys.* 7 (1885), 561 and *Compt. Rend.* **109** (1889), 102.
253. Reported in great detail by Perrin in his *Brownian Movement and the Molecular Reality*, trans. F. Soddy (London: Taylor and Francis, 1910).
254. Planck calculated $N = 61 \times 10^{22}$ from the spectrum of dark bodies; Lorentz obtained $N = 77 \times 10^{22}$. See also Einstein papers of 1905 [*Ann. Phys.* **17** (1905), 549] proving that bodies of microscopic size suspended in a liquid perform movements that can be observed on account of the molecular motion of heat.

255. See Perrin, op. cit., p. 91.
256. 'The stoichiometric laws will continue to exist in chemistry even a long time after the atoms will be found only in the dust of the libraries'. Ostwald, *Vorlesungen*, No. 10, 'Die Warme' (Leipzig: Von Leitz, 1905), p. 211.

CHAPTER SIX

THE MAJOR CHEMICAL ESSAYS (1821)

Two major chemical essays were read by Avogadro before the Royal Academy of Sciences of Turin in February and December 1821. The first, 'Nouvelles considérations sur la théorie des proportions déterminées', dealt mainly with inorganic compounds. The second, 'Mémoire sur la manière de ramener les composés organiques aux lois ordinaires des proportions déterminées', attempted to harmonize the composition of organic substances, as recently revealed by analytical means, with the laws of definite and multiple proportions. The most striking aspect of these two very extensive articles – when published in the *Memorie della Reale Accademia*, they filled over 200 pages – is that in them Avogadro relied exclusively on laboratory findings of other chemists of the time. He reported no experimental results of his own. This purely speculative approach had very little or no impact upon others who practiced chemistry in the 1820s.

He conceded that the subject of the February 1821 essay was outside his prevailing interest in physics. He would consider inorganic compounds because, firstly, one could not 'confide blindly' in the opinion of several chemists, and secondly the determination of the masses of the molecules of different bodies had '. . . une étroite connexion avec plusieurs points appartenans à la physique . . .'[1] He trusted fully in the power of physics and mathematics (the latter he used extensively in the December 1821 essay) to illuminate points which had remained clouded to the chemists.

The 'Nouvelles considérations' reveals Avogadro eagerly seeking to reaffirm his views and to dovetail them with those of the more influential chemists of his time. The essay clearly indicates once again that he realized fully the validity and importance of his gas hypothesis, and at the same time feared that others might not correctly interpret it. The possibility that his ideas might be unacceptable to the majority of his contemporaries then eluded Avogadro. He felt that, among others, Berzelius, Mollet, and Ampère, had adopted the gas hypothesis; clearly, however, this was not true at least for Berzelius.[2] In the 'Nouvelles considérations' Avogadro criticized Berzelius for applying the volumes theory only to hydrogen and oxygen, and argued that evidence existed indicating the elementary nature of both nitrogen and chlorine as well.[3] He also reproved Dalton for doubting Gay-Lussac's 1809 generalization of the regular ratios of combining volumes and, accordingly, for denying that a correlation could exist between that generalization and the law of definite proportions. As for Davy, T. Thomson, and Gay-Lussac, Avogadro pointed out that through their equivocal use of terms such as atom, molecule, proportional, and definite number, and their differentiation between the number of particles entering into combination and the volumes of combining gases, they certainly had not clarified the chemistry of combinations.[4]

'Nouvelles considérations', a 162-page essay, is divided into two parts. The first part has an introductory tone and offers remarkable insights into Avogadro's opinions of the major achievements made by his fellow physicists and chemists in the decade following the statement of the molecular hypothesis. The second part of the essay contains a detailed review of about thirty elements and their binary compounds, based on the latest findings of the leading chemists of the time.

In the opening pages, Avogadro, after clearly restating his 1811 gas hypothesis, defined an 'integral molecule' and its relation to 'partial' and 'total' molecules.[5] He dealt here for the first time with a critical issue left vague in previous essays, namely, the number of 'partial molecules' forming the integral molecule. For some elemental gases, one or two, for others, 'probably' four 'partial molecules' could combine, while in solids and liquids as many as eight or sixteen 'partial molecules' could combine.[6]

Again – yet more lucidly than in his previous writings – Avogadro recounted his logic in establishing the notion of structured elementary particles. He recalled Gay-Lussac's 1809 generalization indicating that, at least for gases, chemical combinations take place by even simple numbers or their fractions, thus leaving out such odd ratios as 1:3, 1:5, 2:3, etc. In addition, the law of definite proportions (to which Avogadro always referred as a 'théorie', although he himself considered it the 'foundation of the entire chemistry')[7] implied that chemical combinations occur by whole numbers. The link which should exist between these two empirical 'laws' Avogadro claimed to have found in the existence of a divisible and, as such, structured molecule. To paraphrase what J. B. Dumas said a few years later, chemistry had the courage to divide the atoms, something that physics did not dare to do.

Once the link between Gay-Lussac's generalization and whole number combinations had been established, contended Avogadro, it was possible to find for any chemical combination a 'proportional number'. Only by adopting his hypothesis, however, would this number not be arbitrary, as that chosen by some chemists[8] and subject to revision every time a new combination between two elements was found. If, by following Avogadro's proposal, the densities relative to hydrogen, for instance, or the weights of equal volumes of gases under comparable conditions were chosen as 'proportional number', then one had at his disposal figures which should not be changed every time a new chemical compound was discovered. The adoption of a set value, such as the relative gaseous density, Avogadro argued, would eliminate the need to refer to fractions of molecules or to the existence of secondary combinations. The gaseous density should, therefore, be used by every chemist as a permanent means to calculate correctly the definite proportions, especially when faced with more than one interpretation of the chemical combination.[9]

Though Gay-Lussac's 1809 generalization of the regularity of combining gases was fundamental, he had not formulated an overall rule to deal with the formation of binary gaseous compounds.[10] Avogadro, on the other hand, had concluded that when equal volumes of two gaseous substances combine chemically, the volume of the compound gas thus obtained is either equal to the sum of the component volumes or is the double of either component. And he claimed now that, with the exception of Berzelius who had expressed it more tentatively in his 'Essai sur la théorie', no one had yet suggested such a generalization.[11] The notion of the division of integral molecules (the term referred to both elements and compounds), which represents undoubtedly the most original and also the most controversial aspect of the molecular hypothesis, was then restated by Avogadro in the 'Nouvelles considérations'. Should the division of integral molecules not take place, he argued, then the volume of a compound gas would equal that of the gases entering in combination. As for the number of molecules of the combining gases, they, too, would equal that of the compound gas. Experience, however, indicated just the opposite.

One must assume then, in Avogadro's words, that 'each compound molecule is divided into two partial molecules which are reciprocally in the same proportion as the whole molecules.'[12] Avogadro recalled here that Berzelius had paid no attention to the concept of division of integral molecules because for him, regardless of whether there was or was not condensation of the combining volumes, the number of 'compound atoms' was always *one-half* the number of 'elementary atoms' present in the same volume before the combination. To explain this the Swedish chemist had hypothesized that an increase in the repelling forces spread the particles further apart in compound gases than they had been in elemental gases, an assumption Avogadro considered more complicated and thus inferior to his own.

Avogadro then cited water as an example of his interpretation of chemical combinations. He considered its molecules to be composed of one molecule of hydrogen and a half molecule of oxygen, and hence to be the product of the combination of two volumes (i.e., two molecules) of hydrogen and one volume (i.e., one molecule) of oxygen. Yet he cautioned that the mass of the molecule of water would not be equal to the sum of the masses of one molecule of oxygen and two molecules of hydrogen.[13]

In these opening pages, then, Avogadro restated very carefully and clearly his molecular hypothesis, and attempted to place it in the context of the chemical ideas of the 1810s.

The next section of the 'Nouvelles considérations' Avogadro devoted to a critical examination of the new explanations offered for the law of definite proportions since the publication of his 1811 and 1814 essays. For binary compounds, a fresh approach had been proposed. Ampère and Gay-Lussac had attempted to apply the law of definite proportions to their researches on nitrogen and iron oxides, respectively. In agreement with this law, they considered each binary compound as formed by two preliminary combinations of the same components. Avogadro recalled that in his 1814 memoir he had assumed, in reference to nitrogen and iron oxides, the possibility of a 'two levels oxidation' involving the combination of oxygen with a pre-existing oxide. He now considered this interpretation 'insuffisante' and 'moins naturelle' because such reactions between compounds and elements are rare.[14]

Avogadro then reviewed Berzelius' generalizations concerning inorganic combinations. The Swede had hypothesized that when a reaction takes place between an acid and a base, both having oxygen as a common component, the amount of oxygen present in the acid is always a simple, even multiple of that of the base. For example, in the case of a base of type RO_2 and of an acid radical of type $R'O_3$, no reaction between them could take place because this would give an odd ratio among atoms of oxygen involved. However, a reaction between compounds of type $R_2O+R'O_3$ would be possible.[15]

Avogadro questioned these generalizations. He contended that no general rule for the composition of salts could be made, although he acknowledged the existence of 'certaines relations' between the oxygen of a base and that of an acid when they chemically combine. As for some ternary and quaternary compounds (such as ammonium sulfate) since their correct vapor densities were unknown, Avogadro conceded that he could not explain their formation.

Another group of ternary compounds included substances formed by the combination of binary compounds and single elements, for instance, the alkaline oxysulfides reported by Gay-Lussac, the oxychlorides and phosgene.[16]

Avogadro speculated that these compounds should be considered as either sulfides or chlorides of a metallic oxide, where sulfur and chlorine act as 'acid' or the electronegative portion of the molecule.

Between 1814, the year of his last major chemical writing, and 1820, five articles had appeared which especially attracted Avogadro's attention. He reviewed them in the 'Nouvelles considérations' and concluded that if their attempt was to simplify the law of definite proportions, then it seemed to him that they all had failed.

In two works published in 1816 and 1818 in the *Annales de Chimie et Physique*, A. J. Frère De Montizon had proposed a correlation between the size of the molecules of metals and their density. Avogadro rejected this as a much too simple view of the complex nature of such correlation.[17]

Next Avogadro questioned William Prout's suggestion, offered in 1815, that the weights of most elements were whole multiples of the weight of hydrogen.[18] The limited accuracy of the experimental results and the extremely small unity taken as a reference (hydrogen) could, claimed Avogadro, account for the even numbers proposed as weights of the elements. Interestingly, two years later, Berzelius raised a similar criticism in his harsh review of Prout and his supporter Thomas Thomson.[19]

Avogadro then examined the relationship between the specific heats of the elements and their atomic weights, as outlined by Dulong and Petit in 1819.[20] This topic specially interested Avogadro because of his extensive studies of thermology (in the preceeding years he had published four papers on specific heats) and because of his deep involvement with the atomic theory and its implications. Although Avogadro considered Dulong's and Petit's essay 'très intéressant', he expressed reservations about the accuracy of their conclusions. His first objection – based on the variation of specific heats with the temperature[21] – would remain for years a serious problem for the generalization of the regularity of atomic heats.[22] His second objection concerned the validity of extending the properties found in particles of solids to the elementary particles of gaseous bodies. Could Dulong–Petit's Law be applied also to gaseous elements? Avogadro would answer this question later on in the 'Nouvelles considérations'.

In an article whose French translation had appeared in 1820 ('Première mémoire sur l'identité de la forme cristalline'), Eilhard Mitscherlich had reported on the connection between chemical composition and crystalline form.[23] Presumably this essay reached Avogadro late in the summer of 1820 while he was completing the manuscript of 'Nouvelles considérations'. In reviewing Mitscherlich's paper, Avogadro cited a previous attempt to relate chemical compounds to their crystalline systems. Seven years earlier in his letter to Claude Berthollet, Ampère had assumed that in any molecule its atoms were actually placed at the vertices of a polyhedron. He had called these polyhedrons the 'forme représentative de la particle', and argued that, through them, crystalline structures eventually took shape.[24] In his comments on these interpretations of the intimate composition of bodies, Avogadro remarked that several compounds appeared to disagree with Ampère's assumption and with the relationship more recently set forth by Mitscherlich. To Avogadro, any generalization of this kind was at least premature. Compounds having very close chemical composition showed, indeed, an analogy in their crystalline structure, but not an identity as the isomorphism laws seemed to assert. He quoted F. Beudant[25] and René Haüy in partial support of his criticism of Mitscherlich.[26] In almost an ironical tone Avogadro concluded: 'il me semble probable qu'il s'y est glissé quelque illusion . . .'[27]

Surprisingly, in reviewing both Prout's hypothesis and Dulong–Petit's Law, Avogadro found himself quite close to the position of Berzelius. He certainly did not, however, share the enthusiasm of Berzelius for Mitscherlich's work, which in 1822 Berzelius had characterized as leading to the '. . . most important discovery made since the doctrine of chemical proportions . . .'.[28]

From an historical point of view, Avogadro's short appraisal of some of the more interesting developments of physical chemistry in those years is quite revealing. Hundreds of articles were published year after year in the journals regularly used by Avogadro. Yet, four of the five that he reviewed in 1821 have left a mark in the development of chemistry. Can we conclude that he was such an uncanny observer that he was able to see 'above the crowd'? Not necessarily, for these works, especially Dulong–Petit's and Ampère's, attracted his attention because they were closely related to his own scientific program. This program, as earlier mentioned, was built around the desire to quantify chemistry, to establish it on a sound mathematical foundation above the uncertainties and human errors inherent in most analytical procedures.

In the second section of the 'Nouvelles considérations', Avogadro thoroughly examined a long series of compounds to determine how his generalizations would apply to them. After separating organic and inorganic compounds, and dividing the latter into binary and ternary combinations, he turned his attention to the compounds of hydrogen with oxygen. By 1820, two competing interpretations of the composition of water existed: Berzelius, Davy, and Wollaston had adopted the two-to-one ratio between hydrogen and oxygen, while Dalton and his followers had stayed with the earlier one-to-one ratio.[29]

Various accurate experimental results (such as the analysis of water and the direct determination of the density of hydrogen) seemed in close agreement,[30] and indicated that oxygen was 16.026 times heavier than hydrogen. Here, for the first time in his chemical essays, Avogadro chose the molecule of oxygen[31] as the reference for determining all other elementary weights. Like Berzelius and Wollaston, he now felt that hydrogen was less reliable as a unit of reference. He then restated that two molecules of water were obtained from the combination of two molecules of hydrogen with one of oxygen or, in other words, that the compound molecule of water was actually formed by one molecule of hydrogen and a half molecule of oxygen. On the basis of the experimental figures available to him, he set the density of the water vapor molecule at 9.013 times that of hydrogen[32] (0.621 relative to air) in close agreement with the values observed by Gay-Lussac.[33]

In 1818, another compound of hydrogen and oxygen had been discovered by Thenard. In Avogadro's view, it contained equal volumes of the two gases.[34] Because of the limited information then available, Avogadro touched only briefly upon the matter. Yet quite pertinent (if one considers his criticism earlier of the oxygen theory of acids) seems his remark that hydrogen peroxide does not present an acid character despite its high content of oxygen.

Avogadro then turned to the compounds of nitrogen and oxygen. On the basis of the experimental data then available, he listed the following nitrogen oxides:[35]

nitrous oxide (protoxide d'azote)	= 1 vol. nitrogen + 0.5 vol. oxygen
nitrous gas (deutoxide d'azote)	= 1 vol. nitrogen + 1.0 vol. oxygen
hyponitrous gas (gaz pernitreux)	= 1 vol. nitrogen + 1.5 vol. oxygen
nitrous acid (acide nitreux)	= 1 vol. nitrogen + 2.0 vol. oxygen
nitric acid (acide nitric)	= 1 vol. nitrogen + 2.5 vol. oxygen

We must recall here the wide conceptual and factual gulf existing in 1820 between the atomists on one side and the volumists on the other. Not only was Dalton still rejecting the law of combining volumes Gay-Lussac had set forth in 1809, but he was so hostile to the notion of volume that in an article written in March 1817, he had used the term only to refer to the work of the French chemists. Shortly thereafter, Thomas Thomson, though one of the earliest and most faithful of the Daltonians, had reached a position unquestionably favorable to Gay-Lussac's findings.[36] In 1820, regarding the determination of the densities of gaseous compounds, Thomson wrote:

I have made a great many experiments upon gaseous bodies with a view to satisfying myself whether this general law (Gay-Lussac's) be entitled to confidence. The result has been the fullest conviction of its precision. I consider it as one of the most important laws hitherto discovered in chemistry.[37]

For nitrogen oxides, Thomson adopted Gay-Lussac's simple volume ratios, and gave his own experimental results in full agreement with the French chemist's interpretation. By 1820 Berzelius, still misled by his assumption of nitrogen as the oxide of *nitricum* (a hypothetical substance in the free state), had discarded Gay-Lussac's suggestion for the composition of nitrous acid,[38] and claimed that this substance was formed by equal volumes of nitrous gas and nitric acid.[39]

The presence of *half volumes* of oxygen in three nitrogen oxides (see above), however could not be easily reconciled with the views of the time, if one considers that the terms atom and volume were often used interchangeably, and that Gay-Lussac's observations on combining gases referred in general to whole volumes. In one of the more revealing sections of the 'Nouvelles considérations' Avogadro cleared this hurdle in a very simple way by using the concept of polyatomic integral molecules to explain the existence of 'demi-molécules'. For example. hyponitrous gas (gas pernitreux) would be formed by one molecule of nitrogen and one-and-a-half molecules of oxygen. In a similar way, nitric acid would be formed by one molecule of nitrogen and two-and-a-half molecules of oxygen. For Avogadro, the formation of both hyponitrous and nitric acids could be accounted for through the combination of pre-existing nitrogen oxides: hyponitrous acid would thus result from nitrous acid and nitrous gas ($NO_2 + NO = N_2O_3$), while nitric acid (N_2O_5) would be a combination half-way between nitrous acid (NO_2) and a higher nitrogen oxide formed by one volume of nitrogen and three volumes of oxygen (NO_3).*

Avogadro's view of ammonia remained unchanged from his 1811 and 1814 essays. Each molecule is formed by a half molecule of nitrogen and one-and-a-half molecules of hydrogen. In fact, in 1808, Gay-Lussac had determined experimentally that two volumes of ammonia are obtained by combining one volume of nitrogen and three of hydrogen. By assuming a division in half of its compound molecule, Avogadro pointed out that the density of ammonia as computed from the densities of its components closely agreed with that found experimentally by Arago and Biot.[40]

* Current chemical symbols are used only to attempt to explain Avogadro's mechanism of combinations. He did not use symbols at this time.

Next Avogadro examined chlorine. He found its density relative to hydrogen, as calculated from the density of hydrochloric acid determined by Biot and Gay-Lussac, as 36.124 or 2.25 that of oxygen.[41] Under the influence of Davy, as he himself conceded, Avogadro had accepted the elementary nature of chlorine before his February 1814 essay, although in the 'Essai d'une manière' (1811) he still used the term oxymuriatic acid,[42] following the prevailing fashion.

In a long footnote in the 'Nouvelles considérations',[43] Avogadro mentioned that Gay-Lussac and Thenard, although challenging Davy's priority on chlorine, had retained their original position on oxymuriatic acid as a compound out of respect ('par déférence') for Claude Berthollet.[44] This casual remark is quite interesting in disclosing Avogadro's awareness of the tight relationships then prevailing among the members of the Arcueil circle. Avogadro also recalled that Berthollet had supported the oxymuriatic theory until 1816.

The next section of the 'Nouvelles considérations' deals with the determination of the relative density of carbon from its gaseous compounds. From the experimental data of Arago and Biot on the density of carbon dioxide, a value of 11.36 had been calculated for carbon in the gaseous state, a figure which Avogadro now corrected to 12.08, in agreement with the updated relative density of hydrogen.[45] In 1814, Berzelius had set the weight of carbon at 12.05 and, of course, Avogadro did not fail to mention this as proof of the fact that Berzelius had thus adopted his 1811 gas hypothesis.[46]

Gay-Lussac's views of the compositions of carbon oxide and dioxide fundamentally disagreed, claimed Avogadro, with both the experimental evidence and the rules generally observed in most binary gaseous combinations. Actually, the French chemist had assumed that if 'carbon were in the gaseous state, it would combine in equal volumes (or in any other proportion) with oxygen, and that the apparent condensation would then be equal to the whole volume of the gaseous carbon',[47] which was not the case. Avogadro concluded that the main obstacle to a logical explanation of these facts resided in Gay-Lussac's failure to recognize the composite nature of the molecules of gaseous elements. For carburetted hydrogen, Avogadro restated his 1814 position that the experimental evidence pointed to a carbon: hydrogen volume ratio of 1:4.[48] Once again, he opposed Gay-Lussac who thought that it contained half as much hydrogen. In Gay-Lussac's opinion, carburetted hydrogen, too, would represent an exception to the rule of volume doubling in relation to carbon, as cyanogen gas does. In 1815 Gay-Lussac had recognized cyanogen as formed by two volumes of carbon and one of nitrogen, but Avogadro thought it was composed of equal volumes of these two elements.[49]

Although Gay-Lussac and Thenard before 1820 had extensively investigated silicon and fluorine, Avogadro did not discuss these researches. Instead he drew from the few other analytical findings available a number of rather accurate conclusions.[50] Avogadro's logic here is worthy of note. In the 1814 'Mémoire sur les masses relatives', he had assumed by analogy with other oxides that silica was formed by one volume of silicon combined with two of oxygen. Now, by considering that for each volume of silicon two volumes of fluosilicic acid (silicon fluoride) are formed and that one volume of fluosilicic acid requires one volume of oxygen in order to precipitate silica, he computed the weight of silicon (31.57) relative to hydrogen. As for fluorine, Avogadro assumed that four volumes of this element combined with one of silicon to give fluosilicic acid. Since he knew the weight of silicon, Avogadro could easily calculate the elemental weight of fluorine which he set at 18 (relative to hydrogen).[51]

Already in 1814, Avogadro had discussed the alkaline and alkaline earth metals. He returned again to them in the 'Nouvelles considérations', updating some of his figures and comparing his conclusions with those of other chemists. Since the evaluation of elementary weights was based mainly upon the composition of their oxides, and since Avogadro agreed with Berzelius on the compositions of sodium, calcium, strontium, and barium oxides, their figures on the atomic weights of these elements were quite close.

Avogadro, however, took a different position when dealing with potassium. Here the experimental evidence[52] seemed to indicate that potassium dioxide[53] contained twice as much oxygen as did sodium dioxide. This led him to infer (as Davy had also done) that in potassium oxide there is only one molecule of oxygen and not two as suggested by Berzelius, and accordingly, that the weight of potassium was 78 (relative to hydrogen) or half that calculated by Berzelius.[54]

In 1814, on the basis of Fourcroy's analytical measurements of its chlorides, Avogadro had proposed two slightly different atomic weights for mercury, the average of which yields exactly 400 (relative to hydrogen) for its weight. Six years later, in the present essay, he fully agreed with Berzelius not only on the elementary weight of mercury (405 relative to hydrogen), but also in his interpretation of the composition of mercury oxides and chlorides.

While reviewing this subject in a long and quite significant footnote,[55] Avogadro recalled that according to the specific heat determined by Dulong and Petit, the weight of mercury's elementary mass should be around 202, or half that calculated by Berzelius and himself. While he conceded that in several cases (e.g., sulfur) the results of Dulong and Petit coincided with his own, this did not mean that the law proposed by the two French chemists in 1819 should apply to all elements. In fact, it may be 'approximativement' true only when there is no difference between the structure of the molecule of an element in the solid, liquid, or gaseous state. As noted above, Avogadro did not unconditionally accept Dulong–Petit's Law of specific heats. Yet he adopted it when it appeared to support his evaluations of the elementary weight of some metals such as, for example, silver.[56]

Avogadro's discussion of iron oxides revealed how confusing the interpretation of chemical combinations had become by 1820. Berzelius had postulated the existence of a black and red oxide (also called oxides at 'minimum and maximum') identified by him as Fe + 2O and Fe + 3O, respectively. Then T. Thomson suggested Fe + O for the lower oxide and, therefore, an atomic weight for iron corresponding to one-half that calculated by Berzelius.[57] In the meantime, his extensive research on ferrocyanic acid and its salts had induced Robert Porret to propose an elementary weight for iron which was still smaller than that determined by Berzelius. Avogadro, after criticizing Thomson's interpretation, seemed to favor the same composition for the iron oxides that had been proposed by Berzelius.[58]

Zinc had not been reviewed by Avogadro in his 1814 essay. By 1821, he concluded that subsequent analysis had indicated the existence of only one oxide that was similar to the higher copper oxide and, as such, was formed by two molecules of oxygen for one of metal. This led Avogadro to calculate an atomic weight for zinc of 129, slightly smaller than that observed for copper.[59] Since the densities of these two metals were comparable, it was difficult to explain why the electropositive nature of zinc could be so much higher than that of copper or even of tin.[60] This characteristic, Avogadro suggested, probably resulted from 'quelque circonstance particulière et inconnue dans la constitution' of zinc. The other alternative, that of doubling the atomic weight of zinc by considering the presence of four molecules of oxygen in its oxide, was implausible for chemical reasons. Besides, as Avogadro pointed out in a footnote, this would yield a figure even less compatible with the Dulong–Petit relationship between atomic weights and specific heats.

In reviewing arsenic and antimony compounds, Avogadro followed Berzelius' results quite closely. Yet he took a very original position by assuming that only half as many oxygen molecules were present in these compounds as Berzelius had held. Avogadro argued that he could now reconcile the composition of arsenic and antimonic acids with that of nitric acid. For nitric acid[61] he had suggested the presence of two-and-a-half molecules of oxygen for each nitrogen, and he now held that the same ratio should be applied to the highest oxides of arsenic and antimony.[62] Consequently, the weights of both elements must be halved. He set them at 75 and 129, respectively.[63]

Avogadro devoted almost two-thirds of the 'Nouvelles considérations', or more than 120 pages, to this very detailed review of the most reliable chemical research conducted since 1814, the year of his previous chemical essay. What conclusions can be drawn from this extensive study of the experimental chemistry of the time? What was Avogadro's purpose in writing this long essay? He clearly indicated that he wanted to determine not only 'the most likely' mass of the molecules of several substances, but also to confirm 'les lois que j'avais établi dans mes mémoires précédents'.[64] Avogadro was quite confident that such 'laws' could be extended to all binary chemical combinations, not only to those in gaseous form. As he had already hinted in the 1814 'Mémoire sur les masses relatives' and also again in the 1821 'Mémoire sur le manière', Avogadro intended to prove that chemistry – both organic and inorganic – could be quantified through nearly identical formulas. In his discussion of iron, for example, he proposed a simple formula which could be used to determine the ratios of iron oxides when combining with each other to give the so-called 'intermediate oxides'.[65] How successful was he in his effort, after completing the 'Nouvelles considérations'?

Of the thirty-one elements discussed here most had already been examined in his earlier chemical essays, the only exceptions being zinc and molybdenum.[66] Sixteen elements known at the time, among them nickel, bismuth, cobalt, and chrome, were not mentioned at all. Avogadro did not offer any explanation for the omissions in the 1821 essay.

Table 4 presents the atomic weights determined by various chemical investigators during the 1810s. They all agree nearly exactly on the weights for the gaseous elements. For boron, calcium, magnesium, barium, strontium, tungsten, tin, and such heavy metals as gold, platinum, and lead Berzelius' findings are closest to the current values. For phosphorus, zinc, and molybdenum, Avogadro did not calculate his own figures, instead, he simply adopted Berzelius' results. Avogadro's elementary weights nearly coincided with those of Berzelius for mercury, iron, and copper, and were more accurate for potassium, silver, manganese, fluorine, silicon, antimony, and arsenic.

Although he often had the same correct experimental figures at his disposal, Avogadro formulated his conclusions independently of those of Berzelius, Davy, and other reputed chemists. When dealing with metallic elements,

TABLE 4. Atomic Weights

	Avogadro 1814	*Avogadro 1821*	*T. Thomson* Ann. Phil., **4** *(1814) 11 ff.*	*T. Thomson* Ann. Phil., **12** *(1818) 84 ff.*	*Berzelius 1819*
Hydrogen	Not revd.[b]	1	0.132	0.125	1
Oxygen	Not revd.[b]	16.026	1.000	1.00	16.026
Nitrogen	Not revd.[b]	14.049	1.803	1.750	14.19
Chlorine	33.61	36.124	4.498	4.5	35.49
Carbon	Not revd.[b]	12.082	–	–	12.052
Sulfur	30.65	32.603	–	–	32.17
Phosphorus	20.00	31.38	2.618	1.500[a]	31.38
Fluorine	10.00	18.00	–	–	12.00
Silicon	33	31.57	–	–	47.42
Boron	60.00	14.73	0.738	0.875	11.14
Potassium	75.78	78.74	–	–	156.8
Sodium	90.00	90.00	–	–	93.08
Calcium	77.3	77.3	–	–	82.00
Magnesium	47.00	101.3	1.577	1.50	51.00
Barium	261	261	–	–	274
Strontium	201	201	–	–	175
Aluminum	34.28	36.2	2.136	2.125	54.84
Mercury	370	400	–	–	405
Silver	206	216	13.71	13.75	432.48
Gold	374	374	–	–	398
Platinum		389	–	–	194.45
Lead	391	391	–	–	414
Iron	106.5	106.5	7.14	3.50	108.5
Copper	123	123	–	–	126.62
Manganese	107.6	107.6	7.115	3.50	114
Tungsten		386	–	–	193.2
Tin	222	235	–	–	235
Zinc		129	4.09	4.125	129
Molybdenum		95.5	–	–	95.5
Antimony	162	129	–	–	258
Arsenic	89	75	6.00	4.75	150
Iodine	–	–	11.160	15.62	–
Nickel	–	–	7.30	3.37	–
Cobalt	–	–	7.32	3.62	–

[a] On the basis of hydrogen phosphide.

[b] Not reviewed since the 1811 'Essai d'une manière'.

Dulong–Petit Ann. Chim. **10** (*1819*) *395*	*Current Value*	
–	1.008	Hydrogen
–	16.000	Oxygen
–	14.008	Nitrogen
–	35.457	Chlorine
–	12.01	Carbon
32.17	32.06	Sulfur
–	30.98	Phosphorus
–	19.00	Fluorine
–	28.06	Silicon
–	10.82	Boron
–	39.09	Potassium
–	22.99	Sodium
–	40.08	Calcium
–	24.32	Magnesium
–	137.36	Barium
–	87.63	Strontium
–	26.97	Aluminum
202.4	200.61	Mercury
108	107.88	Silver
198.9	197.2	Gold
178.56	195.2	Platinum
207.20	207.21	Lead
54.24	55.85	Iron
63.3	63.54	Copper
–	54.93	Manganese
–	183.92	Tungsten
117.6	118.7	Tin
4.03 X 16	65.38	Zinc
–	95.95	Molybdenum
–	121.76	Antimony
–	74.91	Arsenic
–	126.9	Iodine
–	58.71	Nickel
–	58.93	Cobalt

Elements are listed according to the order followed by Avogadro in the 'Nouvelles considérations'. Berzelius' figures are from the 'Théorie des proportions chimiques' (1819).

Avogadro relied mainly on chemical analogies, and on some physical considerations he thought significant. Strictly adhering to his central ideas, Avogadro made great efforts to correlate the elementary weights of both metals and nonmetals and their densities. Whenever this attempt failed, he explained the apparent inconsistency in terms of oxygenicity, the concept he had introduced in 1809.

Very little has been written on the 'Nouvelles considérations' since its publication. Even in Italy the most exhaustive of Avogadro's chemical essays remained essentially unnoticed until Icilio Guareschi reprinted and commented on it ninety years later.[67] It may be interesting to raise the question of whether or not the same fate would have befallen its author had the content of the 'Nouvelles considérations' been widely known. Did Avogadro express his ideas more clearly and consistently in the 1821 article than in his previous two dealing with the same matter?

In the introductory section of the 'Nouvelles considérations', Avogadro certainly expressed the statement of his gas hypothesis with no ambiguities.[68] He stressed once again the immediate and relevant conclusions which could be drawn from his generalization: a gas density indicated the mass of its integral molecule, and the relation among combining volumes represented the number of integral molecules forming the compound. It would be generous, however, to conclude that the 'Nouvelles considérations' helped dispel the confusion in terminology prevailing at the time among physical scientists. In this essay Avogadro defined 'integral' molecule as that body formed by 'partial' molecules and characterized by a much tighter structure and greater mutual attraction than was present in 'total' molecules. While the latter term did not appear in his earlier writings, in the 1811 'Essai d'une manière' he had used 'integral' only in reference to molecules of compounds as distinguished from 'constituent molecule' characterizing the molecules of elements.

Rather puzzling is the fact that, where referring to other authors, Avogadro used the unqualified terms 'molecules' and 'atoms' quite indifferently. The most striking example of this in the 'Nouvelles considérations' is when he discussed Dulong and Petit's 1819 law of specific heats. He described it as 'the product of the specific heats of bodies . . . by the mass of their *molecule* is the same for all those bodies . . .'.[69] Therefore, '. . . the specific heat of each atom should be the same for all . . .', and '. . . one could obtain the *size of an atom* (italics mine) of a substance by dividing . . .'. Dulong and Petit, on the other hand, had stated, '. . . the atoms of all substances have exactly the same capacity for heat . . .' and '. . . the products which represent the heat capacities of different kinds of atoms approximate so nearly . . .'. In

1819 they never used molecule in Avogadro's meaning. Later on, in the 'Nouvelles considérations', Avogadro recalled Dulong's work on phosphorous anhydrides and wrote '. . . this makes the composition of those two acids, by considering *atoms as gaseous volumes* (italics mine), analogous to that of pernitrous and nitric acids . . . .'[70] Are these atoms, without further qualification, 'simple atoms', what Avogadro called 'partial molecules'? And, if so, why did he not use the latter term, rather than the generic 'molecule'?

It had not escaped Avogadro's attention that Berzelius was 'inclined to recognize' that under the same temperature and pressure equal volumes of gaseous elements contain equal numbers of *atoms*.[71] By 'atoms' Berzelius meant elementary, indivisible particles, as distinguished from what he called 'compound atoms'.[72] This distinction may have failed Avogadro when he read Berzelius' 1819 'Essai sur la théorie des proportions'. In the first section of the 'Nouvelles considérations', Avogadro wrote, for instance: 'M. Berzelius a bien vu aussi la necessité de ramener la théorie des atomes ou des molécules à celle des volumes . . . .'[73] One is tempted to conclude that by then he defined as 'integral molecules' what Berzelius called atoms. However, these integral molecules – as seen from the content of the 'Nouvelles considérations' – were all polyatomic, regardless of the physical state of the elements to which they referred. Being polyatomic, they could divide when entering into chemical combination, something that, needless to say, Berzelius' atoms could not do. Hence, there arose the substantial misapprehension between Avogadro and Berzelius, Dulong and other chemists of his time when referring to elementary particles. The Swedish chemist's harsh criticism of Dumas a few years later would only bring this situation more into the open.[74]

Organic Compounds and Definite Proportions (1821)

In several ways the 'Mémoire sur la manière de ramener les composés organiques aux lois ordinaires des proportions determinées', read on 9 December 1821,[75] by Avogadro at the Royal Academy of Sciences of Turin, may be considered a continuation of the 'Nouvelles considérations', the extensive essay published earlier in the same year. In this second major chemical work, Avogadro was walking on shifting ground, to say the least. Here, once again, one sees the scarcely known Avogadro presenting an extremely elaborate and detailed discussion of the composition of substances belonging to a domain of chemistry in which he had little training and even less first-hand experience.

At the end of the 1810s, organic chemistry was virgin territory in which everyone was exploring and discovering something new. Through still unreliable analytical procedures, identification of new natural substances was boldly attempted, and whenever feasible their empirical formulas were established. Many of these investigators were men of the old chemical school, mostly born in the last quarter of the eighteenth century: Prout, Berzelius, Meinecke, Serturner, Theodore De Saussure, Berard, Gay-Lussac, Thenard, Houton, Robiquet, Pelletier, and Chevreul. In 1821 when Avogadro was drafting his 'Mémoire sur la manière', the founders of the organic chemical revolution – Dumas, Laurent, Gerhardt, Wöhler, and Liebig – had not yet appeared on stage. Wöhler's synthesis of urea was eight years away, Liebig was still working as an apprentice to an apothecary in Darmstadt, and Dumas was in a similar position in Geneva with Prevost.

Thanks to the contributions of Gay-Lussac, Thenard, and Berzelius, qualitative and quantitative identification of simple organic compounds had vastly expanded, but appeared inadequate to deal with the ever-growing complexity of the substances which came under investigation. Not surprisingly, these chemists often reached incorrect, misleading, and, unlike those observed in inorganic compounds, inconsistent conclusions.

In addition to the experimental obstacles and the perplexity which arose when findings indicated similar compositions for quite different substances, a very serious philosophical substratum developed in the 1820s, especially outside France. This substratum claimed that an active principle, or a vital force, foreign to sensible matter could be present in organic substances and would make impossible their reproduction in the laboratory. Such a notion, shared by Berzelius and other chemists despite diminishing experimental evidence for vital forces, remained prevalent throughout the first half of the century and had a considerable effect on the approach adopted by those studying organic products.[76]

Less important philosophically, but certainly more relevant to Avogadro's 1821 'Mémoire sur la manière', was the position taken by Berzelius on the way chemical elements gather to form organic compounds. Here, too, Berzelius indicated that on first sight organic combinations seem to take place without limitations in their proportions. 'There is almost an infinite number of different ways that these three or four elements (carbon, hydrogen, oxygen, and nitrogen) may combine', he wrote in 1814.[77] And even more elements are present (even four and five had been detected) in organic compounds, and the 'laws of chemical proportions' governing such compounds became more complex. Berzelius, in fact, pointed out that between the laws concerning the organic compounds and those for inorganic compounds there was as much difference as between the four basic arithmetic operations and algebraic calculations.[78] Yet, this complexity did not rule out the possibility that some 'lois des proportions chimiques' would apply to organic substances. For Berzelius, these substances should be considered as ternary or quaternary oxides in which no element would be present with only one atom (or as unity) and for which, therefore, the number of combinations would be 'almost infinite'.[79] His feelings are understandable, for a constant stream of new analytical findings had revealed to him how similar in their composition were organic compounds which often appeared so different in their properties and origin.

Avogadro in the 'Mémoire sur la manière' rose to challenge these concepts in the most straightforward way. Basing his argument on mathematical considerations, Avogadro sought to demonstrate that the chemical composition of organic substances could be rearranged so that it would obey 'aux lois des composés inorganiques' and consequently that in this respect organic and inorganic compounds were alike.[80] It is only fair to say that while the procedure he followed was relatively easy to grasp in the case of binary compounds, it became most elaborate, involved and, one might add, even bold when applied to the more complex ternary and quaternary organic compounds.

Earlier in the 'Nouvelles considérations', while recalling that Berzelius had distinguished inorganic and organic compounds, Avogadro clearly stated that to him it seemed 'peu probable' that organic substances could be completely different from inorganic ones. Their difference, he argued, may consist in 'certain proportions which do not take place in ordinary (inorganic) combinations, and that only *the operation of vegetation and animalization* (italics mine) can form, though still remaining subject to the same general laws'.[81] What then should be established was the 'way and order' of these organic combinations. When dealing with ternary organic compounds, for instance, should one assume that compound *abc* is formed by the combination of *ac* + *bc* or, alternatively, be *ab* + *c* or *ac* + *b*?

Avogadro replied that, although the method of combination is unknown and although it is not apparent that one element behaves differently from any other, still it is 'natural' to assume that organic combinations take place in a general and simple way. Thus, carbon, or hydrogen, joins the element, e.g., oxygen, which is taken as a 'basis' to form a binary compound. Similarly, ternary and quaternary compounds result from the combination of one further element to a pre-existing binary and ternary compound.[82] Avogadro emphasized here that the choice of this pivotal, or basic, element was arbitrary. However, by analogy with his own views concerning inorganic binary compounds, he favored as basis element the one entering into combination in the least amount.[83]

Of over one hundred organic substances which had been chemically analyzed by the 1820s, Avogadro decided to examine in the 'Mémoire sur la manière' less than fifteen, some of animal, some of vegetable origin, some binary, others ternary or quaternary. He did not explain the reasons behind his choices. Quite likely, he avoided some of the other substances because they were either too complex to be simplified through his mathematical expressions, or the results of their analyses he still considered too unreliable. From Berzelius, Theodore De Saussure,[84] Gay-Lussac, Thenard, Houton, and Chevreul, among others, Avogadro gathered the percentage composition, the densities, and the volume composition of most of these substances. He intended to use these data to relate the organic compounds to the laws of multiple proportions, and thereby to refute further the notion of a chasm between organic and inorganic substances. Once again, Avogadro's program directly challenged that of the leading chemist of the day – Berzelius.

It is noteworthy that in 'Mémoire sur la manière' Avogadro constantly equated molecules and volumes to the extent that in his mathematical calculations there was no limit to his use of fractionary molecules.[85] Since, as noted earlier, the terms 'molecule' and 'atom' often were interchanged at the time, Avogadro's very liberal use of fractions of molecules certainly did not improve the credibility or the degree of acceptance of his work among fellow physical scientists.

In 'Mémoire sur la manière', the first group of substances Avogadro discussed was that of binary compounds formed by the combination of carbon with hydrogen. To illustrate Avogadro's approach, turpentine oil will be examined. Analyses[86] of this volatile oil indicated a ratio by volume between carbon (assumed in gaseous from) and hydrogen of 1 to 1.6, and a vapor density of 5.013 (relative to air). From those figures and, of course, by knowing the density of carbon gas and hydrogen, Avogadro was able to compute the volumes of carbon and hydrogen which, he thought, combined together to form turpentine oil. The results of his calculations indicated that turpentine oil would be formed by twenty-one volumes of carbon as gas and thirty-four volumes of hydrogen. In this way, the C:H ratio would be very close to 1:1.6, as required by the elemental analysis. In order, however, to match the value for the experimental density, Avogadro had to assume that in the course of the combination of carbon with hydrogen, the volume would quadruple; the 'total molecule' of turpentine would then divide by four.[87]

From this, Avogadro developed a generalized mathematical approach for organic compounds. In binary compounds in which proportion of elements by volume may be indicated with 1 volume and $M + p/q$ volumes, one may assume the combination of $q - p$ volumes of the compound in proportion $1 : m$, and p volume of a compound in proportion $1 : (m + 1)$. This works only when $q - p$ is multiple of p or vice versa, or as long as p/q reduces to $q - p$ or $p = 1$. Applying this to the composition of turpentine oil, where the analysis gives $C : H = 21 : 34 = 1 : 1\frac{13}{21}$ and thus $p = 13, q = 21, q - p = 8$. And if a = carbon and b = hydrogen, then turpentine oil is $1a : 1\frac{13}{21}b$ which may also be written:

$$8(1a + 1b) \qquad 13(1a + 2b) \tag{I}$$

meaning 8 molecules a + 8 molecules b, 13 molecules a + 26 molecules b. Since 13 is not a multiple of 8, (I) should be further simplified to $8 : 13 = 1 : 1\frac{5}{8}$; then it is possible to write:

$$3[(1a + 1b) + (1a + 2b)] \qquad 5[(1a + 1b) + 2\,(1a + 2b)] \tag{II}$$

Here again, 5 is not a multiple of 3 and further simplification indicates that the ratio $3 : 5$ is as $1 : 1\frac{2}{3}$, and 5 is not a multiple of 2. Thus it is necessary to further reduce the composition in the following way:

$$\begin{matrix} 1[1\{(a + b) + (a + 2b)\} + 1\{(a + b) + 2(a + 2b)\}] \\ 2[1\{(a + b) + (a + 2b)\} + 2\{(a + b) + 2(a + 2b)\}] \end{matrix} \tag{III}$$

Now by adding all the as and bs one obtains $21a\ 34b$ as the chemical analysis indicated. Through this procedure of simplification, every binary organic compound, regardless of its complexity, can be explained in terms of the law of multiple proportions!

For ternary compounds (containing carbon, hydrogen, and oxygen), Avogadro set aside those having two of their components multiples of a third element from those displaying a more complex composition. Among the former he listed ethyl alcohol and ethyl ether. In 1813 De Saussure[88] and, almost simultaneously, Gay-Lussac[89] had reached quite similar conclusions on the composition of these substances. One volume of ethyl alcohol passing through a red-hot porcelain tube formed one volume of water vapor and one volume of carburetted hydrogen or olefiant gas. As for ethyl ether,[90] Gay-Lussac thought that it was the result of the combination of one volume of water vapor and two volumes of olefiant gas, while De Saussure considered it formed by one hundred parts of olefiant gas and twenty-five parts of water vapor by weight.

On the basis of its vapor density, Avogadro assigned to ethyl alcohol a composition by volume of 1C 3H ½O or, 'autrement,' 2C 6H 1O.[91] He objected to the assumptions made earlier by De Saussure and Gay-Lussac concerning the way such a compound is formed from water and olefiant gas. First, he recalled that the German chemist, Johann Meinecke, had proposed that ethyl alcohol was the result of the union of three volumes of olefiant gas with one volume of carbon dioxide.[92] Next, he noted that any combination of carbon, hydrogen, and oxygen could form ethyl alcohol, as long as it agreed with the percent composition of the alcohol. In addition, Avogadro argued that ethyl alcohol had properties quite different from those of olefiant gas, while hydrates generally were similar to their corresponding anhydrous compounds. Finally, he recalled the impossibility of obtaining alcohol from the direct combination of steam and olefiant gas. In harmony with his general views earlier stated, he concluded that ethyl alcohol was the product of the union of two elements – carbon and hydrogen – with the one, oxygen, the basis element which is present in the least amount by volume.

For ethyl ether, Avogadro agreed with Gay-Lussac that this substance would form by the thermal decomposition of one volume of steam and two of olefiant gas, thus resulting in a compound formed by two volumes of carbon, five of hydrogen, and one-half oxygen.[93] It is worthy of note to mention that by referring to the same arguments earlier used for alcohol, Avogadro concluded that ether could not be considered the product of combination of water and olefiant gas. The transformation of alcohol into ether then occurred because the former loses half of its oxygen and, accordingly, both carbon and hydrogen double their volumes in relation to oxygen; this seems to indicate that Avogadro recognized the formation of ether from alcohol through the loss of water.

The next section of the 'Mémoire sur la manière' is devoted to the ternary compounds, whose complexity makes more difficult their adaptation to the law of multiple proportions. In fact, from their compositions it appears that the other two elements are not always direct multiples by volume of the element entering in the least proportion. However, Avogadro claimed it was possible to arrive at a 'formule générale' which might be applied to all ternary compounds. As he had done for turpentine oil, Avogadro attempted to determine this formula algebraically,

Given a, b, and c in a ternary compound ($a = \mathrm{O}, b = \mathrm{C}, c = \mathrm{H}$), Avogadro's procedure was first to reduce separately to the multiple proportions the combination of *a* with *b* and that of *c* with *b*, and then to take as many molecules of partial compounds of *a* with *b* as necessary in order that the resulting number of molecules of *b* may be separated into the partial compounds of *b* with *c*.

He used cane sugar as a first example, whose composition Berzelius in 1815[94] had given as 12C 21H 10 O. For the same substance, Gay-Lussac had suggested 1C 2H 1O. According to Berzelius, the C:O ratio in cane sugar was such as $\frac{b}{a} = \frac{12}{10} = \frac{6}{5} = 1\frac{1}{5} : 1$.

From here, by reducing this ratio to the theory of multiple proportions, Avogadro argued the existence of two partial compounds as:

4 O 4C (I) 1 O 2C (II)

The H:C ratio was such as $\frac{c}{b} = \frac{21}{12} = \frac{7}{4} = 1\frac{3}{4} : 1$. Hence the two partial compounds 1C 1H, and 3C 6H for the H: C ratio.

The next step was to multiply the two amounts of carbon of compounds (I) and (II) by 3 + 1 in order to separate them into two parts having a 1 to 3 ratio and, finally, to allocate among them oxygen and hydrogen in the necessary proportions. Thus,

$$\text{carbon } 4 \cdot 4 = 16 \quad \begin{cases} 4\text{C} \; 4\text{ O} \; 4\text{H} \\ 12\text{C} \; 12\text{ O} \; 24\text{H} \end{cases}$$

$$\text{carbon } 2 \cdot 4 = 8 \quad \begin{cases} 2\text{C} \; 1\text{O} \; 2\text{H} \\ 6\text{C} \; 3\text{O} \; 12\text{H} \end{cases} \qquad \text{(III)}$$

Summing all the parts gives 24C 20 O 42H, which is equal to Berzelius' formula 12C 10 O 21H.

To put this in more general form, Avogadro, by assuming that in a ternary compound *abc*, $b/a = m + p/q$ and $c/b = m + p'/q'$, suggested the following system, indicating the proportions of the different substances which should be taken:

	b	*a*	*c*
substance b	$(q-p)\,m'(q'-p')$	$(q-p)(q'-p')$	$(q-p)m(q'-p')m'$
$(q-p)\,mq'$	$(q-p)\,mp'$	$(q-p)\,p'$	$(q-p)mp'\,(m'+1)$
			(IV)
substance b	$p\,(m+1)(q'-p')$	$p\,(q'-p')$	$p\,(m+1)(q'-p')m'$
$(m+1)\,q'$	$p\,(m+1)\,p'$	$p \cdot p'$	$p\,(m+1)\,p'\,(m'+1)$

For instance, in cane sugar, $a = 10$, $b = 12$, $c = 21$, and from the expressions seen for C:O and H:C one derives $m = 1$, $p = 1$, $q = 5$, $m' = 1$, $p' = 3$, $q' = 4$. By introducing these values into (IV), one would obtain the same results as indicated in (III).

The preceding calculations clearly indicate that Avogadro relied on a purely algebraic approach, and ignored all chemical considerations. In a further discussion of how to adapt the definite proportions to cane sugar he referred to one-third or one-fourth of a molecule of carbon,[95] and later on, in his review of ethyl ether, he used one-eight and one-twelfth of a molecule of oxygen. This very unrestrained use of fractional molecules emerges here more than in previous chemical essays and reflects once again the special significance that Avogadro gave to the term molecule. As previously noted, he considered molecules as aggregates of smaller parts, although he seemed very prudent with regard to their actual number.

The final sections of the memoir were devoted to the quaternary organic compounds containing nitrogen along with hydrogen, oxygen, and carbon. For Avogadro, the conversion of these substances into a form which met the requirements of the laws of definite proportions was quite simple: add a fourth element to a ternary compound in the same way as a ternary compound resulted from the union of a binary compound with a third element. For example, if *abcd* are the smallest whole numbers through which it is possible to express the relative amounts by weight of four elements, then if a number large enough is chosen for *a*, this element will unite with a number of molecules of element *b* which, in turn, will join with an appropriate number of molecules of *c*, and so on. In this way it will be possible to find out how many molecules of each of these elements are necessary in order to obtain the required final composition.

Worthy of note are Avogadro's comments on the composition of another quaternary compound, urea. The analysis conducted by Berard[96] indicated the presence by volume of 1C 2N 6H 1O.[97] According to the interpretation of Gay-Lussac, this suggested the formation of urea from the union of ammonia and carbon oxide. At the same time, the analytical findings reported by Prout[98] showed for urea a composition by volume of 1C 2N 4H 1O. Avogadro pointed out that, as in Gay-Lussac's compound, nitrogen and hydrogen were in an even ratio, and whole multiples of the other two elements.[99] Hence, urea met the requirements of the laws of definite proportions.

While Guareschi in his eulogistic study[100] suggested that Avogadro was the first to use the correct empirical formula for urea, Avogadro himself recognized that only 'further experiments' could decide which of the two interpretations, Prout's or Gay-Lussac's, was correct. Guareschi, eager to accentuate Avogadro's successes wherever possible, also pointed out the priority of Avogadro in proposing the currently adopted empirical formulas for ethanol, ethyl ether, and ethyl chloride.[101] Most historians have not awarded Avogadro this priority; however, Graebe, Partington, and Cappelletti[102] did do so, probably following the lead of Guareschi.

As Graebe (who became aware of the 'Mémoire sur la manière' through its German translation in Georg Kahlbaum's *Memorien zur Geschichte der Chemie*) emphasized, the two voluminous 1821 chemical essays by Avogadro had no influence on the development of chemistry, for the simple reason that they escaped the attention of most chemists beyond the Alps.[103] Yet, as it has been said correctly, even if contemporary chemists had read the 1821 essays, Avogadro's words would have fallen on deaf ears.[104] Had the essays appeared, say, in the influential *Annales de Chimie* rather than in the little known *Memorie della Reale Accademia delle Scienze di Torino*, one may wonder what would have been the reaction to such a long series of speculations. It should be mentioned here that between 1814 and 1824 Avogadro's work had no access to foreign journals. He published in Italian periodicals and mostly in the 'house organ', the *Memorie della Reale Accademia di Torino*. Although these articles were written in French, due to their length and tone, it is doubtful that they attracted a wide readership in other European countries.

Fifteen years after Avogadro's 'Mémoire sur la manière', a paper by Auguste Laurent appeared in the *Annales de Chimie*.[105] In it, the then young French chemist suggested that atoms and molecules, when they combine to form organic compounds, follow simple laws such as those established for the gases or, in other words, that one, two, three, or four atoms, molecules, or volumes of a simple or compound body combine with one, two, three, or four atoms, molecules or volumes of another simple or compound body, a view very close to that set forth by Avogadro in 1821. When Berzelius reviewed Laurent's paper in his *Jahresbericht* of 1838, he described its content as 'hasty legislation' and added: 'I think it useless that my reports shall deal in future with such theories.'[106]

Hence, one must say that in 1821 the chances for Avogadro to find any reception for his mathematical approach to organic chemistry were slim at best, and not only because of Berzelius' antipathies. As mentioned earlier, until the late 1840s, chemists had more immediate obstacles to overcome, such as the refinement of analytical procedures, the isolation, identification, and classification of natural organic substances, and the synthesis of new ones. They were simply not inclined and were too busy to pay any attention to extravagant computations or conjectures which seemed to have little to do with their benchwork.

Nevertheless, in retrospect, the historical relevance of the views advanced by Avogadro in 'Mémoire sur la manière' cannot be overlooked. He tried to simplify the experimental data at his disposal, thereby seeking to prove that a dualism of organic/inorganic chemistry was not necessary, and that a substantial unity existed in chemistry. This science was undivisible for him. If one recalls the confusion of ideas then prevalent on the classification and chemical composition of organic compounds, then Avogadro's attempts to order organic compounds may become more understandable.

To see the unitary system earnestly defended, one had to wait a quarter of a century for Gerhardt and his *Introduction à l'étude de Chimie*, published by mere coincidence in 1848, the year of Berzelius' death. When Gerhardt wrote in this book, 'the molecule is an edifice, an unique system formed by the assembly, in a *determinate* but unknown order, of infinitely small particles called atoms . . .', he was only stating, twenty-seven years later, the ideas of Avogadro.[107] In another way, too, Gerhardt had been anticipated by Avogadro. In 1845,[108] Gerhardt suggested that the sum of the vapor densities of the elements present in an organic substance could be used to establish agreement between the composition given by the analysis and the formula assumed for that particular substance. In the 'Mémoire sur la manière' Avogadro had on several occasions used this same procedure as a corollary of his gas hypothesis to determine the composition of a number of organic compounds and in his 1811 and 1814 essays had frequently applied it to ordinary gaseous substances.[109]

One final comment must be added regarding Avogadro and Berzelius' dualistic theory. In his attempt to explain the formation of organic compounds, Avogadro avoided any discussion of the electrical properties of the elements involved. As earlier mentioned, only in the February 1821 memoir did he suggest that, when a compound contains two, three, or four elements in the same amount, then the element taken as a basis will be the most electropositive. How then could chlorine enter into and stay in combination with carbon and hydrogen, as in ethyl chloride, to cite one of the compounds discussed in the 1821 essay? This was not explained. One may wonder whether Avogadro purposely avoided this issue on which, of course, Berzelius' position was founded. In the 'Mémoire sur la manière', at any rate, the Swede's dualistic doctrine was ignored.

Between 1821 and 1830, with the exception of two minor articles,[110] Avogadro focused his studies on physical questions, where his main line of scientific pursuit lay. He probably felt that for the time being there was little that he could contribute to the development even of theoretical chemistry. This decade coincides approximately with his exile from the chair of physics at Turin, following the political events of the spring of 1821.

NOTES

1. 'Nouvelles considérations sur la théorie des proportions determinées dans les combinaisons et sur la détermination des masses des molécules des corps', *Memorie Reale Accademia Scienze Torino*, **26** (1821), 1–162, page 3 cited. (Hereafter referred to as *MRAST*.)
2. As Avogadro himself recognized later on in *Fisica dei corpi ponderabili*, vol. II, (Torino: Stamperia Reale, 1838), p. 868.
3. Since 1811, Avogadro had perceived the elemental nature of chlorine.
4. Unfortunately, Avogadro himself sometimes contributed to this confusion.
5. *MRAST*, **26** (1821), 11–12.
6. Ibid., 14, 23.
7. Ibid., 3.
8. He refers here to Dalton and his followers without naming them.
9. In support of this claim, Avogadro quoted Berzelius' assertion, presumably from his 'Essai sur la théorie des proportions chimiques et sur l'influence chimique de l'électricité', Paris (1819), that atoms and volumes are nothing more than two different ways of representing the combining elements. Berzelius had expressed this idea several years earlier in his 'On the cause of chemical proportions', *Ann. Phil.* **2** (1813), 450.
10. On 'Sur la combinaison des substances gazeuses', *Mem. Soc. Arcueil*, **2** (1809), Gay-Lussac said: 'We must admit as a truth . . . that the condensation of the molecules of two combining substances, in particular of two gases, has no immediate relation to the condensation of the volume' (trans. *Alembic Club Reprint*, no. 4). There is no further elaboration of the matter in this essay.
11. *MRAST*, **26** (1821), 15.
12. Ibid., 16.
13. As he explained in the 'Essai d'une manière' (p. 62), the actual mass of the water molecule is one-half the sum of the masses of its elements, because there is a division in half of such molecule. This applies also to other gaseous compounds, such as nitrogen oxide, nitrous oxide, carbon oxide, carbon dioxide, etc.
14. *MRAST*, **26** (1821), 22.
15. There are, of course, limitations to these rules. For instance, nitric acid, whose salts have a 2.5: 1 oxygen ratio between acid and base, and phosphoric acid, where five of oxygen combine with one phosphorous. See Berzelius, *Ann. Phys.* **53** (1816), 393.
16. See *Ann. Chim.* (2) **6**, (1817), 321. 'Mémoire sur les combinaisons du soufre avec les alcalis'.
17. *MRAST*, **26** (1821), 41. This was also criticized by Thomson in *Ann. Phil.* **7** (1818), 12.
18. *Ann. Phil.* **6** (1815), 321.

19. *Jahresbericht*, **2** (1823), 39.
20. *Ann. Chim.* (2) **10**, (1819), 395. 'Recherches sur quelques points importants de la théorie de la chaleur.'
21. The specific heats are also affected, Avogadro added, by the different heats of fusion and heats of expansion displayed by some metals. Generally neglected by historians is Dalton's odd criticism of Dulong–Petit's generalization. He observed that since they had shown the different capacity for heat of the solid elements, it would be *impossible* to obtain a constant product of their atomic weights, times their specific heats. See Dalton, *A New System of Chemical Philosophy*, Vol. II (London: Wilson, 1827), p. 293.
22. Regnault arrived at very similar conclusions nineteen years later. See *Ann. Chim.* **73** (1840), 5.
23. *Ann. Chim.* **14** (1820), 172. The original communication was submitted by Mitscherlich to the Berlin Academy in December, 1819.
24. It should be recalled that Ampère termed as 'particules' the constituent and integral molecules of Avogadro.
25. French mineralogist, author of the *Traité Elémentaire de Mineralogie*, 1824.
26. While Avogadro was writing this essay, Mitscherlich, working in Berzelius' laboratory, was gathering the information that would enable him to publish his fundamental paper in 1822 ('Om Forhallandet einellan Chemiska . . .') on the crystalline systems of alkali phosphates and arsenates. In 1820 Avogadro was not aware that a more restricted interpretation of the isomorphism laws would be made by Mitscherlich in that essay.
27. *MRAST*, **26** (1821), 44.
28. *Jahresbericht*, **1** (1822), 67. These were actually Berzelius' own words, since he edited the *Jahresberichte*.
29. See *Ann. Phil.* **4** (1814), 84.
30. These are the data of Berzelius and Dulong from *Ann. Chim.* **15** (1820), 386, whose density of hydrogen relative to air was at 0.0688.
31. In both the 1811 and 1814 papers he had adopted hydrogen as term of reference.
32. Thus obtained: [16.026 (oxygen) + 2 (hydrogen)] / 2 = 9.013.
33. This is the same figure quoted by Avogadro in the 1811 'Essai d'une manière'.
34. For Thenard it was HO_2, or hydrogen bioxide (O = 8). See *Ann. Chim.* **9** (1818), 441, 'Nouvelles recherches sur l'eau oxigenée'.
35. It is interesting to compare these with the compositions given only a few years earlier by Dalton, quoted from *Ann. Phil.* **9** (March 1817), 188:

    One vol. of nitrous oxide from 1 vol. nitrogen and 0.62 vol. oxygen.
    Two vol. of nitrous gas from 1 vol. nitrogen and 1.24 vol. oxygen.
    Subnitrous acid (volume not defined) from 1 vol. nitrogen and 1.86 vol. oxygen.
    One vol. nitrous acid from 1 vol. nitrogen and 2.48 vol. oxygen.
    Nitric acid (volume not defined) from 1 vol. nitrogen and 3.10 vol. oxygen.

    In the same paper Dalton, after recalling that 'the greatest difference amongst us (he means among himself, Gay-Lussac, and the "volumists") is in regard to the absolute weights of the elements Azote and Oxygen', lists the series of nitrogen oxides according to the findings of Gay-Lussac.
36. These findings were originally submitted by Gay-Lussac to the Academy of Paris in 1816, appeared in the *Annales de Chimie* of that year, and, quite interestingly, when published in the *Ann. Phil.* **8**, (1816), 71, T. Thomson gave their 'translation' in terms of the Daltonian theory:

	Atoms of nitrogen	Atoms of oxygen
Nitrous oxide	one	one

Nitrous gas	one	two
Nitrous acid	one	three
Nitrous vapor	one	four
Nitric acid	one	five

When Dalton published, *Ann. Phil.* **9** (1817), 186, his own conclusions concerning nitrogen oxides, he reported that by volume they contained in general 24% more oxygen than the analyses of Gay-Lussac indicated. He also disagreed with Gay-Lussac in the definition of some of these oxides (for instance, by calling nitrous acid what Gay-Lussac, Avogadro, and others called nitrous vapor), thus further increasing the confusion on an already entangled situation. (Davy, too, sided with Gay-Lussac on this issue.)

37. *Ann. Phil.* **16** (1820), 170.
38. 'Über die Bestimmung der relativen Anzahl', *Ann. Phys.* **7** (1826), 71, 397; **8** (1827), 177.
39. Accordingly, for Berzelius the nitrogen oxides' series was the following: 2N + O, N + O, 2N + 3O, 2N + 5O.
40. Although there is no specific reference for this, Avogadro probably, in quoting Biot and Arago, was referring to their 1806 essay in *Mêmoires de l'Institut*, I, no. 7, p. 302.
41. The modern value is 35.457.
42. *J. Phys.* **73** (1811), 60.
43. *MRAST*, **26** (1821), 63.
44. Berthollet had accepted only in 1816 the elemental nature of chlorine.
45. The experimental figure (Berzelius') was then 12.02.
46. *Ann. Phil.* **3** (1814), 51. Also *Ann. Chim.* **92** (1814), footnote pp. 142–43, where the density of carbon assumed in gaseous form is calculated from carbon oxide as Avogadro had done in 1811.
47. From the 'Mémoire sur les combinaisons', *Alembic Club Reprint*, No. 4, trans., p. 12.
48. At the end of 1814, Davy analyzed this gas from samples collected in northern Italy and reached the same conclusion, giving the composition by weight. *Phil. Trans.* **104** (1814), 557, 'Some experiments on the combustion of the diamonds'.
49. *Ann. Chim.* **95** (1815), 136, 'Recherches sur l'acide prussique'.
50. Avogadro did not quote his sources, but he had access through the *Journal Phys.* **79** (1814), 259, to French abstracts of *Phil. Trans.* **103** (1813) and **104** (1814) for Humphry Davy. John Davy's work on fluorine compounds was also reported in the 1814 abstract published in the *Journal de Physique*. Analysis indicated that 61.4 parts of silica were formed by treating 100 parts of fluosilicic acid (silicon fluoride) with ammonia. By basing his calculations on combination weights and gaseous densities, whenever available, he calculated the composition of silica first and from there the relative density of silicon as a gas. Then from fluosilicic acid he arrived at the relative density of fluorine.
51. A figure which he found in pretty close agreement with that obtained through Berzelius' analytical results on calcium fluoride. It is interesting to note that for the Swedish chemist hydrofluoric acid at that time was approximately composed of $\frac{3}{4}$ oxygen and $\frac{1}{4}$ fluorine. *Ann. Chim.* (2) **11** (1819), 120, 'Suite des expériences pour déterminer la composition des plusieurs combinaisons inorganiques'.
52. Berzelius, *Ann. Phil.* **3** (1814), 360.
53. Called potassium peroxide (Po + 6O) by Berzelius; sodium peroxide he indicated as So + 3O. For reference, see preceding note.
54. Thomson in his *System of Chemistry* (1817) had proposed K_2O and K_2O_3.

55. *MRAST,* **26** (1821), 128.
56. This was in contrast with other physical scientists of the time. See Berzelius, *Jahresbericht,* **1** (1822), 19, and R. Fox, *Brit. J. Hist. Sci.* **4** (1968), 13.
57. Which was 108.5. Avogadro did not refer specifically to Thomson here. In *Ann. Phil.* **5** (1815), 10, Thomson indicated two iron oxides having the same composition as the oxides of nickel and cobalt – Fe + 2O; Fe + 3O – like those adopted by Berzelius.
58. This meant in Berzelius' symbols 2Fe + 6O (black) and 2Fe + 4O (red).
59. In 1814 Berzelius had reported an atomic weight of 122 for zinc, in *Ann. Phil.* **3** (1814), 358.
60. Tin has almost the same specific gravity and almost twice the atomic weight of zinc.
61. Now called nitric anhydride.
62. It should be recalled that Berzelius' formulas at the time were As + 5O and Sb + 5O.
63. These weights are in fairly close agreement with the modern values.
64. *MRAST,* **26** (1821), 45.
65. In current terms, Fe_3O_4, or magnetite (now a mixture of FeO and Fe_2O_3), was then considered an 'intermediate' oxide between the red and the brown iron oxide.
66. For molybdenum Avogadro said that he had nothing to add to the conclusions reached by Berzelius.
67. See I. Guareschi, *Opere scelte* (Torino, 1911).
68. He not only thought that molecules of elements could divide, but also hypothesized that the intermediate integral molecules could also split into two, four parts, etc. The resulting number of molecules in the final volumes would remain the same.
69. *MRAST,* **26** (1841), 41.
70. Ibid., 93.
71. Ibid., 12.
72. As earlier discussed, Berzelius later used the expression 'group of atoms' when referring to certain gaseous elements, but he never adopted 'molecule' for this meaning.
73. *MRAST,* **26** (1821), 10.
74. Berzelius' reaction to Dumas' 1826 article has been examined in a previous section of this work (pp. 163–4).
75. *MRAST,* **26** (1821), 440–506. At this time he is listed in the notes of the Royal Academy of Turin as *Professor Emerito*, a title that he maintained after his reinstatement at the University of Turin.
76. Liebig in his *Letters on Chemistry* (1844) supported the view that, besides heat, chemical affinity, forces of cohesion, and crystallization, there is a fourth intangible present in living bodies that he called *Lebenskraft*.
77. *Ann. Chim.* **92** (1814), 141, 'Essai sur les proportions déterminées dans les quelles se trouvent . . . '.
78. Ibid., 147.
79. Ibid., 149.
80. *MRAST,* **26** (1821), 440.
81. 'Nouvelles considérations', *MRAST,* **26** (1821), 37.
82. This may be explained schematically by $a + b = ab$, then $ab + c = abc$, then $abc + d = abcd$, and so on.
83. The purpose of this was to avoid odd ratios in the compound.
84. Theodore De Saussure (1767–1845) was the author of the *Recherches chimiques sur la végétation* (1804) in which one of the first investigations was conducted on the conversion in plants of CO_2 into carbohydrates. He was Genevian, as was his father, Horace, the geologist. Most of his 1821 essay was published in *Bibliothèque Universelle*.

85. On the degree of submolecularity of molecules, according to Avogadro, see B. W. Mundy, *Chymia*, **12** (1967), 151. There is no evidence that Avogadro clearly defined the number of atoms present in his molecules.
86. Conducted by Jacques Houton De Labillardière and Theodore De Saussure gave C = 87.78%, H = 11.64%, and N = 0.56%. The density was determined by Gay-Lussac from data collected between 1817 and 1820. For them turpentine was formed by 2 volumes C plus 4 volumes olefiant gas (1C 2H) and thus was equal to one volume of 6C 8H, corresponding to a density of 6.03 (relative to air).
87. To visualize all this in current terms, one should write: $21C + 34H \rightarrow 4(C_{21}H_{34})$ and $(C_{21}H_{34})/4 = C_{5\frac{1}{4}}H_{8\frac{1}{2}}$. In fact, for the calculation of density (relative to air): 5¼ · 0.843 (density of carbon assumed in gaseous state) + 8½ · 0.073 (density of hydrogen relative to air) = 5.00 vs. 5.01 (experimental figure). This means that we should assign C_4 and H_4 to have the equation balanced. It would help to recall here that in the case of carbon dioxide (and by assuming again tetratomic molecules) Avogadro had explained its formation in the following terms: $C_4 + 2O_4 \rightarrow 2(C_2O_4)$ duplication of volume in formation of 'total molecule' followed by $(C_2O_4)/2 = CO_2$ division of 'total molecule' by two.
88. *Ann. Chim.* **89** (1814), 273.
89. Ibid., **95** (1815), 311.
90. Or sulfuric ether as it was commonly called then, quite improperly Avogadro added.
91. *MRAST*, **26** (1821), 450. For the vapor of ethyl alcohol, density is 1.613 (relative to air). For Avogadro, by defining ethyl alcohol as 1C 3H 1/2O, then the calculated density (relative to air) would be: $0.843 + \frac{1}{2}(1.10) + 3.\ 0.073 = 1.613$, as the experimental figure (where 0.073 is density of hydrogen, 1.10 that of oxygen, and 0.843 the density of carbon in vapor state (all relative to air = 1).
92. 2C 2H as methane was considered then.
93. In current terms, this would be the equation for Gay-Lussac's reaction: $2CH_2 + HO = C_2H_5O$. For the French chemist, water was HO in the Daltonian interpretation. In general, Avogadro aimed to avoid the use of fractionary volumes or molecules in 'total' compounds as he did instead with 'partial' compounds, 'Expériences pour déterminer les proportions . . . '.
94. *Ann. Chim.* **94** (1815), 5. C = 44.2%, H = 6.80%, O = 49.0%. Also, *Ann. Phil.* **5** (1815), 265.
95. *MRAST*, **26** (1821), 475.
96. In his thesis at Montpellier, 9 July 1817.
97. Thenard, *Traité de Chimie*, vol. IV, gave only the percent composition and not the empirical formula for urea which is: C = 19.4%, N = 43.4%, O = 26.4%, H = 10.8%.
98. *Ann. Phil.* **11** (1818), 352, C = 19.9%, H = 6.6%, O = 26.6%, N = 46.6%.
99. Berzelius in *Lehrbuch*, **4** (1831), 356, assigned to urea the empirical formula N_2CH_4O. Prout gave to urea this composition in volumes: $\frac{1}{2}$O 1N 1C 2H. Actually, this interpretation of the composition of urea was correct.
100. *Amedeo Avogadro e la teoria moleculare* (Torino, UTET, 1911).
101. Berzelius seemed still uncertain whether ethyl alcohol was $C_4H_{12}O_2$ or one half of this formula as late as 1840.
102. Graebe, 'Der Entwicklungsgang der Avogadroschen Theorie', *J. Prakt. Chem.* **87** (1912), 153; J. R. Partington, *A History of Chemistry*, vol. IV (London, 1962), V. Cappelletti, *Diz. Biograf. Italiano*, vol. IV (Ist. Enc. Ital., 1962), p. 695.
103. Graebe: 'Entwicklungsgang', op. cit., p. 154. A 16-line review appeared on the *Edin. Phil. J.* **10** (1824), 190, where the author's name was misspelled as Avogrado.
104. N. G. Coley, *Ann. Sci.* **20** (1964), 210.
105. Title, "Théorie de composées organiques', *Ann. Chim.* **61** (1836), 125.
106. The tone of this comment is not surprising because it is neither the first nor the last

occasion on which the great legislator of chemistry in the first half of the nineteenth century cast his stones against a view which did not fit his own conceptual system. The dualistic theory, one of his keystones throughout his life, had been attacked by a man new to the scientific arena (Laurent was only twenty-eight at the time) and with very limited experimental evidence to back his ideas. Berzelius immediately set them aside as unfounded.

107. Page 7 ff.

108. *Précis de Chimie Organique*, p. 57.

109. See, for instance, note 3, p. 8 and note 1, p. 7.

110. 'Sur la densité de corps solides et liquides comparée avec la grosseur de leurs molécules et avec leur nombre affinitaires', *Giorn. Fisica, Chimica e Storia Naturale*, **31** (1827), 1–41, and 'Note additionelle au mémoire précédente', ibid., 42–94.

CHAPTER SEVEN

AVOGADRO'S OPUS MAGNUM: **FISICA DEI CORPI PONDERABILI** (1837–1841)

The chair of theoretical physics at the University of Turin was reinstated in 1832, and for the following two years Augustin Cauchy, in exile from Louis Philippe's France, occupied it. When the famous French mathematician left Turin for a more rewarding office in Prague, the Sardinian Government reappointed Amedeo Avogadro to his previous position. This event took place in November 1834; Avogadro had remained outside the academic world for over twelve years.

The considerable amount of time at his disposal in the period preceding the end of his enforced retirement presented a favorable opportunity for him to plan and carry out the writing of a huge treatise, which appeared in four volumes published from 1837 to 1841.[1] A work of such magnitude (almost 4000 pages) on physics had never before been published in Italy. In this sense, it represents a landmark for the historian of Italian science. Yet, seen in the overall context of European physics, Avogadro's treatise was untimely. In the late 1830s, new fundamental theories were unfolding, and as a result, the conceptual foundations on which most of his book rested were obsolete even before the printing had been completed.

In *Fisica dei corpi ponderabili*, Avogadro continued to support the caloric theory of heat and a distinctive interpretation of the nature of molecular forces, which had been essential components of the Laplacian program of physics. This program had found in Laplace and Berthollet, and in their associates Biot, Gay-Lussac, Thenard, and Poisson, its most prominent representatives, and had reached its peak in the Napoleonic period, a quarter of a century before the appearance of *Fisica*.[2]

There is little doubt that Laplace and Berthollet had a strong enduring influence on Avogadro in his formative years, as did Biot and Poisson in the maturity of his scientific thought. It is also true that Avogadro was critical of those working outside the Laplacian program, like Ampère, and the younger generation of French scientists represented by men such as Dumas, Dulong, Petit, and Gaudin.[3]

The *Fisica dei corpi ponderabili* was dedicated to Charles Albert, then King of Sardinia: 'To show Your Majesty that it is my wish to reciprocate the benefits Your Majesty bestowed on me, particularly by awarding me the distinguished Order established by You, to reward merits of the scientists in these magnificent regions placed under the very happy rule of Your Majesty and with such great benefit to our own country's studies.'[4]

What lay behind the facade of this courtly and turgid prose? Although it is difficult to prove his loyalty to the Savoyard dynasty, Avogadro had, nevertheless, ample reason to be grateful to his sovereign. In fact, shortly before, Charles Albert had knighted him with the Civilian Order of Savoy, a recently created middle-rank honor. More importantly, three years earlier the King, very likely upon the advice of the marquis Cesare Alfieri,[5] had reappointed Avogadro to the chair of theoretical physics at the University of Turin.

The purpose of the *Fisica* was also clearly outlined in the dedication:

> In my studies I especially pursued that section of physics dealing with the general constitution of ponderable bodies and, accordingly, with the quality of their *component molecules*, with the forces by which these molecules are stimulated, with the capacity of different bodies for caloric, and with the density and elastic forces of their vapors . . .[6]

Thus, in the tradition of early nineteenth-century science, Avogadro hoped to investigate yet again the correlation between ponderable and imponderable bodies. He stressed that the study of these two kinds of bodies formed the subject of what he defined as 'physics itself'.

In the introduction, he also attempted to classify the sciences, as Ampère had done a few years earlier on a much different scale in his 'Essai sur la philosophie des sciences'.[7] Avogadro emphasized both molecular geometry and crystallography as part of the mathematical sciences, and classified chemistry as an offshoot of general physics. Another branch of general physics was 'physics itself' (fisica propriamente detta) which included the physics of ponderable bodies (the title of his book). This physics dealt with bodies formed by tightly joined molecules having a sensible mass. A further branch of physics had to do with the tenuous imponderable fluids (corpi imponderabili) which were diffused everywhere in space, surrounding molecules of ponderable bodies and subject to vibrational movements.[8] These fluids were of two types. The first type included those fluids which moved around the molecules of sensible bodies, adhered to their surfaces, caused them to move, but did not have any influence upon their intimate constitution. Magnetism and electricity, quite analogous to each other, belonged to this class of fluids. Their properties, but not their essence, were known.[9] Caloric and light comprised the second class of imponderables. They were thought to be either single molecules, which gathered around ponderable molecules, but did not stir any movement in their masses, or fluids which were diffused everywhere in space, but were possibly subject to vibrational movement extending from one ponderable particle to the next.

The *Fisica dei corpi ponderabili* was divided into four volumes. The first volume (1837) covered the constitution of bodies at a given temperature. There Avogadro set forth his ideas concerning molecular structure (see Table 5): molecules and their types, the nature of molecular forces and their relation to the physical state of the bodies. He then discussed mathematical and mechanical problems involving the equilibrium of molecular forces with external forces. His analysis of bodies subject to special strains (elongation, flexion, vibration, and oscillation) relied heavily on Poisson's[10] and on Cauchy's treatment of stress. For example, in discussing molecular attraction, Avogadro referred to Poisson's exponential expression linking the force holding molecules together to its intensity and to the molecular distance.[11]

Table 5. Nomenclature adopted by Avogadro in the *Fisica dei corpi ponderabili*

= *Elementary molecules* also defined as *primitive molecules* or *partial* or *constituent molecules* are present in elements or simple substances.

= *Integral molecules or compound molecules*: the term integral applies to both chemical elements and chemical compounds. By gathering these molecules together, one obtains a sensible mass.

= *Total molecules*: are formed by partial molecules and they can be alike in chemical elements or of different kind in chemical compounds.

= *Physical atom or gaseous atom*: used after 1830 in the same context as integral molecule earlier. They are formed by several *partial atoms*.

= *Chemical atom or solid atom*: used after 1830 in the same context as constituent molecule or elementary molecule earlier.

= *True atom or simple atom or partial atom*: used as equivalent to chemical atom.

= *Compound atom*: used as equivalent to compound molecule or integral molecule.

Later on, Poisson's theory of the attractive and repulsive forces acting upon molecules, as advanced in his 'Mémoire sur l'equilibre' of 1828, would be used by Avogadro to examine the mutual action between two adjacent molecules and the caloric molecules surrounding them. The importance of this discussion cannot be over-emphasized. It provided a detailed exposition of

Avogadro's system of molecular forces and actions, as functions of a molecule's distance, form, and number. In the treatment of this subject, he again followed the classical mechanics of Poisson. Unfortunately, his reasoning, encumbered by the intricacies of the caloric theory, was anything but clear.

In dealing with solids, Avogadro devoted almost two-thirds of the first volume to the study of crystallization. The conjectures of Wollaston, Ampère, and Mitscherlich, correlating crystal forms to atomic and molecular constitution of bodies, were extensively discussed. He then examined isomerism and dimorphism and Persoz's interpretation of these properties.

In 1836, Laurent had presented his theory of fundamental radicals found in organic compounds. From these 'derivative radicals' could be obtained compounds which still retained an 'analogous atomic' constitution. Avogadro wondered whether this same mechanism could be extended to other binary compounds besides those containing only carbon and hydrogen. He then speculated on the correlation between the density of solid elements and their compounds, and the size and distance of their molecules. He defined the density of a body, regardless of its state, as directly proportional to the mass of its integral molecules and inversely proportional to the cube of the distance existing between their molecular centers. Hence, by comparing the molecular mass of a body[12] with its density at a given temperature, it would be possible to calculate how mass and molecular distances could be related.

The similarity observed among most metals suggested to Avogadro that an equal number of elementary atoms might form their integral molecules and, therefore, that the ratio of the masses of the atoms of different metals was the same as the ratio of the masses of their molecules. The polyatomicity of all metals was an important assumption which appeared in his other writings, and was re-emphasized in *Fisica dei corpi ponderabili*. Much later, both Laurent in the *Méthod de Chimie* (1854) and Gerhardt in the *Traité de Chimie Organique* (1854) recognized the diatomic nature of most elements. This concept, scarcely orthodox at the time, later inspired Cannizzaro, who reached a similar conclusion for a number of metals in the elaboration of his 'Sunto' in the late 1850s.[13]

In the 1820s, Dumas had examined the problem of the atomic volumes of solid elements and had reported his views in the *Journal de Physique*[14] and in the first volume of his *Traité de Chimie* (1828). The same problem had also attracted Avogadro's attention, as indicated by his detailed review in *Fisica dei corpi ponderabili* of Dumas's work on atomic volumes. Between 1846 and 1852, Avogadro published four very lengthy essays on this subject and on the relationship between other physical properties and atomic volumes.[15]

In the second book of Volume One of *Fisica*, Avogadro turned to the study of crystallization. He realized the importance of establishing the molecules' positions in solids, and examined how the crystalline forms could be related to the structure of the elementary molecules. Avogadro's interpretation of the crystalline forms rested on his theory of the constitution of matter. While he designated as integral molecules the smallest portions in which a substance can be subdivided and 'still keep its nature',[16] he identified such integral molecules with the crystalline form typical of that substance. For example, a cube would be the form of sodium chloride. He re-iterated that integral molecules were made by constituent or elementary molecules, very close to each other, and placed at the corners of the crystalline forms.

The shape of the elementary molecules, unlike that of the integral molecules (a term applied to both compounds and elements) was not specified by Avogadro, although he probably thought that they were all hard spheres. The mutual position of the elementary molecules within the integral molecule was affected by both the repulsive and attractive forces acting upon them.

When a change of state occurred, for instance in cooling a solution, the attractive forces acting upon the molecules of the solute could be related to their shape. As a result of this relationship – not clearly defined by Avogadro – the molecules gathered together in accordance with the laws of aggregation, and crystals were thus formed. Therefore, the study of crystallization was to be extended to include the causes behind the shape of the integral molecules, and not to be limited solely to either a geometrical or physical interpretation of the crystalline structure.

Earlier, in discussing molecular forces in solids, Avogadro had argued that the equilibrium between attractive and repulsive forces among molecules determined their mutual distance.[17] In solids, unlike liquids, only a very large outside force might change the relative position of the molecules. He further assumed that an 'extrinsic force', no better qualified, was required to maintain the liquids in their particular state. This force would act in addition to the ordinary attractive forces.

Avogadro's consideration of molecular equilibrium as a function of the molecular distance deserves to be further examined. He supposed that at very small distances, both attractive and repulsive forces decreased very quickly, while at sensible distances they became insensible. However, this behavior, as he pointed out, was not easy to reconcile with the inverse square law governing the intensity of forces between contiguous molecules, unless one considered that forces did not act directly upon ponderable molecules, but rather upon such molecules and those of a weightless fluid (caloric or ether) forming atmospheres around them which superimpose and mutually penetrate.

Therefore, the attractive and repulsive forces among the ponderable molecules were subordinate to the distribution and to the position of the molecules of the interposing fluid. While the molecules of fluid repelled each other, they attracted the ponderable molecules. In discussing attractive and repulsive forces among ponderable molecules, Avogadro introduced Ottaviano Mossotti's 1836 theory of molecular forces.[18] This theory assumed the existence of essentially two molecular forces involved in the constitution of bodies. Only one force, however, the repulsive, was similar to the forces proposed by Avogadro, and it decreased very rapidly with increasing molecular distance. The other force, the attractive, instead followed the inverse square law of Newtonian gravitation. For Mossotti, this gravitational force was represented by the small difference that he assumed to exist between the attractive force of the material molecules for the ether and the repulsive force between the material molecules themselves, which was also in inverse proportion to the square of the molecular distance. This kind of force, Avogadro pointed out, had been proposed by Aepinus in his attempt to explain the distribution of a single electric fluid on the surface of electrified bodies. Now Mossotti's application of such a repulsive force to the molecular level was not easy to accept, Avogadro contended in his single criticism of the theory.[19] He clearly missed the central idea of Mossotti's 1836 essay, which shortly after its publication drew enthusiastic response from Faraday,[20] i.e., the existence of very close links among all natural forces.

In his discussion of the molecular forces involved in the solid state of matter, Avogadro never considered the behavior of such forces when subjected to the passage of an electric current. This is hard to explain, because by the 1830s, i.e., the time he was gathering material for *Fisica*, a substantial amount of experimental and theoretical work had been done by Ritter, Davy, Ampère, and Berzelius, among others,[21] and Avogadro was certainly aware of the conclusions reached in those years by them.

In this section of *Fisica* there was also no mention of the conclusions Faraday had drawn after observing the electrochemical decomposition of solids in solutions. Although Faraday avoided any corpuscular interpretation of matter, he believed by 1836 that chemical and electrical action were the same.[22]

Avogadro gave no indication in *Fisica* of his reactions to these new ideas and facts linking matter and energy, chemistry and electricity.[23] This section of his treatise strongly suggests that he remained faithful to a model of molecular structure for solids and liquids that excluded all electrical considerations, while it rested instead on a system of mechanical and thermal forces.

The second volume of the *Fisica dei corpi ponderabili* appeared in 1838 and contained two main sections, the first devoted to the constitution of liquids, the second to the constitution of gases and air-like fluids.

A review of the theory of capillary action, with special emphasis on molecular forces, comprised most of the first section. The peculiar attractions and repulsions which occurred between the 'poles' of the solid molecules were not present in liquids, Avogadro pointed out. He also observed that molecular attraction takes place not only among molecules of the same kind, but also among heterogeneous molecules, and, accordingly, the molecules of a liquid act upon the molecules of another liquid or those of a solid with which they are in contact. The unusual behavior observed whenever solid and liquid molecules interacted was generally described as capillarity, while the term 'capillary action' denoted all the forces responsible for such unusual behavior.[24]

The mathematical theory of capillary forces, in Avogadro's view, had reached its most advanced development in Poisson's *Nouvelle théorie de l'action capillaire* (1831).[25] He also mentioned Gay-Lussac, Ivory, Link, Dutrochet, and naturally Laplace, but there is no doubt that Avogadro favored Poisson's interpretation of capillarity.[26] Shortly before the publication of *Fisica*, Avogadro had studied this subject. In 1838 he had conducted and later published the results of a series of experiments to determine the rising of mercury in a capillary tube made from a metal capable of forming an amalgam with mercury.

The next section of the second volume of *Fisica* deals with gases and air-like fluids. It begins with the study of air and barometric pressure, followed by a description of the experimental methods adopted by Boyle and Mariotte which led to the formulation of their famous law. The discussion of this generalization introduced Avogadro's explanation of the behavior of gaseous substances, and strikingly illustrated the conceptual path by which, in 1811, he had formulated his hypothesis.

In the gaseous state, he argued, the distance between molecules was such that their mutual attraction ceased to exist. This explained why the relation of pressure to volume remained constant for all gases. Furthermore, as Boyle and Mariotte had observed, the attractive force exerted by the molecules of caloric upon neighboring molecules was slight in the gaseous state because of the high degree of dispersion observed there.

Avogadro then focused his attention on the density of fluids, and this led in turn to a re-introduction of his 1811 gas hypothesis. Avogadro devoted almost forty pages to the subject, some of them remarkable for a lucidity often missing in his previous writings.

'The fundamental principle . . . , that I believe I have been the first one to state', he opened, 'relies on a general fact that Gay-Lussac has observed first . . .'.[27] It was the same principle that Dalton had formulated after studying chemical combinations and observing the definite ratios with which elements combined to form chemical compounds.

Not only were there simple ratios of combinations, as chemistry indicated, but also the number of combining molecules was small. If such were not the case, one would face integral molecules of 'enormous size', something that 'cannot be assumed in nature'. Avogadro had made the same observation in the 'Essai d'une manière' a quarter of a century earlier, and apparently saw no reason to change it now. After explaining how the molecular mass[28] could be determined from the combining weights, Avogadro discussed the law of combining volumes as stated by Gay-Lussac in 1808; he observed that if the combining weights represented molecules, then the corresponding volumes indicated the number of molecules. Therefore, some whole number relationship had to exist between the combining volumes and the molecular masses. Two alternatives were possible. Equal volumes of different gases, at the same temperature and pressure, contained: (a) an equal number of molecules, or (b) a multiple number. The determining factor was the distance between molecules. Since there was no reason to believe that the size or the nature of the molecule could influence it, this distance should then be considered constant for all gases under the same physical conditions. Therefore, only the first alternative seemed to be 'probably' correct. By further assuming that at the same temperature and pressure the inter-molecular distance was invariable for all gases, the mass of their molecules was proportional only to their density. Thus was inferred the most important corollary of the 1811 hypothesis.[29]

The question debated by historians on the limits Avogadro set for the division of integral molecules[30] may be answered in his own words: 'We must then conclude that the integral molecules of gases are formed by several *single* molecules, although we cannot indicate their number nor establish whether it is constant in all elementary gases.'[31] This indeterminacy, while leaving many speculative avenues open to those few contemporaries who accepted the molecular hypothesis, probably did not help Avogadro's ideas gain wider acceptance.

The next chapter of Volume Two stressed once more that only by adopting the molecular hypothesis would it be possible 'to remove that arbitrariness' in determining weights (both atomic and molecular) and compositions which plagued chemistry for so many years.[32]

The division and recombination of molecules during a substance's change of physical state, a cornerstone of Avogadro's theory of molecular constitution, was discussed extensively in the fourth section of the second volume of *Fisica*. Several pages of this section also critically analyzed experimental findings of Dumas and Mitscherlich. The experiments were conducted between 1826 and 1833 and yielded results explainable, only barely, in terms of the molecular hypothesis. It is quite interesting to see how Avogadro stretched his imagination in the attempt to harmonize Dumas's and Mitscherlich's findings with his own interpretation of the molecular structure.

Although Avogadro mentioned the conclusions Dumas drew from his researches, and specifically that the vapor densities of the 'gaseous' or 'physical atom' sometimes could not be considered as multiple of the vapor densities of the 'chemical atom' (as in the case of mercury), he did not disclose his own thoughts. In Germany, in 1833, Mitscherlich, who already enjoyed an international reputation for his studies on isomorphism, published a series of articles on vapor densities.[33] His work, which was conducted at temperatures up to 700°C, had also appeared in the *Annales de Chimie*[34] and did not escape Avogadro's attention. Mitscherlich reported data on five non-gaseous elements and on a number of their compounds. For bromine as for iodine, there was no problem in reconciling their 'gaseous atoms' with their 'chemical atoms', since both hydrobromic and hydroiodic acids behave as hydrochloric acid, with the division of its molecule by two upon formation of the compound gas.

In Mitscherlich's determinations of the vapor density of phosphorus trichloride, however, some puzzling conclusions became apparent. In the formation of this compound, phosphorus's 'gaseous atom' would divide by six.[35] Such an anomaly could not be explained even by assuming that the 'true atom' of phosphorus (he defined this as a molecule composed by the same number of indivisible atoms as that of the element of reference, i.e., oxygen) was one-half that given by the vapor density of this element. For Avogadro, then, there were only two possible interpretations: an experimental error or the possible 'decomposition' of phosphorus chloride when it vaporized.[36] Needless to say, the latter interpretation would eventually prove to be correct, but the significance of this idea failed to impress Avogadro at the time. Once more, he revealed his natural inclination to uphold a preconceived hypothesis rather than to accept empirical data.

The concluding section of Volume Two of *Fisica dei corpi ponderabili* contained both a review of the principal essays on molecular constitution which had appeared after 1811, and a number of considerations on the possibility of applying to organic compounds the generalizations which, in Avogadro's view, guided the behavior of gaseous compounds during their formation. The determination of the molecular weight of complex organic substances through their vapor densities was an objective that Avogadro may or may not have discerned, although he had hinted at it earlier.[37] He correctly understood, however, that even the constitution of organic compounds could be better explained through the application of his equal volumes–equal numbers generalization.

Avogadro's analysis of the essays of previous decades which dealt with the constitution of gaseous bodies[38] is very revealing. It was quite a thorough review, although few words were devoted to Ampère and his letter to Berthollet in 1814. This might be explained by noting the more extensive examination of Ampère's elaborate system in Volume One of the *Fisica dei corpi ponderabili* where Avogadro discussed crystallography. Dalton, Prout, and William Charles Henry[39] were also mentioned in the review. Quite significantly Avogadro contrasted the British scientists with their French counterparts who, in his view, in general had more favorably received his ideas.

The third volume of *Fisica* treated the influence of temperature variations on the constitution of bodies. This volume, published in 1840, was divided into two books. The first examined the concepts of caloric and temperature and their relationship. It also reviewed the specific heats of gaseous and non-gaseous bodies and their variation with the change of temperature. Avogadro discussed here the experimental findings of Berard, Delaroche, Marcet, and de La Rive, among others, and evaluated their conclusions. He further elaborated the possible link between the specific heats of compound gases and their components.

The second book of Volume Three looked into the effect of temperature on volumetric changes of bodies, and the laws of expansion and condensation of solids and liquids. Avogadro also discussed the expansion and compression of gases, and the relation between their specific heats and changes of pressure.[40] He also devoted a great deal of space to Laplace's[41] explanation of the effect of caloric on gases in terms of their molecular forces.

After mentioning Fourier's renowned contribution to the theory of the transmission of heat and the related experimental work of Pictet, Rumford, and Leslie, Avogadro re-emphasized that light, caloric, and ether were all imponderable fluids, and as such, their study lay outside the subject of his textbook.[42] Yet, he considered the effect of caloric upon the constitution of ponderable bodies.

In the opening chapter of Volume Three, caloric was defined as 'the cause of heat'. It was also noted that some physicists explained heat as motion internal to bodies stimulated by temperature and present in both the particles of bodies and in those of the 'other fluid' dispersed around them. Heat, thus interpreted, would be the result of the *vis viva* of bodies, caused by the vibrations of their particles, and its total amount would remain constant even when bodies changed their physical state.

In his discussion of the relationship between specific heats and other properties of bodies, Avogadro began by recalling Dulong and Petit's experiments of 1819 and their important conclusions:[43] an almost constant product could be obtained by multiplying the atomic weights of some elements (or 'masses of their molecules', as Avogadro added) by their specific heats. He then argued that since the specific heat of one atom was the amount of caloric required to raise its temperature by one degree, then by dividing the specific heats of Dulong and Petit by the number of atoms comprising the same weight of each element, one should obtain the specific heat of *one atom*.

In Avogadro's view, Dulong and Petit's generalization was only 'approximate' for two reasons: (a) the variation in specific heat with change in temperature, and (b) the difference existing between the affinity for caloric and the molecular attraction of different substances.[44] Furthermore, the link that Avogadro had hypothesized between the affinity for caloric and the specific heat – which after 1816 was one of the foundations of his heat theory – appeared to contradict Dulong and Petit's conclusions. Yet, Avogadro conceded that their generalization could still be valid with respect to the elements used in their experiments (all metals, except sulfur).

Avogadro noticed, of course, that the atomic weights derived from Dulong and Petit's generalization were generally one-half of those found by Berzelius. In fact, only for sulfur and platinum did the new and old figures coincide. In Avogadro's view, if one assumed as correct the atomic weights calculated by the two Frenchmen, then, since each element observed had the same capacity for heat, one could draw the conclusion that each integral molecule was formed by the same number of 'partial simple molecules'.[45] He suggested, however, that in the solid state a different number of such molecules might sometimes gather to form an integral molecule. It was not yet settled, for him, whether in the case of sulfur, 2.011 represented the weight of a 'simple atom' or the weight of a molecule formed by the same number of 'simple atoms' as that present in a molecule of oxygen in the gaseous state (i.e., two).[46]

Clearly, Avogadro here sought not only to reconcile the experimental data with the corrections he wished to bring to Dulong and Petit's Law, but also to explain the differences between the atomic weights determined by chemical procedures and those obtained from vapor densities. To account for this variance between 'chemical' atoms and 'thermal' (physical) atoms, Avogadro in 1834 assigned to all integral molecules – and not only to those of gaseous elements – a degree of polyatomicity which changed from element to element. Only for those elements displaying similar behavior – for example, the four halogens – might the polyatomicity be similar.

The investigation of molecular structure in all three states of matter remained unquestionably at the center of Avogadro's speculations. Around it he explained both chemical and physical behavior. However, established chemical combinations were sacrificed whenever they disagreed with physical experiments. Avogadro's lack of concern for chemical facts greatly contributed to the skepticism he roused among contemporary chemists.

In the next section of Volume Three, Avogadro examined within the framework of his gas hypothesis the equality of the specific heats of elemental gases, established by Dulong in 1829.[47] Once it is assumed, Avogadro reasoned, that under equal physical conditions the densities of gases are proportional to the masses of their integral molecules and that, accordingly, under equal volumes, all gases contain the same number of molecules, then they must have equal specific heats as well. Thus, everything could be explained in terms of molecular structure: densities, masses, and specific heats. However, in his discussion of molecular constitution and specific heats for gaseous compounds, the equality of specific heats does not apply. Avogadro pointed out that he had been able to establish a relationship between the specific heats of gaseous compounds and those of their chemical components. Before Dulong's experiments of 1829, this correlation had to do with the specific heats of gases under constant pressure; now it was made 'much simpler' by Dulong's conclusions. Essentially it stated that, under equal volume, the square root of the sum of the whole (or fractionary) number of integral molecules of simple gases forming a compound gas was equal to the specific heat of that compound, or put in another way, the square of the specific heat of a compound gas was equal to the sum of the number of integral molecules of its components. In 1830, this equation predicted results so close to the experimental data for a few compounds that Avogadro referred to it again eight years later in *Fisica*.[48]

In the early 1830s Avogadro extended to nongaseous bodies the ideas he felt he had earlier applied successfully to the theoretical determination of the specific heats of gaseous compounds. The general equation he had proposed in 1833 indicated the specific heat of a solid or liquid compound as a function of the square root of its *constituent number*, multiplied by 0.1875 (one-half the Dulong and Petit's constant coefficient) and divided by its molecular weight. Avogadro assumed that solid compounds of similar chemical composition would have the same constituent number. By this he meant the whole or fractionary number of 'atoms or parts of simple atoms' forming that particular compound.[49] The abundance of experimental data then available on the specific heats of sulfides, oxides, hydrates, sulfates, etc., allowed him to verify the proposed equation.

Probably one of the more remarkable examples of Avogadro's calculations of the specific heats of solid compounds concerned the metal oxides, such as alumina, with a general chemical composition of 1Me 1½O. In order to reconcile the experimental figure with the calculated one, he assumed first that this group of oxides should contain only three-quarters of an oxygen atom for each atom of metal, (1Me$\frac{3}{4}$O); secondly, that there is a division by two of the oxide molecule. As a result, alumina would be composed of $\frac{1}{2}$ Al $\frac{3}{8}$O, and the square root of its constituent number ($\sqrt{\frac{1}{2}+\frac{3}{8}} = \sqrt{\frac{7}{8}} = 0.935$), a figure in agreement with the observed specific heat.

No less revealing of Avogadro's approach was his attempt to extend similar concepts to organic compounds – also called by him 'ternary compounds'.[50] He discussed two of them, ethyl alcohol and ether. By assuming 1C 3H$\frac{1}{2}$O for the molecular composition of the former in vapor state, the calculated specific heat was 0.274, very far from the experimental value (0.62). However, if for liquid alcohol the constituents were further divided by four, then the introduction of the new values into the general equation gave calculated specific heats only 'slightly' lower than those found.[51] Thus, for Avogadro the molecule of alcohol in a liquid state would contain $\frac{1}{4}$C$\frac{3}{4}$H$\frac{1}{8}$O rather than 2C 1H$\frac{1}{2}$O, although even in this case the discrepancy between the experimental and the calculated figures seemed scarcely acceptable.

Avogadro then returned to the problem of the specific heats of elements. He speculated again that not only compound atoms upon their formation, but also the atoms of simple bodies, regardless of their physical state, underwent division.[52] From this it followed that Dulong and Petit's generalization could not be used for all elements, and that some of the atomic weight, which they modified as a result of their generalization, had to be re-evaluated. Avogadro examined sulfur to test the validity of his argument. Using his general equation and assuming its atom in solid state to be one-quarter that in gaseous form, he calculated the specific heat of sulfur: $\sqrt{(\frac{1}{4} \times 0.1875)}/\frac{1}{4} = 0.375$, which was approximately twice the specific heat obtained by dividing the Dulong and Petit constant coefficient (atomic heat 0.378) by Berzelius's atomic weight of sulfur (2.011). In fact, one has to recall that Avogadro had assigned to sulfur one-half the atomic weight obtained from its chemical combinations.[53]

Despite these considerations, there is no doubt that Avogadro realized too many uncertainties remained before an acceptable correlation could be established between chemical atomic weights and those deduced from specific heats. This was revealed, for instance, by his hope that a correct experimental determination of the specific heats of both elements and compounds in gaseous state would be feasible. If such were the case, then one could ascertain the mass of their integral molecules from their density.

The final sections of Chapter 2 dealt with a series of general considerations. First, Avogadro attempted to explain why, according to Dulong and Petit's Law, atoms of different weight but the same constitution had the same specific atomic heat regardless of their physical state, and why a variation in the atomic constitution of a substance caused a change in its specific heat. As a conditional and possible interpretation,[54] Avogadro re-introduced the concepts of *affinity* for caloric and *attractive power* for caloric in relation to the atomic weight. Since the attractive power, he wrote, seemed to be the same for the molecules of all bodies, it could be assumed that elementary gases, whose molecules have the same constitution, under equal temperature and pressure have equal space available to caloric for its diffusion and, therefore, the same attractive power for caloric and the same specific heat.

This attractive power appeared to be an additive property in compound gases, although the experimental results suggested that the specific heat of each molecule was not directly proportional to its attractive power for caloric, but to the square root of the same. On the other hand, when a division occurred in the molecule of a gaseous compound upon its formation, then the number of simple molecules decreased, and the attractive power for caloric and the specific heat of that compound declined proportionally.

Similar considerations, Avogadro added without further elaboration, might be applied 'approximately' to both solid and liquid bodies.

The question of how the caloric increased or decreased and distributed itself in the intermolecular space could not be answered 'a priori',[55] Avogadro suggested. The validity of the equation he had proposed in 1816 ($c^2 = n'c'^2 + n''c''^2$ + etc.) for the attractive power of a gaseous compound (naturally nullified when the specific heats c, c', c'' of the components were found to have the same numerical value) was now limited, in Avogadro's view, to showing the indirect proportion which exists between the affinities for caloric of different substances and their atomic weights.

This section of the *Fisica dei corpi ponderabili* did not mention the attempts made in 1820 to link the refractive power of gases with their affinity for caloric.[56] But Avogadro did devote a few lines to his extensive studies between 1823 and 1825, in which he thought that he had succeeded in establishing that relationship between the electrochemical series and the affinity for caloric of a number of substances.[57] Although he treated this subject only in passing, it had occupied a great deal of his time during his forced retirement.

The concluding paragraph of the chapter briefly mentioned the vibrational theory of heat which, in Avogadro's view, did not satisfactorily explain the mutual effect between the molecular weight of a substance and its affinity for the 'ether' surrounding it.[58] Actually, this theory made it even more difficult to understand the same effect.

In the section devoted to the characteristics and uses of different kinds of thermometers, Avogadro offered his views on absolute zero. He recalled that some physicists had denied the existence of absolute zero, or had argued that temperature could not be defined by a finite number of degrees, beginning with a state in which a body was absolutely deprived of 'free caloric'. In his opinion, this position had no foundation if one understood temperature as the force exerted by caloric in leaving the body. Since this was a finite force at any given temperature, Avogadro argued, then any part of the force would contain a finite number of degrees of temperature. Eventually, in decreasing a finite number of degrees from a given temperature, this force would be wholly expended.[59]

Clapeyron's article 'Sur la puissance motrice du chaleur' (1834), as well as the earlier 'Réflexions' by Sadi Carnot, often quoted by Clapeyron in the same article, did not escape Avogadro's attention.[60] Yet, it is evident from the context that Avogadro had failed to recognize the significance of these works, as had most physicists of the time.[61]

The fourth volume of the *Fisica dei corpi ponderabili* appeared in 1841, and carried further Avogadro's series of observations on the influence of temperature on the constitution of bodies.

After examining the phenomena caused by variations of caloric, such as changes of volumes in liquids and solids, specific heats, and absolute zero in Volume Three, Avogadro turned to the theory of evaporation and condensation. In particular, he discussed the densities of vapors and their relationship to the densities of the liquids producing them; and also the absorption and the evolution of heat observed in the dissolution of solids. The specific heats of solids were again investigated and compared with those of the same substances in the liquid state. As usual, Avogadro devoted a section to instruments, giving a detailed account of different types of hydrometers and their use in measuring the humidity of air and gases.

Avogadro also surveyed the experimental work of Watt, Despretz, and Clement on latent heats, as well as the investigations of Dumas, Mitscherlich, and Persoz on the comparative volumes of vapors, liquids, and solid. Dalton's law of partial pressures was recalled, along with his findings concerning the vapor pressure of water and his theory of the mixture of vapors and permanent gases.

The section of Volume Four devoted to the relationship between specific heats of liquids and those of their vapors deserves some attention, because here Avogadro emphasized again one of the key points of his physical theory. He had calculated the ratios between the specific heats of water, alcohol, ether, and turpentine oil, and the specific heats of their vapors. From these calculations he now concluded that the integral molecule is divided by four whenever there is a change from the vapor to the liquid state.[62] This conclusion merely restated what had already been discussed in Volume Three. Although not explicitly mentioned, the converse process seemed equally clear in Avogadro's mind; that during the passage from the liquid to the vapor state in water, alcohol, and ether, four 'partial molecules' of these substances combined into one integral molecule.

Since this transition usually occurred when the temperature of a substance was increased, it was surprising that an 'association' of separate 'partial molecules' might take place at all. Thus, it may be interesting to follow Avogadro's reasoning here in some detail. Through a series of calculations, and by assuming that the specific heat of water vapor at constant volume was equal to that of carbon dioxide under the same conditions, he figured that, in agreement with Dulong's experiments, the specific heats of water vapor at constant volume and at constant pressure were respectively one-third and one-half those of the same weight of liquid water.[63] In 1840 Avogadro had discussed this subject.[64] It was, however, well known that water vapor expanded more than water and that the specific heats of both liquids and gases usually increased with their expansion. To account for this inconsistency, he offered two explanations: first, that less caloric was required to heat a vapor at atmospheric pressure than to heat the liquid producing it; second, that the specific heat of the vapor would be less than that expected if one considered its integral molecule to be formed by a larger number of simple molecules than that present in the same substance in liquid form.

Thus, Avogadro emphasized here that *molecular construction* affected specific heat more than the change of physical state.[65] When water evaporated, a grouping of its molecules – four of them – occurred and, therefore, the specific heat of its vapor decreased considerably. Naturally, he concluded, the opposite occurred when water vapor condensed.

In 1826, in a seventy-page essay which appeared in Turin, Avogadro had speculated that a direct correlation between the affinity for caloric and the intermolecular distance could exist in solids at ordinary temperature.[66] Hence, it appeared probable that the same correlation would also be valid for liquids.

More than fourteen years had passed since he had published these ideas. In *Fisica* Avogadro now contended that it was 'quite likely' that the characteristic attraction for caloric exhibited by the different molecules in solids and liquids, influenced the mutual distance of the molecules. Founded on this belief was the recognition that a constant ratio, whose average was set at around 135, might be established between the vapor density, at 100° and 760 mm, of any substance and its density in the liquid state, provided that no division or association of molecules took place during the change of state. Should such division or association have occurred, then the ratios should be one-half, one-fourth, one-eighth, or double, quadruple, etc. of 135. To establish to what extent this happened, one had to compare the experimentally determined density with that obtained through calculation, by assigning the atomic weight of an element to its integral molecule. In lead, for example, two integral molecules joined together when this element changed from the solid to the vapor state. There is no doubt that the relation of atomic weights to densities and atomic volumes, which Avogadro extensively elaborated in a later series of essays,[67] remained at the center of Avogadro's scientific interest for over twenty-five years.

After this review of the most salient sections of the *Fisica dei corpi ponderabili*, it may be appropriate to summarize some of the themes upon which this treatise rests. In his extensive study of the constitution of matter and in his theory of heat, Avogadro reflected the attitude of the more conservative circles of the French school of physics. Anyone familiar with the content of J. B. Biot's *Traité de Physique* published twenty years earlier, and with C. Pouillet's *Eléments de physique* published in 1832, will find them quite consonant with the line of reasoning followed in the *Fisica*. By way of contrast, one may look at Avogadro's treatise in the background of a major textbook of physics which had been published in France at the same time. Antoine-César Becquerel's *Traité de Physique* appeared in 1842, when its author, the founder of a dynasty of famous physicists, was professor at the *Musée d'Histoire Naturelle*. To say the least, the two books have very little in common; in fact, they seem to be written a generation apart. The old distinction between ponderable and imponderable, still so dear to Avogadro, was ignored by Becquerel who devoted full sections to acoustics, magnetism, and electricity. The close analogy of heat with light was emphasized again and again. Heat, which could be caused by mechanical, chemical, or electrical action, was for Becquerel a radiation moving from molecule to molecule whenever a warmer body was in contact with a cooler one. In the *Physique* it was deplored that *chaleur* and *calorique* were still confused. Heat, electricity, attraction, and affinity had to be taken into consideration in discussing molecular forces. For Avogadro, on the other hand, affinity was still the keystone. Thus, while physics studied only 'aggregation affinity', chemistry dealt with 'composition affinity'. The *Fisica* mainly examined 'mechanics applied to insensible molecules', and, accordingly, mechanical considerations fully permeated its discussion of molecular forces.[68]

Avogadro's molecules were composite even for the elements; they were extremely small, invariable in mass and shape, and did not touch. The last characteristic was supported by the traditional arguments of the time – i.e., macroscopic properties such as porosity, transparency, and the ability to expand and to contract. That an empty space exists among molecules remained an important element in Avogadro's theory of the constitution of bodies.

Molecular distances could vary greatly, as indicated by the wide range of densities. Closely related to molecular distance was the force, defined as molecular attraction, which acted among adjacent molecules. Although Avogadro introduced Poisson's mathematical elaboration correlating the force and distance of mutually attracting points, he conceded that the true function indicating the variation of molecular attraction was still unknown.[69] Avogadro, however, recognized the non-Newtonian character of this force and shared this view with Davy, although it had not been accepted earlier in France by Fourcroy and Guyton, among others.[70]

While little was known about the forces presiding over molecular attractions, the opposite force – repulsive in nature and without whose existence there would be no porosity of matter – found its ultimate cause for Avogadro either in caloric or in a vibrational movement of the molecules and the ether surrounding them.[71]

In view of the extensive treatments of such fields of inquiry as specific heats, density, molecular forces, and changes of state, heat was probably the most discussed subject in *Fisica*. As earlier noted, Avogadro was a calorist and remained so even after the caloric theory began to decline. To understand his loyalty to and reliance upon such a waning theory, one must remember that his famous 1811 generalization linking the densities of gases to the weights of their elementary and compound molecules, could be deduced equally well through the caloric theory as within the framework of another novel explanation of heat. In the 1830s, while Avogadro was gathering the material for *Fisica*, this novel interpretation – the wave theory of heat – had already been proposed, and was in fact occasionally mentioned by Avogadro. Throughout his treatise, however, Avogadro never questioned the existence of caloric itself. Instead, he asked only whether caloric should be considered as an imponderable fluid or as the manifestation of a movement of the particles of the same fluid.

It has been generally contended that, after the establishment in the 1840s[72] of the doctrine of the convertibility of heat into other forms of energy, the caloric theory very quickly disappeared from the scene. Actually, this was not the case. The transition from the caloric theory to thermodynamics occurred gradually.[73] The wave theory of heat, which attracted several supporters after 1830, emerged as the natural link between the caloric and the dynamical theory of heat. In fact, in the years which saw the publication of the *Fisica*, the wave theory incorporated an essential element of the caloric theory – the existence of an ether – vague in its nature, and yet responsible for the transmission of vibrational motion from one particle of a body to another. The wave theory of heat, supported by the extensive experiments of William Herschel and Macedonio Melloni, was recognized favorably in Britain and in Italy.[74] But it was not until the publication of two articles by Ampère in 1832 and 1835 that the wave theory found new significance.[75] These articles are important here because they attracted Avogadro's attention, as is clearly evident from some pages of *Fisica*.[76]

In the 'Note sur la chaleur' (1835), Ampère proposed a system of attractive and repulsive forces among atoms, much stronger than those acting within the molecules. Three dynamic components were present: an attractive force directly proportional to the mass of the molecules and inversely proportional to the square of the distance between the molecules; a repulsive force caused by the vibrational movement of the interposed ether; and attractive and repulsive forces between the atoms within the molecules. In the same essay, physical forces were distinguished from chemical forces, the latter alone being responsible for the division of molecules in the course of chemical combinations. This last assumption, certainly not original in 1835,[77] Avogadro found very close to his own views, formulated twenty years earlier.

Ampère's conjecture that the vibrations of the particles of a body were transmitted by vibrations of the surrounding ether attracted Avogadro's attention. He mentioned it in the *Fisica* whenever he outlined a possible alternative to the caloric theory.[78] It is apparent then that for Avogadro, as well as for other physicists after 1830, the replacement of caloric with ether no longer represented a serious conceptual obstacle. Yet, the presence of a weightless fluid remained fundamental to his system.[79]

It is interesting to note that Avogadro, in over 900 pages of Volume Three of *Fisica*, devoted only a few paragraphs to discuss emission and wave 'hypotheses' on the nature of heat. In citing the physicists to his time who had studied radiant heat, he omitted Leslie[80] and William Herschel. On the other hand, he recalled Melloni[81] only for his great 'success in the study' of diathermancy and for his observation that this characteristic property of both solids and liquids transmitted radiant heat.[82] Pierre Prevost received special attention for his description in 1791 of temperature equilibrium as a dynamical effect caused by two opposite but equal caloric radiations.[83]

Especially significant are the paragraphs of the first chapter of Volume One of the *Fisica* where Avogadro discussed the shortcomings of the vibrational theory of heat. He argued that the acceptance of this theory would imply, in the case of molecular vibrations, a change in the permanent volume of a body, which was something that could not be conceived. Thus, the permanence of size and volume of the ultimate components of matter symbolized a fundamental conceptual hurdle for Avogadro's acceptance of an alternative theory of heat.

The problem of determining the specific heats of non-gaseous compounds had also attracted Avogadro's attention since the early 1830s. In *Fisica* he remained substantially in agreement with the ideas previously elaborated. He relied more on the chemical compositions than on the physical properties of compounds for the determination of specific heats.

In his study of specific heats in the third volume of *Fisica*, there is another consideration worth mentioning. In 1824 and three years earlier, in his discussion of the specific heats of gases, Avogadro had assumed the existence, in both elemental and compound gases, of an 'affinity for caloric' that he considered, together with oxygenicity, density, etc., as a characteristic property of each substance. After 1828, when the evidence indicated that all elementary gases had the same specific heat at constant volume, Avogadro set aside as useless the notion of attractive power for caloric.[84] Yet, more than a decade later, in the *Fisica*, this concept reappeared. The question involved was how to explain specific atomic heats, atomic constitution and mass, and what might be their relationship. Avogadro recognized now that no one knew in what way the 'laws' governing the distribution of caloric in the intermolecular space changed as a function of temperature.[85] Attractive power, nevertheless, was still present in Avogadro's reasoning as a property of the molecule of each substance and proportional to its mass. The affinity for caloric was assumed to be the same in all gaseous substances.[86] This meant that when the molecular mass increased, the attractive power for caloric decreased and vice versa.

While he explicitly stated that he intended to follow the mechanical approach chosen by Poisson and earlier by Laplace in the study of molecular forces, he did not mention in *Fisica* other interpretations of the structure of matter. Thus, Boscovich's *Philosophiae Naturalis Theoria*, whose impact in the early part of the nineteenth century had been rather limited on the Continent, was never cited by Avogadro. He only hinted at the possible existence of contiguous 'mathematical elements' as an alternative to a structure of solid atoms and molecules.[87] One cannot fail to notice also the neglect in *Fisica* of the atomic theory, which he probably considered to be simply a chemical theory and as such unrelated to the main subject of the book. Avogadro appeared unfamiliar as well with Daniel Bernoulli's 1738 theory of gas pressure.

As mentioned earlier, in *Fisica* there was no discussion of the interdependence between chemical, thermal, and electrical phenomena. Very little chemistry appeared in this treatise and even less electrochemistry, although its developments by the late 1830s certainly had not escaped Avogadro's attention. Of course, electrochemistry also involved electricity, an 'imponderable', but this entire approach scarcely seems reconcilable with the scientific philosophy of a man who had always tried to link together different facets of the physical sciences. The fact remains, however, that, even in his chemical essays after 1820, Avogadro avoided any consideration of the possible electrical nature of the forces keeping chemical combinations together, in contrast with Berzelius's approach in those same years.

Scores of pages in *Fisica* dealt with density.[88] In Avogadro's system, this property was closely tied with the caloric present in a substance and with the molecular distances. He attempted again and again to rationalize the discrepancies observed between the density of an element and its atomic weight determined through chemical analysis. In fact, the rift between 'chemical' and 'physical' atoms sometimes appeared to be so wide that Avogadro had to summon all his mathematical ingenuity to overcome it. Avogadro's insistence in this endeavor seems almost to indicate that by then he was fully aware that those few cases of anomalous vapor densities would – and with hindsight we know they did – represent one of the most serious hindrances to the general validity and acceptance of his theory of the constitution of bodies.

One last comment should be made on the content of Avogadro's major physical treatise. Like Ampère, Dumas, Cauchy, and Biot, Avogadro was a very devout member of the Catholic Church. Nevertheless, throughout the entire *Fisica*, including the introduction which contains some glimpses into his scientific philosophy, there was not even a hint of his metaphysics or of his personal religious convictions. In this he was again within the mainstream of the French cultural tradition of the early part of the century, which, in general, sharply separated science and religion.[89] One has to wait until the late 1830s to observe the effects of Auguste Comte's writings on the younger generation of chemists and physicists.

NOTES

1. It is fair to assume that Avogadro devoted a great portion of the previous years to gathering material and preparing the draft for his treatise. This may be inferred from the fact that he, hitherto a prolific writer, did not publish at all in 1833, wrote only a short essay in 1834, nothing again in 1835, and only in 1836 did an important article, 'Nouvelles recherches sur le pouvoir neutralisant . . . ', appear in *Memorie della Reale Accademia della Scienze di Torino*, vol. 39, p. 57.
2. See R. Fox, *Hist. Stud. Phys. Sci.* 4 (1975), 89; and M. Crosland, *The Society of Arcueil* (Harvard Universiry Press, 1967) and *Gay-Lussac: Scientist and Bourgeois* (Cambridge University Press, 1978).
3. This may seem ironical since the position of these four men was closer to Avogadro's than to that of the Laplacian School.
4. *Fisica*, Introduction, p. i.
5. Cesare Alfieri di Sostegno became the first Minister of Public Instruction of the Sardinian Kingdom, appointed with such a title, in 1831. Previously, the functions of this Ministry were the responsibility of 'Il magistrato della riforma' who had

ruled over educational institutions, including the two universities of Turin and Genoa.

6. *Fisica*, Introduction, p. ii. The qualification of 'component', if considered in opposition to 'compound' molecule, may be considered as synonymous with 'constituent'. Avogadro's nomenclature in the *Fisica* will be examined below. See Table 5, p. 305.
7. The first volume was published in 1834 in Paris; the second, revised by Jean Jacques Ampère, appeared posthumously in 1843.
8. *Fisica*, Introduction, pp. viii ff.
9. By contrast, Berzelius, in discussing light, heat, and electricity, in Volume I of his *Traité de Chimie* (Firmin Didot, 1845), p. 13, qualified as 'a real absurdity' their definition as imponderables, because these agents were deprived of the essential character of matter. For their analogy with forces, Berzelius assigned them the collective name of 'Dynamids' or forms of force. Shortly afterwards, he stated that a science cannot be a 'tissue' of conjectures, but a system of proved realities and, of course, in so stating, invalidated both the atomic and molecular theories.
10. S. D. Poisson, *Traité de Mécanique* (Paris, 1833).
11. S. D. Poisson, *Traité de Mécanique*, vol. I (Bachelier: Paris, 1833), pp. 174 ff. This treatise was of fundamental importance for a whole generation of students of mechanics.
12. By which is meant its atomic weight.
13. Laurent said that each molecule of a simple body is at least divisible into two parts. See also Cannizzaro's 'Sunto', *Nuovo Cimento,* 7 (1858), 345 and 353.
14. The June 1821, 'Essai sur le volume des atomes des corps', was written in collaboration with Royer.
15. They will be discussed in detail in Chapter 8.
16. *Fisica*, vol. 1, p. 333. Characteristically, he did not define this nature as chemical.
17. Ibid., sec. 1, p. 54.
18. 'Sur les forces qui regissent la constitution intérieure des corps . . . ' (Imprimerie Royal: Turin, 1836), (incidentally, the same printing house which published the *Fisica*).
19. *Fisica*, vol. 1, p. 203.
20. Mossotti's memoir was translated and published in Taylor's *Scientific Memoirs* in 1837 upon the strong recommendatons of Faraday himself, and later reported in *Phil. Mag.* **10** (1837), 320.
21. The first nine series of Faraday's *Experimental Researches on Electricity* had been published before 1836 in the *Philosophical Transactions*. Berzelius's electrochemical theory, which supported an atomic polarity such that each atom carried both a positive and a negative charge, appeared after 1811 and in detail in his *Lehrbuch der Chemie*. In 1821 Ampère had offered his own contribution based upon the specific charge carried by each atom neutralized by an opposite charge on its outer layer. In this way he explained how atoms were held together in a compound through the neutralization of their outer layers and through their permanent opposite charges.
22. Series VIII, *Experimental Researches on Electricity* (1836).
23. Yet, more than thirty years earlier, he had formulated a theory of molecular polarization in nonconductors. As mentioned in Chapter 2 of this book, these two essays (1806 and 1807) represent the closest Avogadro ever came to establishing a correlation between the physical constitution of bodies and their chemical and electrical behavior.
24. *Fisica*, vol. 2, pp. 2 ff.
25. In 1832 Poisson received the Copley Medal for this work.

26. Quite surprisingly, in his extensive review of capillarity in the *Fisica*, Avogadro neglected to mention T. Young's 1805 essay on the 'Cohesion of Fluids', remarkable for its consideration of surface tension. Many of these ideas were later developed by Laplace. Avogadro only very briefly recalled Gauss's fundamental *Principia generalia theoriae fluidorum*, which appeared almost at the same time as Poisson's *Nouvelle Théorie*; his only comment was that the same objections raised to Laplace's theory of capillarity applied also to Gauss's.
27. *Fisica*, vol. 2, p. 843.
28. Avogadro never adopted the expression atomic or molecular weight in his writing, but always used molecular mass.
29. *Fisica*, vol. 2, p. 845.
30. S. H. Mauskopf, 'The atomic structural theories of Ampère and Gaudin', *Isis*, **60** (1969), 61; and B. W. Mundy, 'Avogadro on the degree of submolecularity of molecules', *Chymia*, **12** (1967), 151.
31. *Fisica*, vol. 2, p. 842. Italics mine.
32. This represents a leitmotif whenever Avogadro argued in favor of the adoption of his gas hypothesis.
33. He then had the chair of chemistry at Berlin. These researches appeared in *Ann. Phys.* **105** (29), (1833), 193.
34. *Ann. Chim.* **55** (1833), 5.
35. That is, its molecule should be hexatomic in modern terms.
36. *Fisica*, vol. 2, p. 862. 'Decomposition' is used by Avogadro to indicate what is currently defined as 'dissociation'.
37. Ibid., vol. 1, p. 879.
38. Ibid., vol. 2, pp. 868 ff.
39. Only M. Donovan's observations on the gas hypothesis in his *Treatise of Chemistry*, 1832, p. 379, were not mentioned by Avogadro.
40. As in the other volumes of the *Fisica*, Avogadro extensively described a number of instruments. In this case, he detailed thermometers, pyrometers, and thermoscopes.
41. Especially in vol. 5 (1823) of the *Mécanique Celeste*.
42. *Fisica*, vol. 3, p. 20.
43. *Ann. Chim.* **10** (1819), 395.
44. *Fisica*, vol. 3, pp. 204–5. For the anticaloric and anti-Irvinistic attack by Dulong and Petit in 1819, see R. Fox, *Brit. J. Hist. Sci.* **4** (1968), 1.
45. *Fisica*, vol. 3, p. 205.
46. At this time Avogadro considered the integral molecule of oxygen to be biatomic, a term which he never used. However, see his 1843 'Proposizione di un nuovo sistema di nomenclatura chimica', *Mem. Mat. Fis. Soc. Ital. Sci.* **23** (1844), 273.
47. *Fisica*, vol. 3, p. 212. Dulong's findings, based upon the velocity of sound, were reported in *Ann. Chim.* **41** (1829), 113.
48. *Fisica*, vol. 3, p. 213.
49. Ibid., p. 218, contained a quite remarkable admission that 'simple atoms', too, could have parts. This appeared also in the 1834 essay on the specific heats of solids ('le nombre constitutif de l'atome, c'est à dire, du nombre entier ou fractionnaire des atomes *simples* qui entrent dans la formation . . . ' *Ann. Chim.* **57** (1834), 125). The fractioning of simple atoms by Avogadro in the 1830s escaped N. Coley's attention in his 1964 article on 'The Physico-Chemical Studies of A. Avogadro', *Ann. Sci.* **20** (1964), 205.
50. *Fisica*, vol. 3, p. 222.
51. Actually, the difference is about 12 percent. This is the calculation: $\sqrt{4.5} = 2.12$ and $(0.1875 \times 2.12)/1.45 = 0.274$.
52. *Fisica*, vol. 3, p. 224.

53. In current symbols, Avogadro seemed to prefer the reactions S + O and S + 1.5O to the reactions S + 2O and S + 3O then generally adopted.
54. *Fisica*, vol. 3, p. 225.
55. Ibid., p. 226.
56. *Mem. Mat. Fis. Soc. Ital. Sci.* **18** (1820), 153.
57. More than 250 pages altogether in three articles.
58. *Fisica*, vol. 3, p. 228.
59. The Irvinist approach to the determination of absolute zero was examined in *Fisica*, vol. 4, pp. 91 ff.
60. *Fisica*, vol. 3, pp. 629 and 759.
61. William Thomson 'rediscovered' Sadi Carnot in the late 1840s. See his paper of June 1848 in the Cambridge Philosophical Society Proceedings. Also see S. Brush, *Brit. J. Hist. Sci.* **5** (1970), 165, who has examined in detail Thomson's responsibility in creating some of the myths on the development of the theories of heat in the first half of the nineteenth century.
62. *Fisica*, vol. 4, p. 605.
63. As noted earlier, this assumption relied on the number of molecules involved in the formation of water and carbon dioxide, which are the same, 1.5 for both, and the square root of this number is 1.225.
64. This formed the subject of a note published in *Biblioth. Universelle*, **29** (1840), 142.
65. *Fisica*, vol. 4, p. 605.
66. 'Mémoire sur la densité des corps solides et liquides . . . ', *Mem. R. Accad. Sci. Torino*, **30** (1826), 80.
67. The first 'Mémoire sur les volumes atomiques . . . ' appeared in 1843, shortly after the completion of the *Fisica dei corpi ponderabili*, the second two years later; the third in 1849; and the last one in 1850, at the time of his definitive retirement from the faculty at Turin. They were abstracted by Avogadro himself in the *Annales de Chimie* in those same years.
68. 'Introduction', vol. 1, p. 29.
69. *Fisica*, vol. 1, p. 10.
70. See T. H. Levere, *Affinity and Matter* (Clarendon Press, 1971), p. 49.
71. *Fisica*, vol. 1, p. 11.
72. J. Mayer's 'Bermerkungen über die Krafte' appeared in 1842.
73. See the well-documented and convincing exposition of this subject by S. G. Brush, 'The wave theory of heat', *Brit. J. Hist. Sci.* **5** (1970), 145.
74. Macedonio Melloni, born in Parma in 1798, worked for eight years in Paris where he enjoyed the personal support of Arago. The importance of his research was soon widely recognized and, among other honors, he received the Rumford Medal in 1835. His best known work (1842) recognized the strict analogy between light, radiant heat, and ultra-violet radiation. Among the early supporters of the wave theory of heat mentioned by S. G. Brush, p. 155, are T. Forbes, D. Lardner, G. Lamé, A. de La Rive and W. Whewell.
75. 'Idées sur la chaleur et sur la lumière', *Bibl. Univ. Geneve*, **49** (1832), 225 and 'Note sur la chaleur et sur la lumière . . . ', *Ann. Chim.* **58** (1835), 432.
76. See the *Fisica dei corpi ponderabili*, vol. 1, p. 12 and vol. 3, p. 22.
77. Avogadro's 1811 hypothesis concerning polyatomic molecules had been clearly restated by M. Gaudin in 1833, but Ampère failed to mention either Gaudin or Avogadro's hypothesis here.
78. See, for instance, *Fisica*, vol. 1, p. 12 and vol. 3, p. 9.
79. See S. G. Brush, 'The wave theory of heat', (fn. 4.73) p. 157. Avogadro's position can be identified in Step 5 of the development given by Brush, p. 147.
80. Leslie was cited, however, on several occasions for other studies. See *Fisica*, vol. 3,

pp. 150 and 600.

81. Ibid., p. 17.
82. M. Melloni had very close relations with both French and British scientists. The scant attention Avogadro gave to his work may be justified by the fact that Melloni's most important contribution to the study of radiant energy, *La Thermochrose*, appeared in 1850, nine years after the *Fisica*.
83. In *Observations sur la Physique,* **38** (1791), 314.
84. See N. G. Coley, *Ann. Sci.* **20** (1964), 202 ff. and R. Fox, *The Caloric Theory of Gases* (Clarendon, 1971), p. 225 ff.
85. *Fisica*, vol. 3, pp. 225, 226, and 227.
86. This derived from Dulong's findings of equal specific heats for all elementary gases.
87. *Fisica*, vol. 1, p. 20. I could not find a copy of Boscovich's main work in the library of the Accademia delle Scienze of Turin.
88. One could say that if weight was a fundamental parameter for Dalton and volume for Gay-Lussac, so density was for Avogadro.
89. According to M. Crosland, *The Society of Arcueil* (Harvard Press, 1967), pp. 90 ff., the presence of Saint Simon at Arcueil and Auteuil in the first decade of the century and his contacts with Laplace and Poisson may have given a 'pre-positivistic' character to French science. Crosland points out that anticlericalism, as well as nationalism, was deeply rooted in France in the early nineteenth century. Among intellectuals, and in general among educated people, this situation was true also in Italy at the time, with few notable exceptions, Avogadro among them.

CHAPTER EIGHT

LATER WORKS

Between 1832 and 1851 Avogadro published five essays covering a wide range of subjects. Two of them dealt extensively with a new field of inquiry for their author: a systematic rearrangement of chemical nomenclature. The topics of two other articles included a mathematical analysis of electric induction and a study of an equation to describe the behavior of gases approaching liquefaction. In another article Avogadro reported one of the few experimental researches he conducted in his scientific career, concerning the determination of the vapor pressure of mercury at temperatures very close to its boiling point.

This essay 'Mémoire sur la force élastique de la vapeur de mercure . . .'[1] was written while Avogadro was still barred from teaching at the University of Turin. Hence, it is not clear whether he conducted his experiments as a guest of the *Gabinetto di Fisica* of the University, or whether he used some private facilities. Rather surprisingly, the 'Mémoire sur la force . . .', written in the form of a very long abstract, was accepted for publication by the *Annales de Chimie* before it appeared in the original form in the memoirs of the Academy of Sciences of Turin.[2] In this article one can detect once more Avogadro's deeply rooted habit of searching for a mathematical notation to relate different physical properties of substances. In this case, he sought to formulate an equation to compute the vapor pressure of mercury at very high temperatures. However, in a remarkable break from his habit, he conceded that no *a priori* mathematical expression could be used for this problem, and that he would be forced to derive empirical formulas from the experimental data at his disposal.[3]

In arranging his simple experimental apparatus (Figure 2), Avogadro relied on the very well-known observation that for a liquid in a closed container connected to an inverted siphon, any increase of vapor pressure caused by a rise in temperature would result in an expansion of the liquid's volume and its rising through the siphon. Thus, he had built a U-shaped glass tube open at one end and having at the other end a small, closed spherical container. Due to the atmospheric pressure, mercury filled the tube and about four-fifths of the container, reaching the same level in the open branch. A metal scale graduated in millimeters, was placed behind the glass tube in order to

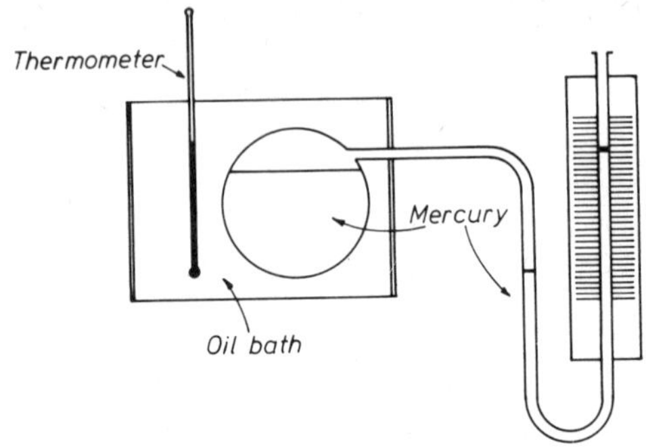

Fig. 2

measure the level of mercury in the open branch. By immersing this modified manometer (as in practice the whole instrument was) in an oil bath whose temperature was slowly raised and measured, Avogadro could then observe the effect of temperature on mercury vapor pressure.

For temperatures ranging from 100°C to 300°C in 10° increments[4] Avogadro measured the level of the mercury by first increasing and then lowering the temperature of the oil bath. By taking into consideration the vapor pressure of the air contained in the glass bulb and the effect of heating on the expansion of mercury in the open side of the tube, Avogadro arrived at the general equation $T = P\ (1-l/L)$. This equation related the mercury pressure T, the total vapor pressure P, and the height L in millimeters, reached by mercury in the open branch of the tube, where l was the height that air alone would have reached at the same pressure and temperature, according to the Gay-Lussac and Boyle-Mariotte Laws, which Avogadro assumed would apply to mercury.

In the range of temperatures between 230°C and 290°C Avogadro found that this equation gave results extremely close to those experimentally observed. However, at lower temperatures and lower vapor pressures – as he himself recognized – the experimental errors became large and the observations no longer agreed with the equation. Accordingly, he tried to formulate a tentative equation which would give 'au moins approximativement' an idea of the behavior of the vapor pressure of mercury at temperatures below those observed. There were indications that only near 0°C does the vapor pressure of mercury drop almost to zero.[5] This meant that the simple function earlier proposed $e = (1 + 0.4548 \times t)^{2.875}$ valid between 230°C and 290°C should be replaced with a three-term function similar to that used by Biot in his

Traité de Physique. This equation $\log e = \log A + at + bt^2 + ct^3$ was modified[6] by Avogadro to

$$\log e = at + bt^2 + ct^3 \qquad \text{(I)}$$

(where e = vapor pressure determined at temperature t, and a, b, c are constant coefficients to be determined experimentally), since $\log A = 0$, by taking, as Avogadro did, the vapor pressure of mercury at 760 mm = 1. Next he proceeded to calculate a, b, c from the mercury vapor pressures determined at 230°C and 290°C, the temperature extremes of his experimental observations. General equation (I) then gave values which closely agreed with the vapor pressures he observed below 230°C.

Yet, equation (I) when applied to low temperatures (for instances 0°C) led to results which did not fully agree with Faraday's findings that, at that temperature, the vapor pressure of mercury should be zero. In fact, equation (I) indicated that, at 0°C, $e = 8.5 \cdot 10^{-7}$ mm, a figure extremely small, but still not equal to zero. Hence, Avogadro speculated that at low temperature perhaps even Faraday's sensitive methods had been insufficient to detect the extremely small pressures. At 100°C equation (I) indicated a vapor pressure for mercury of about 0.03 mm, again almost unmeasurable.[7]

These results gave Avogadro enough confidence to use his general equation to calculate all vapor pressures for mercury from 100°C to 360°C, its boiling point. He trusted that his figures would be useful to physicists, and that through his findings one could also estimate the exact amount of mercury present in the air above the container.[8]

As earlier discussed,[9] Avogadro's article of 1833 on mercury vapor pressure was reviewed in the *Jahresbericht* of 1834 in the condescending tone which often characterized Berzelius' comments towards other scientists' work. In this case he turned his criticism mainly to the rather simplistic procedure adopted by Avogadro leading, in his view, only to approximate results.

The early 1840s saw a return of Avogadro to a field of physics – electrostatics – which had not attracted his attention for over thirty-five years. Faraday's experimental researches on induction published in the XIth Series in 1837 and 1838 certainly sparked the renewed interest displayed by Avogadro in the subject. By 1842 Avogadro felt that the time was propitious to restate and to update a series of theoretical considerations he had much earlier developed on the behavior of an insulated body when placed between plates charged with opposite electricity.[10] In that year he submitted to de La Rive's *Archives de l'électricité* (a supplement to the *Bibliothèque Universelle*) a short 'Note sur la nature de la charge électrique'.[11] Almost at the same time he sent to an Italian journal, *Memorie di Matematica e Fisica* (the official organ of the Societa' Italiana di Scienze in Modena) another article on the same subject. This was published after an unusual delay of over one year.[12] At the time Avogadro forwarded a copy of the article to Faraday, who showed it to the then twenty-one-year old William Thomson. Thomson, then at St. Peters in Cambridge, mailed back to Faraday in August 1845 a short but very accurate review of Avogadro's essay.[13] As Thomson himself mentioned in his letter, the distribution of electricity on two insulated spherical bodies had been studied experimentally by Coulomb almost half a century earlier.[14] Avogadro, however, indicated in the opening section of his memoir that these studies and those previously conducted by Poisson and by others rested on the theory of action-at-a-distance which the more recent findings of Faraday had set aside.[15]

Avogadro began the 1844 memoir by assuming that the presence of a given amount of electricity on the surface of a body required an equal amount of electricity of opposite charge on another surface separated from the former by an insulator. Avogadro, aware that Faraday had detected an interaction of opposite charges in curved lines, understood the difficulty of mathematically treating curves, and limited himself to the study of the 'density' of electricity when opposite charges acted on a straight line.[16] For his analysis, Avogadro used as a model two equal spheres, both insulated and touching each other. These spheres were imagined to be located at the center of a spherical surface placed at such a distance that it could not affect them. The experimental evidence indicated that the electrical density was zero at the contact point of the two spheres, and that at opposite sides they contained equal and uniform electricity.

Avogadro now aimed at a mathematical expression for the electrical density at any point on the surface of two spheres. Avogadro employed both differential and trigonometric functions to determine the surfaces of a series of spherical segments of the two touching bodies, and thereby to evaluate the electrical density that such segments would acquire at a given point. He clearly pointed out, however, that for such evaluation one should also consider the effect of the 'obliquity' of each surface element when related to the other sphere.[17] In fact, Coulomb's findings showed that the influence of the curved inductive effect was such that electrical density was less at 90° than 180° from the contact point of the two spheres.

By introducing the values of the angles above the contact point of the two spheres, Avogadro finally reached a quite complex mathematical expression for the electrical density of any given point of the surface of one of the spheres. This was:

$$\frac{2\cos\theta-1}{\sqrt{\zeta-4\cos\theta^3}}\left[\frac{1}{2}\pi+\text{arc}\left(\tan=\frac{2\cos\theta-1}{\sqrt{[(\zeta-4\cos\theta)\cos\theta]}}\right)\right]-$$

$$-4\,\frac{(1-\cos\theta\sqrt{\cos\theta})}{\zeta-4\cos\theta}+\text{arc}\left(\tan=\sqrt{\cos\theta}\right)$$

where π = electrical density
θ = angle of contact point over the line joining the two touching spheres

At 90° from the contact point, or when $\theta = 90°$, the whole trigonometric expression is equal to zero and, therefore, the electrical density is at its maximum. The opposite is true at point of contact between the spheres, where there is no free surface and where electrical density is zero. For other points on the surface, Avogadro compared his calculations with the experimental results determined earlier by Coulomb.[18]

The results of these comparisons, it is only fair to say, are not encouraging. In fact, Coulomb's figures indicated a greater electrical density at 90° from the contact point than the calculations showed. At 30°, the difference between the set of figures appeared to be even larger. By graphing these variations, Avogadro inferred that the decrease of electrical density from 90° to 0° as given by his formula was generally slower than that found by Coulomb. He thought that this might be explained by assuming that inductive action occurs in curved as well as in straight lines. He further argued that a kind of shielding effect would be greater for the surface points closer to the contact point between the two spheres (Figure 3), where *r, s, t* are tangential planes, 0 = contact point, G = point of higher electrical density (according to Coulomb), H = point at 90° from 0, C = point at 30° from 0. Accordingly, the electrical density decreased in those points more than it would if the inductive action took place only in straight lines. While he recognized that he could not mathematically evaluate the effect of curving on induction, the existence of this effect, as Faraday's experiments proved beyond doubt, was recognized by Avogadro.

This highly elaborate analysis of induction not only became known to Faraday and Thomson, but also attracted Ottaviano Mossotti's attention, when in August 1846 he published an article in *Memorie di Matematica e*

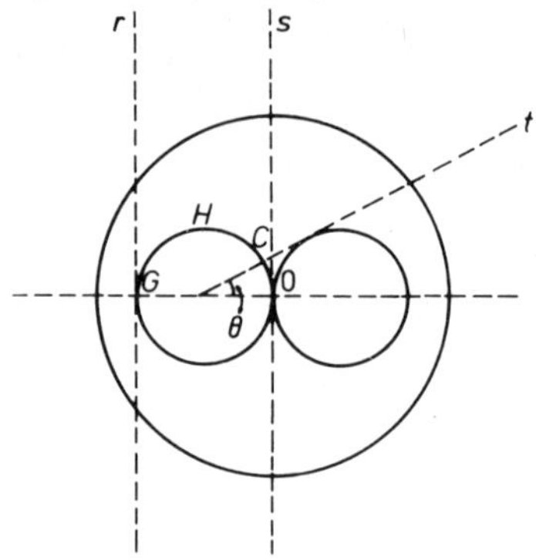

Fig. 3

Fisica which further developed some of Avogadro's ideas.[19] Mossotti[20] was not remiss in recognizing the significance of Avogadro's theory of molecular polarization of an insulator placed between oppositely charged bodies, a theory which was by then almost thirty years old.

As has been stressed earlier, Avogadro was neither a practicing chemist nor did he claim to be one. Thus, one may wonder about the motives behind his involvement in a subject which seems quite remote from his main interest.

There is supporting evidence here that Avogadro's proposal of a new system of chemical nomenclature was more than just a whim. First of all, this was an extensive work; in fact, the two essays dealing with this subject cover more than ninety pages. It should also be recalled that while he was writing them, Avogadro had returned to his full responsibilities as a professor of physics at Turin, was finishing the last volume of his *Fisica dei corpi ponderabili*, and was preparing the material for four more articles. Furthermore, Avogadro's insistence on having the two essays published indicates his eagerness to have his ideas brought into the open. A partial record exists of the early attempts for publication: on 19 July 1842, Avogadro mailed from Turin to Auguste de La Rive in Geneva the manuscript of his 'Proposizione d'un nuovo sistema di nomenclatura chimica'. He asked for its publication in the *Bibliothèque Universelle.*[21] He recognized that the article in its original form might be too long to appear in one issue, and suggested that it be divided into two parts. He also asked for the return of the manuscript in case de La Rive decided not to publish it. No reply came from Geneva, and six months later Avogadro again wrote de La Rive asking for a decision. He also feared that the manuscript had been lost in the mail. To be sure that this time his letter safely reached de La Rive, Avogadro had it personally delivered through the Sardinian Consul in Geneva.[22] A month later still no reply; at this point, Avogadro became quite concerned about his manuscript and wrote to the editor that he was 'annoyed' (fâché) by the whole situation. It is definitely surprising that, after more than seven months, de La Rive had failed to reply. After all, he and Avogadro knew each other personally,[23] and Avogadro was not a newcomer in the scientific world. The only plausible explanation is that de La Rive may have been abroad or too ill to keep up his correspondence at the time.

At any rate, four months later Avogadro submitted the manuscript of his article, now in the Italian version, to *Memorie di Matematica e Fisica* which published the complete text in 1844.[24] A reading of this essay may help to understand Avogadro's concern for its fate. Here he had good occasion to restate some of his ideas concerning chemical compositions, and he took advantage of this opportunity.

Avogadro's fundamental argument was that the chemical nomenclature, as it had been proposed by Fourcroy and Lavoisier in 1789, was no longer adequate even to describe the composition of simple salts. Furthermore, it was unsuitable for the very complex organic combinations that chemists had recently identified. Subsequent modifications of the nomenclature had not met general acceptance. Thus, it had become necessary to invent special names to avoid lengthy descriptions of new compounds, names which, in some instances, did not accurately express the nature of the compounds. As Avogadro pointed out, this profusion of names would create chaos in nomenclature. In his chemical studies, he had faced the drawbacks of the current terminology; this had convinced him of the need for a new method of defining chemical compounds.[25]

For the chemical elements, Avogadro suggested that only minor modifications would be in order. For example, *azoto* and *potassio*[26] should respectively replace the terms *nitrogeno* and *calio* (or kalium) which existed in the terminology proposed by Berzelius and other chemists, especially those outside France. For both organic and inorganic compounds Avogadro suggested a classification based upon taxonomic criteria similar to those adopted by biological scientists: a generic name to indicate the elements present in a compound, and a specific term to represent the proportions in which those elements combined. However, unlike botany and zoology, chemistry's generic names should be such that they could be used also to define compounds which do not yet exist in nature or have not been prepared in the laboratory. Avogadro criticized such terms already adopted by some organic chemists as benzamide, aldehyde, etc., because, although they might indicate chemical composition, they followed no common or general classification. What Avogadro advocated was a name which would immediately denote the presence of all the components of a given substance or, in other words, would provide a literary translation of the chemical formulas. He referred to his new names as 'formule parlate' (spoken formulas).

When writing these spoken formulas, electronegative elements were to precede the electropositive ones. Thus, in oxides, chlorides and, sulfides, respectively, oxygen, chlorine, and sulfur were to be mentioned before the metals. As for the spelling of the new names, Avogadro recommended the adoption of the Latin language which, in his view, could easily have been adopted by the French, the Italians, and also by the Germans with some slight modifications.[27] Some samples illustrate the new nomenclature: the oxides of lead, copper, and sodium would be *oxyplumbum, oxycuprum, oxysodium*. For ternary compounds, *ocarpotium* would mean a combination of oxygen, carbon, and potassium.

Avogadro's discussion of the second component of the new nomenclature, i.e., the specific name,[28] is especially interesting to the historian. Avogadro had suggested that in addition to the generic name, a specific name should signify the proportions of each element in a given compound. However, a question at once arises. Should these proportions refer to the number of *equivalents* or to the number of *atoms*?

Avogadro's answer is quite revealing. While some chemists might still disagree on the notion itself of atom, almost all agreed on the numerical value to be assigned to the equivalents. As specific examples, O and S can denote either the equivalent or the atom of oxygen and sulfur, but H_2 and Cl_2 are one equivalent (or 'double atoms', the term he used) of hydrogen and chlorine, etc. For Avogadro the equivalents and what he called 'Berzelius' atoms' coincided for most metals.[29]

Yet, still unsettled in his view was the situation of nitrogen and carbon, on whose weights chemists did not agree. For carbon, Avogadro speculated that atom and equivalent would coincide. This meant, as he himself pointed out, that CO and CO_2 (rather than C_2O and CO) would be the correct formulae for carbon monoxide and dioxide, respectively. For nitrogen, he shared the prevailing view that its equivalent was the double of Berzelius' atom, or in modern symbols, N_2. The same, he wrote, would hold for phosphorous and arsenic. For silicon, Avogadro repeated his earlier conclusion that this element combines with two atoms of oxygen to give silica, analogous to the formation of carbon dioxide from carbon. Berzelius opposed this interpretation, as Avogadro himself pointed out.

Once established, the smallest whole number of chemical equivalents present in a given compound would be used to indicate the specific composition of that substance. These numbers, for example, would be three and one for oxygen and sulfur respectively in sulfuric acid (oxide). Among the examples given by Avogadro, he cited water whose 'relative equivalents' for him were *one* hydrogen and *one* oxygen.[30] In binary compounds, such terms as simplex, duplum, triplum, quadruplum, etc., indicated the number of equivalents, for instance, of oxygen combining with the metallic element. As examples of ternary compounds, terms as semi-simplex, tertio-simplex, tertio-duplum. respectively referred to 1-2-2, 1-3-3, and 2-3-3, equivalents. Such a system of nomenclature, relatively simple for small numbers, becomes awkward for more complex compounds. For example, vigiplum meant twenty and dogitriplum meant thirty-two.

Very significant here is Avogadro's emphasis on the nature of a compound (oxazotum, for example, to indicate oxygen and nitrogen) and on the proportion of its components rather than on its chemical properties. Accordingly, to designate a substance as an 'acid' was superfluous, Avogadro argued; in fact, acid had been introduced by Lavoisier and Fourcroy only to indicate a 'higher degree of oxidation' in certain substances. For thirty-five years, Avogadro had not changed his mind on this topic.[31]

By 1850, at the very eve of his retirement from active teaching at Turin, Avogadro published a second memoir on chemical nomenclature in the *Memorie di Matematica e Fisica.*[32] He followed nearly the same outline of his previous 1844 article, restating that his nomenclature had no relation to the 'constitutional or rational formula' of the compounds. In the six years since the first article, however, at least two obstacles had risen which complicated the application of Avogadro's nomenclature. One was the ever-increasing complexity of the new organic compounds being identified; the second arose from the introduction of the type theory and its growing acceptance by many chemists in the 1840s. This had resulted in lengthy and complicated names for many new compounds. Consequently, simplicity and easy pronunciation of their names, two of Avogadro's goals in his proposal, seemed no longer within easy reach, as he conceded in the 1850 memoir. According to the type theory, Avogadro argued, one, two, or more equivalents in a given radical were replaced by a 'group' of equivalents. The resulting new compound was then to be considered as a 'variety' of the original radical or 'primitive type'. This variety also required a qualification, such as a special adjective, to be added to the generic name of the original compound.

For instance, the qualifying prefix *uniclorico, biclorico*, or *triclorico* must be affixed to the original radical to indicate the replacement of one, two, or three equivalents of hydrogen by the same number of chlorine equivalents. In case of two elements replacing hydrogen in a radical, then both should appear: biclo-tribromico, meaning two chlorine and three bromine equivalents entering the radical. As for radicals themselves, Avogadro now proposed for the sake of simplicity to replace their full name with an abbreviation. In this way, the radical *etereno* C_4H_4 – whose name should be changed to *lampeno* as more appropriate, according to Avogadro[33] – was to be written $\overline{L}$, where the line above the letter means that such a group of equivalents is a radical.

Interestingly, neither of these two essays indicate that Avogadro was aware of similar work on chemical nomenclature being conducted in other countries. In the 1844 article, he cited some of the terms for napthalene derivatives from a series of reports published by Laurent in the *Comptes Rendus* of the Paris Academy. Yet, Avogadro found these terms impractical, too complicated, and confusing. Avogadro seemed unfamiliar with Gerhardt's *Précis de chimie organique*, which had been published between Avogadro's first and second essay on nomenclature and which contained a classification by genera and species similar to that which Avogadro had proposed. In 1848, Leopold Gmelin, too, in his famous *Handbuch der Chemie* had tried to put some order in the names of organic compounds by indicating in the word which elements and how many of them would combine to form the compound. In 1850, Avogadro failed to cite this work which shared a number of his proposals.

To place his attempt for a new nomenclature in its proper historical setting, it is only fair to recall that as long as the debates continued between the 'equivalentists' and those who supported Berzelius' atomic weights, all efforts to reach a quantitative nomenclature would remain futile.[34]

One more comment should be made concerning these essays on chemical nomenclature. In almost one hundred pages of writing about chemical compounds, Avogadro never used the term molecule. This is quite surprising, because early in 1849 he had published a short note on the need to distinguish between integral molecules and equivalents.[35] One can only speculate about such a surprising omission in Avogadro's terminology. Very likely, the omission was intentional and may have to do with the perception that molecules, holding in Avogadro's view more physical than chemical meaning, did not belong properly in a work dealing with chemical nomenclature.[36] Alternatively – but this is a less plausible explanation – Avogadro may have realized that in the chemical world of the 1840s there was really no interest in a term, molecule, which then seemed obsolete to some, confusing or meaningless to others.[37] Now, in 1850, he used 'atoms' when discussing oxygen and he referred to 'double atoms' when referring to hydrogen and nitrogen. It is in some way bewildering that in his last article the author of the molecular hypothesis failed even to mention those 'entities' which had formed so great a part of his whole conceptual framework.

A series of experimental investigations on the properties of gases under compression conducted by Henri Regnault a few years earlier induced Avogadro to work on this topic in which he had not previously researched. Regnault with his thoroughness had determined the ratio of pressure to density, near the liquefaction point, for air, nitrogen, hydrogen, and carbon dioxide.[38] His conclusions were that, contrary to prevailing opinion, at ordinary temperature, gases – with the exception of hydrogen – could be compressed more than the Boyle–Mariotte Law would allow.[39]

Three years after Regnault's article, Marcelin Berthelot, then only twenty-three had established that, for certain gases, pressure alone was insufficient to cause a change in their physical state.[40]

By 1850, Avogadro had enough information to draft his paper which he read before the Royal Academy of Sciences of Turin in June 1851; it was published in French in the Memoirs of the same Academy three years before his death in 1856.[41] This article is one more example of the type of work Avogadro had been conducting in theoretical physics for most of his life: a purely mathematical exercise based upon somebody else's experimental findings. As stated earlier, the subject was new to him; yet, one can better understand his interest in it by recalling that, beginning with his celebrated 1811 essay, he had closely studied the relationship between gas densities and molecular distances. Forty years later, molecular forces – now divested of the caloric theory – still represented a focus of attention for Avogadro.

Now, the formulation of a mathematical expression to replace the Boyle–Mariotte Law for some gases was Avogadro's aim in the 1851 memoir. He was convinced by Regnault's findings that a 'loi générale de la compressibilité des gaz parfaits' could be found. To this effect he found Regnault's equations 'arbitrary and complicated', and he proposed to replace them with a simpler exponential equation involving only one term. This mathematical expression would specify the density at which a given gas liquefied, and the pressure under constant temperature at which the liquefaction occurred. Avogadro admitted that it would not be feasible to establish a general relation solely on the basis of theoretical considerations to replace the Boyle–Mariotte Law. Yet, an equation 'empirique' might be formulated by drawing from Regnault's results. Such an equation would predict the approximate behavior of perfect gases.

As a further aim of this study, Avogadro sought to determine the variations required in the general equation when applied to other gases examined by Regnault. In the first part of his article he directed his attention to hydrogen, carbon dioxide, nitrogen, and oxygen. In the second part he attempted to correlate the general equation of the compressibility of gases which he had earlier proposed with the pressure under which they liquefied.

The law of compressibility for hydrogen, within the range of its densities as determined by Regnault, appeared to yield values quite close to those observed for the perfect gases. In fact, r/m (pressure/density) for hydrogen at 3°C deviated only slightly from the value given by the Boyle–Mariotte Law. The equation in second degree of m proposed by Regnault to calculate r/m and which introduced two coefficients (experimentally determined) gave, in Avogadro's view, unacceptable results. For instance, for $m = 0$, $r/m = 0.999$.[42] In fact, the variations of r/m were just small fractions for $m = 2, 4, 8, 16$, and the decreasing values of m between 1 and 0 were sensitive to small decreases of r/m. Accordingly, even for $m = 0$ the value of r/m would still be slightly smaller than 1 and the value of $(r/m) - 1$ would be equal to a small negative fraction instead of being $r/m = 0$ or $(r/m) - 1 = -1$. This convinced Avogadro to use an exponential equation such as

$$a^{(r/m)-1} = m \tag{I}$$

where a is a constant determined experimentally. In this way, an algebraic function of $(r/m) - 1$ will change according to an arithmetic progression, while m changes in geometric progression. By adopting this equation, $(r/m) - 1 = 0$ and $r/m = 1$.[43]

The next step for Avogadro was to replace $(r/m) - 1$ in equation (I) with another exponential term. To do this he used a fractionary power of $(r/m) - 1$ and determined its exponent as well as the value of its base a. Equation (I) thus became $a^{((r/m)-1)^x} = m$ and, therefore,

$$\left(\frac{r}{m} - 1\right)^x = \frac{1}{\log a} \log m = A \log m \tag{II}$$

where $A = 1/\log a$. Avogadro, by introducing the values for m and r found by Regnault, thus proceeded to calculate x and A in equation (II). Although the experimental findings for the density of the gas gave values for x ranging from 0.33 to 0.54, he still thought he could adopt as exponent $x = \frac{1}{3}$.[44]

For A Avogadro also determined four different values, and decided (through quite unconvincing logic) to adopt for this coefficient not their average, but the highest value.

The final exponential equation thus became:

$$\left(\frac{r}{m} - 1\right)^{1/3} = 0.250\,73 \cdot \log m \quad \text{or} \quad 9735\ \sqrt[3]{((r/m) - 1)} = m \qquad \text{(III)}$$

which was not influenced by the logarithmic system as equation (II) was. By adopting equation (III) Avogadro indicated that r/m should go to zero even when m was not zero, an important condition for the law of compressibility of gases. In fact, according to (III), when $r/m = 0$, the density of the gas would be approximately 0.00010. At this very small density, there was no sensible repulsive force among the integral molecules of the gas.[45] When he checked r/m, determined by using his equation against Regnault's experimental values, Avogadro noticed that the difference between the two sets of figures increased almost ten times, and in direct proportion to the increased density of the gas. To account for this difference, he suggested that approaching the point of liquefaction would make the gas more and more compressible.

In discussing a mathematical expression for determining r/m for carbon dioxide, Avogadro made the only reference to his 1811 hypothesis. He found 'remarquable' that the density of carbon dioxide at the atmospheric pressure was almost identical to its 'poids atomique'.[46] This density of 1.5202 (relative to air) gave in fact for carbon dioxide a weight twenty-two times that of hydrogen, a figure close to that calculated by Avogadro as early as 1811 in his 'Essai d'une manière'. For carbon dioxide, Avogadro formulated another equation relating pressure to density. The 'Formule de compressibilité' was quite similar to that he had adopted for hydrogen, except that the numerical value assigned to the coefficient of the term indicating the 'tendency' of the gas to liquefy was eight times the same coefficient for hydrogen. In doing this, and in relying on Regnault's experimental data, Avogadro made some questionable assumptions, including the claim that, by interpolation, the density of carbon dioxide would not further decrease at pressures lower than 190 mm of mercury. Yet, Avogadro noted how close his values and Regnault's results were for carbon dioxide, and he felt quite confident about the validity of his general equation for gases.

Next he extended his calculations to nitrogen and oxygen. For nitrogen, he noted that it could not be liquefied even at very high pressures and low temperatures, and he assumed (quite safely) that the influence of the 'tendency' to liquefy could not be felt at a density greater than that which existed under 1 meter of mercury of pressure. For oxygen, Avogadro, as Regnault had done earlier, had to use indirect numerical values drawn from those of air and nitrogen. However, he found himself in a quandary: how are nitrogen and oxygen distributed in the air and how is air pressure divided between the two component gases? Avogadro recalled on this occasion Dalton's generalization, by then over half a century old. In a gas mixture, each component gas would expand throughout it as if the other component did not exist. Accordingly, nitrogen and oxygen would be subjected to partial pressures proportional to their densities, or respectively 79% and 21% of the atmospheric pressure. The alternative was to consider each gas in a mixture as divided into an indefinite number of small particles scattered throughout similar small particles of the other gas. In this case, each gas will behave as a separate horizontal layer, with each layer subjected to the atmospheric pressure.

Avogadro defined this second interpretation as 'l'idée la plus simple',[47] while recognizing at once that his calculations leading to a general law of compressibility of gases would differ depending on which of the theories of gas mixture would be adopted.[48] When he applied Dalton's theory of gases to Regnault's experimental results on air, using the equation of the compressibility of nitrogen he had earlier established, Avogadro found 'unacceptable results'. For instance, the ratio r/m for oxygen would decrease too quickly with regard to m to be considered valid in a gas which, at the pressures experimentally determined, still seemed to be far away from its liquefaction point. Therefore, Avogadro decided to adopt the layers theory of gas mixtures as more congenial to his mathematical interpretation of Regnault's data. He then assumed that an equation similar to that previously set for the other gases could also be used for the compressibility of oxygen if new values would be introduced into the coefficients.[49]

In determining the coefficients, he conceded that one value was quite anomalous with respect to their average. To justify this, Avogadro argued that in a mixture of gases like air, the molecular distance is no longer related to the density that each gas would have if it were in a separate layer.[50] Eventually, after discarding another anomalous value in the computation of coefficients, Avogadro arrived at another 'formule générale' for oxygen. He again used this equation to calculate values of m and r which he compared with those determined by Regnault. The two sets of figures agreed rather closely, and Avogadro felt quite confident about the validity of his equation for oxygen. Yet, he admitted that the layers' theory of gas mixtures on which it was based was not 'perhaps free from objections'.

In the concluding section of the article, Avogadro discussed the application of a modified equation with three terms to evaluate the pressure-to-density ratio of gases earlier examined. For hydrogen, for instance, $r/m = 1 + 0.000\,412\,38(m-1) + 0.000\,036\,53(m-1)^2 - 0.000\,001\,2632(m-1)^3$. With this equation, he concluded that the maximum value of r/m was for $m = 25$, and that above this density, r/m began to decrease and was less than one when the gas approached its liquefaction point. At that point, the equation of compressibility no longer applied.

Avogadro, although aware that the experimental value for the liquefaction point of hydrogen was not yet available, still attempted to obtain this figure by extrapolation. He concluded that at a temperature 'above 0°C' (but not better defined) hydrogen would liquefy under a pressure of 285 atmospheres, and that its density would then be about 470 times that at one atmosphere. However, he did not fail to mention that these were 'approximate figures'.[51]

A review of Avogadro's 1851 article indicates that he had not been impressed by Regnault's assumption, six years earlier, that there was a range of temperatures inside of which gases followed Boyle–Mariotte's Law. By the 1840s, experiments conducted on permanent gases clearly indicated that, even at extremely high pressures, gases could not be liquefied at ordinary temperatures.[52] Consistent with one of his main conceptions, Avogadro, in formulating his equations on the behavior of certain gases when they ceased to be perfect, had focused his attention on the relationships between pressures and densities, but always within a narrow range of temperatures (around 3–4°C). As Regnault had done, Avogadro took these temperatures almost as a constant in his calculations, while he speculated on what the densities of these gases would be when they began to liquefy. The ratio of these densities he found to be approximately 16: 8: 4: 2 for hydrogen, nitrogen, oxygen, and carbon dioxide, a surprising regularity that he, however, failed to notice. In fact, he avoided any interpretation of the influence of temperature on liquefaction.

Avogadro recognized – as Regnault and others before him had done – the Boyle–Mariotte Law as a 'loi limitê' that would be valid only within definite limits of density and pressure of the fluid under consideration. And he concluded that such law should be modified when a gas approached its liquefaction point.

The far flung and yet distant association that Avogadro had for most of his life with the French physicists had begun as early as 1816 with men such as Berard and Delaroche; it was still clearly detectable forty years later when he devoted this long memoir, his last, to an attempt to 'refine' some general conclusions on the behavior of gases set forth by one of the most reputable experimentalists of the time. In almost half a century of scientific production Avogadro's style had changed very little: with a good control of the mathematical instrument he went through a long series of elaborate computations to reach his conclusions. To do this he relied on the experimentalists's figures; sometimes he accepted them, most of the time he tried to change them to fit what in his view was a better theory or a new generalization. In 1851 Avogadro considered Regnault's 'formules tout-à-fait arbitraires' and labored for seventy pages to demonstrate that his own path would lead to nothing less than to a law replacing Boyle–Mariotte's. For a seventy-five-year-old man at the periphery of the scientific world the task seems ambitious to say the least. Was then Avogadro conceited to the extent to lose the perspective of the quality of his work? Probably it would be fairer to say that he was above all the victim of a cultural involution which had its roots in a bygone age, and of which he himself seemed unaware.

NOTES

1. *Mem. R. Acc. Sci. Torino,* **36** (1833), 215.
2. *Ann. Chim.* **49** (1833), 369.
3. As Avogadro himself recalled, in 1819 he thought he had found a mathematical function relating the expansion of water to its vapor pressure. *Giornale di Fisica,* **2** (1819), 187. Now, at the conclusion of the 1833 essay, he recognized that the equation could not be applied to mercury vapors.
4. 300°C was the upper limit of most mercury thermometers then used.
5. Avogadro cited Faraday's work here.
6. *A* is vapor pressure at boiling temperature of a given liquid.
7. Just for reference, current values of the vapor pressure of mercury are 0.27 mm at 100°C and $1.8 \cdot 10^{-4}$ mm at 0°C.
8. This indicates that people working in laboratories with this metal were becoming increasingly aware of the health hazard involved.
9. See Chapter 4 of this work on the reaction to the molecular hypothesis after 1811.
10. Reference is made here to the theoretical work conducted by Avogadro between 1806 and 1807 and examined in the second chapter of this book.
11. This also has been discussed in Chapter 2.
12. *Mem. Mat. Fis. Soc. Ital. Sci.* **23** (1844), 156. It is surprising that Avogadro, as a full-fledged member, did not receive priority over other contributors to the *Memorie*. There were only forty members of the Society at any given time.
13. See *The Selected Correspondence of Michael Faraday*, vol. 1, ed. L. P. Williams (Cambridge University Press, 1971), p. 458.

14. *Mem. de l'Institut,* **12** (1811), n. 37.
15. *Mem. Mat. Fis. Soc. Ital. Sci.* **23** (1844), 157.
16. This was the only explanation offered by the action-at-a-distance theory, as presented by Poisson and Coulomb.
17. *Mem. Mat. Fis. Soc. Ital. Sci.* **23** (1844), 167.
18. Avogadro quoted the 'Mémoire sur l'électricité' which Coulomb published in 1787; for these experiments Coulomb had used small rings of gold foil to detect the electric charges on the surface of two spherical bodies.
19. *Mem. Mat. Fis. Soc. Ital. Sci.* **24**, s. 2 (1846), 49.
20. Whose 1836 essay 'Sur les forces qui regissent la constitution intérieure des corps' had immediately aroused Faraday's interest.
21. 'I could have sent it to the editors of the *Annales de Chimie et Physique*', he wrote to de La Rive, 'but it seemed doubtful that the chemists of Paris would undertake to make known the work of a foreigner on a subject in which they took a pledge ... to give themselves the law. ...' Quite a revealing statement of the feelings that Avogadro had for the French chemical circles of the time. This letter, with others mentioned below in the text, belongs to a small collection found in de La Rive archives in the Bibliothèque Publique et Universitaire of Geneva.
22. See letter from Turin dated 9 December 1842. According to the stamped date on the envelopes, the mail took two days from Turin to Geneva.
23. They had met in Turin during the second national congress of the Italian scientists in 1840, and certainly were familiar with each other's work. In 1843 de La Rive had been reappointed Rector of the Académie of Geneva, which may have kept him very busy at the time.
24. *Mem. Mat. Fis. Soc. Ital. Sci.* **23** (1844), 260.
25. Ibid., 261.
26. Azoto and potassio are the terms still used by Italian chemists, for instance, ossidi di azoto, cloruro di azoto, while they say acido nitrico, nitrobenzolo, etc.
27. Nothing is said about English.
28. *Mem. Mat. Fis. Soc. Ital. Sci.* **23** (1844), 273.
29. Ibid., 273.
30. His earlier interpretation of water in terms of the relative combining volumes of hydrogen to oxygen seems forgotten here.
31. In the *J. Phys.* **69** (1809), 142, he had for the first time argued that acidity and alkalinity were only relative properties.
32. *Mem. Mat. Fis. Soc. Ital. Sci.* **24** (1850), 166.
33. Because the olefiant gas contained in ethylene was used for lamps lighting streets.
34. This has been examined in detail by M. Crosland in the fifth section of his *Historical Studies in the Language of Chemistry* (Cambridge, Mass., 1962), pp. 317 ff.
35. *Bibl. Univ. Genève,* **11** (1849), 285, 'Note sur la necessité de distinguer les molecules intégrantes des corps de leur equivalents chimiques ... '.
36. When defining atoms, Avogadro quite characteristically qualified them as 'chemical atoms' or alternatively as Berzelius' atoms.
37. See for instance, A. Laurent's *Méthode de Chimie* (written in the late 1840s, published in 1854) – now considered a classic by historians – where the term molecule appears only when the author refers to 'molecular groups' or complex organic substances, while he uses 'atoms' to refer to compounds such as water or to gaseous elements (hydrogen, chlorine, etc.).
38. See 'Mémoire sur la compressibilité des fluids elastiques', *Mem. Acad. Sci. Paris,* **21** (1847), 329–428.
39. Avogadro, following the French tradition, always spoke of Mariotte's rather than Boyle's Law.

40. In 1850, Berthelot did not succeed in the liquefaction of oxygen even by subjecting it at 780 atm. See *Ann. Chim.* **30** (1850), 237–24.
41. *Mem. Reale Accademia Scienze Torino*, s. 2 (1853), 171. From here referred to as *MRAST*.
42. For Avogadro r/m becomes zero when density decreases even before $m = 0$ or for a definite value of density. If one makes $m = 0$ in Regnault's equation $r/m = 1 + A(m-1) + B(m-1)^2$ one finds that $r/m = 0.999$, slightly less than one, or the value that r/m should have according to Mariotte–Boyle's Law. This does not reconcile with r/m becoming zero before m becomes zero, according to Regnault's equation; on the contrary, r/m is almost equal to one even when density is zero. See *MRAST*, s. 2 (1853), p. 179.
43. Ibid., 180.
44. Avogadro argued that in $(r/m) - 1 = A^2 (\log m)^2$ and $(r/m) - 1 = A^3 (\log m)^3$; $(\log m)^3 < (\log m)^2$ and coefficient $A^3 > A^2$ and, therefore, if $x = 1/3$, then $(r/m) - 1$ has a faster rate of increase in relation to m.
45. By his exponential equation, Avogadro calculated that at 760 mm of mercury, the density of hydrogen would be $m = 0.76002$ (and $m = 1$ under one meter of mercury of pressure) which was slightly higher than that expected from the Boyle–Mariotte Law. He found this value in agreement with his assumption that m should decrease in relation to pressure r but less than that law would require. *MRAST*, s. 2 (1858), 189.
46. Ibid., 180.
47. Ibid., 213.
48. In fact, Avogadro calculated that by adopting Dalton's theory of gases, at a pressure of one meter of mercury, nitrogen's density would be 79 percent of that observed at that pressure. In the other theory, under the same pressure of one meter of mercury, both nitrogen and oxygen would retain the same density of one, and each gas would contribute to a partial pressure in proportion to its percentage in air.
49. It may be interesting to note that in calculating one of these coefficients for different densities of oxygen, Avogadro had arrived at the values 1.191, 1.131, and 0.883; he then considered it legitimate to average these figures to 1.068, despite their range of over 30 percent.
50. Density, for elementary gases, in Avogadro's interpretation, has always been closely linked to the molecular distances.
51. Actually, the ratio of density of liquid hydrogen at –239°C to that of hydrogen at 0°C and 760 mm, is 674.
52. After 1844, J. Netterer reached pressures up to 2800 atmospheres without changing the physical state of the gas involved.

CONCLUDING REMARKS

A prolific writer – he published scores of articles and a massive treatise on theoretical physics – Amedeo Avogadro remains known, 125 years after his death, only for two statements he made almost at the outset of his scientific life. With few exceptions, the rest of his work has been ignored.[1] The most acceptable explanation for this treatment by historians is that his writings may seem as irrelevant today as they were when they first appeared.

A first aim of this book has been to provide a reasonable historical background to the scientific biography of this little known figure. Here I have suggested that Amedeo Avogadro cannot be considered an Italian scientist in the sense in which we consider Dalton an Englishman or Lavoisier a Frenchman. In his times, Italy was a geographical expression, and Avogadro, Count of Quaregna, was a Piedmontese, a citizen of the Sardinian States, for most of his life.[2] He can be seen as the product of a close, narrow-minded society which bore little resemblance to the cosmopolitanism of the Florentines and the vitality of the Milanese. Turin, the capital of the Sardinian States, where Avogadro was born and spent most of his life, was a small, provincial town, full of barracks and churches, and very French in its outlook and its lifestyle.

A further purpose of this book has been to offer a wider group of readers the opportunity to become familiar with the content of all the major studies conducted by Avogadro in a secluded corner of Europe through half a century; studies which, needless to say, have to be examined within the perspective of the time and circumstances in which they were pursued. The main theme emerging from Avogadro's writings is his persistence in linking together the physical and chemical properties of substances. In his vision there is a web which interrelates such diverse characteristics as vapor densities, molecular sizes, electrochemical potentials, specific heats, weights of chemical elements, and affinities for caloric.

This associative trend is already present in the early essays on electricity and electrochemistry, and can still be detected more than forty years later in the series of memoirs devoted to the atomic volumes. When, in 1811, Avogadro set forth his gas hypothesis in the 'Essai sur la manière' he was consistent with his main conceptual system, by again establishing a link between molecular sizes, densities, weights of elements, and their compounds; he also related directly combining volumes and the number of combining molecules. This was not a causal attempt, but, as he claimed, he saw, through the application of his hypothesis, a method to remove the uncertainties and the confusion then prevailing among those who were trying to establish the correct weights of the chemical elements and compounds.

Yet Avogadro failed to convince his fellow scientists of the validity of the ideas deriving from his 1811 generalization. The molecular hypothesis remains an example of those scientific assumptions which experienced long delays before their general acceptance. The reasons for the belated adoption of Avogadro's hypothesis are complex and multifold, and have been analyzed in detail in Chapter 3. Here I will sketch briefly the shifts in appreciation of this important scientific statement since its formulation.

After being announced in 1811, the molecular hypothèsis was ignored, rejected, or misunderstood for almost fifty years. It was then debated for another quarter century before being rapidly recognized and restated as a major physical law. At present, it is undergoing a new process of critical analysis by historians and philosophers of science.

The understanding of Avogadro's main contribution may be evaluated along three lines: the traditional historiography, the orthodox textbook interpretation,[3] and the current historical-philosophical reassessment. The current trend has been to question a number of myths which, as a result of the conventional historical approach, had found deep roots in the textbook tradition.

Historians of chemistry, beginning in the 1860s, created many legends on the subject. One of the most misleading of these is that Cannizzaro, as a result of his presentation at the Karlsruhe Congress in 1860, convinced his fellow scientists of the crucial importance of Avogadro's contribution. That this was not the case was demonstrated by Cannizzaro himself in May 1872, when he gave the Faraday Lecture to the Chemical Society in London. This event came six years after the atomic debates at the Chemical Society and represented a remarkable attempt to provide a full conceptual and historical background to the atomic-molecular theories, as discussed in Chapter 5. Had Cannizzaro firmly established his position at Karlsruhe, as many historians maintain, the London lecture would have been an historical curiosity rather than a vigorous statement in a continuing scientific debate.

Cannizzaro is generally considered the key figure in rediscovering Avogadro's hypothesis. A careful reading of his famous 'Sunto di un corso'

(1858) suggests that he went beyond simple rediscovery to the creation of a fundamental myth: that Avogadro had proposed that the molecules of gaseous elements were diatomic.[4] Not only did Cannizzaro seem to ignore Avogadro's assertion that all elements, in the gaseous state, including the metals, were polyatomic, but he also seemed totally unaware of the fact that Avogadro had extended his gas generalization to solid elements and their compounds.

In my view, Cannizzaro misread Avogadro because of the vast conceptual gulf between these two men which extended quite beyond their fifty-year age difference. As Cannizzaro himself conceded in his 'Sunto', after the work of Berthelot, Wurtz, Hofmann, Williamson, Frankland, Gerhardt, and Clausius, he had at his disposal a wealth of information earlier unavailable. Cannizzaro then invested the notion of molecules with what has become our present meaning, something which it was not in Avogadro's power to do. Thus, it is fair to say that Cannizzaro did not merely 'resurrect' the original hypothesis, but by translating it into its modern terms, made it more acceptable to physical scientists of his time. While Avogadro had visualized a two-tier system of particles in his system of constitution of matter, Cannizzaro interpreted this as a clear distinction between atoms and molecules.

As indicated in Chapter 5, Lothar Meyer fully accepted Cannizzaro's version of Avogadro, as did Mendeleev. Both their textbooks, first published in 1864 and 1868, respectively, were extremely successful. In 1869, Alexander Naumann, a physical chemist, published in the *Chemische Berichte* a short clear note, where the equal volumes–equal numbers generalization was qualified, probably for the first time, as 'Avogadro's Law'. By the 1880s, the gas hypothesis had been elevated to the present status of universal recognition.[5] Subsequently, the post-Sunto interpretation of the molecular hypothesis was kept alive through three generations of historians, beginning with Wurtz and Hoefer and including Kopp, Pattison-Muir, Ostwald, Thorpe, Meyer, Graebe, Guareschi, Giua, Moore, Partington, Mason, and many others to our own day.[6]

From this deep-rooted historical tradition began what I call the orthodox textbook account of Avogadro's hypothesis. And through explanations that have changed little in the past century, multitudes of students have learned in their introductory chemistry courses whatever they know about Avogadro, Dalton, Gay-Lussac, their theories and their laws. These textbooks give the impression that for their authors the temptation of weaving history into science is too strong to resist; unfortunately, the result of these efforts is not always successful. The most common error found is that Avogadro proposed a diatomic structure for all elementary gaseous molecules, which is, of course, Cannizzaro's version.[7] Another common error of all textbooks is that they never mention that Avogadro failed to adopt (even twenty years after Berzelius introduced them) chemical symbols and numerical subscripts which would have made his theory more understandable.[8] While some textbooks almost shun the historical issues, others are good examples of whiggish history (such as to fault John Dalton for the fact that by 1890 some chemists were still reluctant to give up HO as the formula for water.)[9] Those textbook writers who attempt to explore the reasons behind the rejection of the molecular hypothesis, generally neglect the towering influence of Berzelius and his ideas at the time and the very negative effect that Dumas's early understanding of the gas generalization had on its acceptance, as discussed in Chapter 4. A common and serious misapprehension, quite misleading for those unfamiliar with the history of the period, arises from the introduction in some books of the molecular hypothesis in the context of the kinetic theory of gases, thus suggesting that Avogadro followed the same path.[10] Yet no mention is made of the caloric theory of heat which actually served as the foundation of his whole conceptual system. And in the textbooks we find many recurring minor blunders, such as that Avogadro was a practicing chemist, a teacher of Cannizzaro (they never met), and a university professor when he proposed his generalization.[11]

It may be appropriate to note at this point that both the writers of textbooks and those who earlier inspired them, the traditional historiographers, are not and were not professional historians. History of science is of course a relatively young field, and for over a century it was principally the preserve of practicing scientists. Scientists often display a factual, pragmatic approach in their historical accounts; in so doing they may lack the in-depth analysis that serious history requires. The influence and interdependence of social, political, economic, and personal events on scientific developments is either neglected or only hinted at.[12]

Another trend has been emerging in reaction to the classical historiographic tradition. Professional historians and philosophers of science have begun to appraise major scientific breakthroughs in a new critical light. Avogadro's contribution is among them. In their analysis, some philosophers of science criticize the inductivist approach of the conventional history of science. In so doing they have reassessed the meaning of Avogadro's hypothesis, now considered as an *ad hoc* scheme to solve an apparent riddle. In this view, Cannizzaro's research represents a definite improvement over previous positions, while Avogadro's work can be seen as deteriorating.[13] For others, Avogadro's generalization is qualified as empirically progressive.[14]

Side by side with the philosophical appraisal, some contemporary historians of science assess Avogadro's work as a collection of wild generalizations, and characterize him as a 'scientific fossil' because in his later years he remained bound to an obsolete scientific tradition.[15] In some cases the overpowering effect of the Laplacian school of physics and of the French physicalist tradition on Avogadro's thought is emphasized, and his position outside the context of the atomic theory is stressed.[16] A recent view argued that practicing chemists could do without the molecular hypothesis, and accordingly they ignored it.[17] The increasing adoption of the concept of equivalents up to and past the middle of the century is seen as confirming the trend. As for organic chemists, until Kekulé, radicals for them were more meaningful than molecular aggregations.[18]

Some of the points raised here have been discussed in the main text.[19] With regard to the influence exerted by the Laplacian school on Avogadro, my main argument is that the molecular hypothesis could not have been proposed by Avogadro had he been a strict disciple of that school. The sense of loyalty among the members of the Berthollet's and Laplace's circle was so deep that Gay-Lussac had not the confidence to propose in 1808 the same concepts that Avogadro would set forth only two years later. It has been pointed out, correctly, that the 1808 memoir on the combination of gases was the 'by-product' of the Arcueil Society and, accordingly, Gay-Lussac had to be extremely cautious in moving over to any consideration which could be considered as not in harmony with that of his confreres.[20]

The debate over the neglect of Avogadro's hypothesis is far from being over, as some very recent and extensive research indicates. A more balanced overall view of the etiology of such neglect is now quite properly advocated.[21]

Let me state now my position vis-à-vis the current historical and philosophical evaluation of the molecular hypothesis. There is no doubt that, as earlier suggested, Cannizzaro's theories were more progressive than Avogadro's and played a major role in developing the atomic program;[22] this despite the fact that the 1858 'Sunto' had limited impact on scientific circles when it appeared. It is also true that, while Avogadro was vague about the degree of submolecularity of chemical elements, Cannizzaro fully grasped this important feature of Avogadro's system and gave it its modern meaning.

This said, I have to part company with many positions, both interpretative and factual, marked out by the current criticism. Were Avogadro's main assumptions based upon unscientific analysis and was his later work completely without value?[23] That his experimental work was almost nil there is no doubt. And his speculations based on the experimental work of others often led him into blind alleys. Yet the validity of his main assumptions is difficult to challenge. The thesis that there was no way to check the proposed existence of polyatomic molecules is refuted by the experimental evidence from the chemical behavior of gases which could not be interpreted otherwise. Some of the new critics do not explain why Berzelius refused to recognize the polyatomic nature of the molecules of the elements. This recognition would have involved – as discussed in Chapter 4 – an unacceptable violation of one of Berzelius' main tenets, the electrical nature of the chemical combinations; something he was not ready to do even twenty years after the gas hypothesis had been proposed.

There seems to be sufficient evidence to support the view that Avogadro – while considering the theory of definite proportions the 'base of the entire modern chemistry' (in his own words) – took over where Gay-Lussac had left off in his 1808 essay on the combination of gases. Thus, the concept of 'condensation' (or 'rapprochement') of volumes in relation to condensation of molecules, as I outlined in Chapter 3, fully escaped Gay-Lussac, but was accounted for by Avogadro. I argue further that neither Gay-Lussac nor Berzelius could offer an explanation of the densities of the compound gases in terms of the density of their components, unless they assumed a splitting of the particles of these gases. The idea of linking densities to weights of the chemical elements was hinted at by both Dalton and Gay-Lussac, but it was explicitly put forward first by Avogadro, as an obvious corollary to his hypothesis. By the 1830s, Dumas and Mitscherlich had developed the techniques to determine the vapor densities of several solid elements and their compounds. This makes irrelevant the objection often raised that Avogadro's generalization could find only a few applications due to the limited number of gaseous substances then known.

I contend also that by 1815 Berzelius was as confused as any other physical scientist of the time about the link between atoms and volumes, and that in his position as the leading chemist of his generation he was more responsible than others in perpetuating to his death this confusion. Then the question arises: if Berzelius' attempt to solve the problem of consistent atomic weights (the so-called 'independence problem'[24]) was better than Avogadro's, why did Cannizzaro build his own system on the latter's hypotheses rather than on the former's empirical data? As Cannizzaro mentions in the 'Sunto', it was not possible for Berzelius to differentiate between atoms and molecules of the elements, and, accordingly, he could only assume that equal volumes of gaseous elements contained equal number of atoms, and naturally when these atoms combined to form a compound, they did it as a whole. On the other hand, Gerhardt who, like Avogadro, was prone to generalizations, took up both of the latter's assumptions: that all molecules of chemical elements were complex and would divide after combining, and that equal volumes of gases contain equal numbers of molecules. For this, Gerhardt received proper credit by Cannizzaro in the 'Sunto'.[25]

Perhaps the greatest asset of Cannizzaro when he wrote the 'Sunto' was his intellectual independence which allowed him to be daring. Not only was he young when he was drafting his famous essay, but he owed no allegiance to any scientific school. A review of his writings clearly indicates this. While he may have acquired from Raffaele Piria, his early mentor at Pisa, his experimental skill in organic chemistry, yet the cluster of concepts that he developed in 1858 and later on proposed at the meeting in Karlsruhe were his own. The dichotomy 'chemical science vs. physical science' still well rooted in the second half of the nineteenth century,[26] did not seem to represent an obstacle for Cannizzaro. And in this reference, it is ironical that Avogadro in his mature years adopted this dichotomy and used it constantly in his writings.[27]

In examining the attitude of those who seemed to agree with Cannizzaro's proposals after the Karlsruhe Congress, one is not only aware of the influence of their age, but also of the fact that they had studied or taught at Heidelberg, where apparently the diffusion of Gerhardt's new chemistry in the 1850s had been the most widespread. This appears to be in sharp contrast with the conservativism (although an exception could be made for Wurtz) of the French delegation led by Dumas.

Now to the query: was there a need for the molecular hypothesis in 1811? My answer is that a similar question can be raised for the atomic theory. Once the empirical validity of the laws of definite and multiple proportions had been demonstrated, then attempts to characterize such invisible entities as atoms and molecules, and to determine their number, shape, and size certainly moved into the realm of fantasy.[28]

In conclusion, it is difficult to foresee what effect, if any, the recent historical and philosophical criticism will have on future evaluations of Avogadro's scientific work. Some of the points raised are valid, while others cannot be endorsed. A purely internalist methodology, for example, remains highly questionable. To set a man apart from the circumstances of his life and his times makes poor history. There is also the possibility that those who have demolished the old myths are creating new ones of their own. Yet it was important to move Avogadro's most celebrated contribution from pure eulogy to reality, and this has been done.

If, in teaching a science, historical considerations must be introduced, then figments of imagination cannot prevail over history. Accordingly, those who practice a science and those who write its history should welcome periodical re-evaluations of important scientific events, such as the molecular hypothesis. To do otherwise would only lead to bad history and to bad science.[29]

I am most grateful to John H. Brooke and Noel G. Coley for their criticism and valuable comments while I was drafting this section.

NOTES

1. Among the exceptions are Noel G. Coley (*Ann. Science*, **20** (1964), 195); John K. Bonner's dissertation (Johns Hopkins, 1974). Robert Fox's chapter on Avogadro in his *The Caloric Theory of Gases* (Oxford: Clarendon, 1971); and Icilio Guareschi's *Amedeo Avogadro e la sua opera scientifica* (Turin: UTET, 1911).
2. A Piedmontese or a Sicilian could not be identified as an Italian in the early nineteenth century, as a Bavarian could not be properly called a German.
3. Orthodox is used here to mean 'conventionally approved'.
4. It should be stressed here that qualifications such as 'triatomic' and 'tetraatomic' never appeared in Avogadro's writings.
5. See, for instance, the review in the *Journal of the Chemical Society*, **54** (1888), 779 of van't Hoff's work on dilute solutions.
6. The conventional attribution to Avogadro of the diatomic nature of all gasesous elements has been continued to the present. See S. F. Mason in his popular *History of the Sciences*, first published in 1956 as *Main Currents of Scientific Thought*, rev. 1962. For current edition, see Collier Books, New York, 1973, p. 453. The same interpretation of Avogadro can be found in the following:

   C. A. Wurtz, *Histoire des doctrines chimiques* (Paris: Hachette, 1868).
   J. C. Hoefer, *Histoire de la chimie* (Paris: Firmin Didot, 1866).
   H. Kopp, *Geschichte der Chemie* (Braunschweig: Vieweg, 1843–1847).
   M. Pattison Muir, *History of chemical theories and laws* (New York: Wiley, 1907).
   W. Ostwald, *Klassiker der Exacten Wissenschaften*, vol. VIII (Leipzig: Akad. Verlag).
   T. E. Thorpe, *Essays in Historical Chemistry* (London: Macmillan, 1902).
   E. von Meyer, *Geschichte der Chemie* (Leipzig, 1889).
   C. Graebe, 'Der Entwicklungsgang der Avogadroschen Theorie', *J. Prakt. Chem.* 87 (1913) 145.
   I. Guareschi, *Amedeo Avogadro e la sua opera scientifica* (Turin: UTET, 1911).
   M. Giua, *Storia della Chimica* (Turin: Chiantore, 1946).
   F. J. Moore, *A History of Chemistry* (New York: Macmillan, 1939).
   J. R. Partington, *A History of Chemistry*, vol. IV (New York: Macmillan, 1964).

7. Among twelve recent textbooks randomly chosen only R. Eastman in his *General Chemistry* (New York: Holt Rinehart, 1970), suggests that for Avogadro hydrogen could be H_4, H_6, etc.

8. The adoption by Avogadro of geometric symbols as those used by M. Gaudin in 1833 would have made his argument more convincing.
9. See, for instance, W. L. Masterton's chapter devoted to the atomic and molecular theories in his *Chemical Principles* (Philadelphia: W. B. Saunders, 1973).
10. Avogadro's number is discussed within the kinetic theory of gases; and most of the times Avogadro is introduced within the context of N when – it should be well known – that he had nothing to do with the formulation of N.
11. For Linus Pauling, *College Chemistry* (San Francisco: W. H. Freeman, 1964), Avogadro was professor at the University of Turin in 1811 (he was not) and Cannizzaro was professor at the University of Geneva (sic) in 1858. The same error appears in L. Fine, *Chemistry* (New York: Appleton, 1972), p. 35. For R. H. Eastman, *General Chemistry* (New York: Van Nostrand, 1974), and for others, Cannizzaro was a student of Avogadro.
12. I have never found, for instance, in any college chemistry book, mention of the fact that it would have been impossible for Germany to wage four years of World War One without having the help of her mighty chemical industry which, in turn, was fully based upon the vast research programs of the previous thirty years.
13. This is the position taken by Martin Frické in his 'The rejection of Avogadro's hypotheses' in *Method and Appraisal in the Physical Sciences* (London: Cambridge University Press, 1976), p. 277.
14. See H. Gay, 'Radicals and types', *Studies Hist. Phil. Sci.* 7 (1976), 17.
15. For this see J. K. Bonner, 'Amedeo Avogadro: a reassessment of his research and its place in early nineteenth century science', Ph. D. dissertation, Johns Hopkins University, 1974, and Robert Fox, *The Caloric Theory of Gases from Lavoisier to Regnault* (Oxford: Clarendon Press, 1971), p. 225 ff.
16. See J. K. Bonner, chap. 1 and 3 *passim*.
17. See N. Fisher, 'Avogadro's hypothesis: why did the chemists ignore it', paper read at the meeting of the British Society for the History of Science, July 1973, kindly conveyed by author. Now published in *Hist Sci.* **20** (1982), 77 and 212.
18. Ibid., p. 13.
19. For my views of the influence of Berthollet and of the French physical school on the formulation of the molecular hypothesis, see Chapter 4.
20. See M. Crosland, *The Society of Arcueil* (Cambridge, Mass.: Harvard Press, 1967), pp. 358 ff., and M. Crosland, *Ann. Science* **17** (1961), 8.
21. In a recent essay, 'Avogadro's hypothesis and its fate: a case-study in the failure of case-studies', *Hist. Sci.* **19** (1981), 235, John H. Brooke examines the issue of the neglect of the molecular hypothesis in latest philosophical studies, and raises several incisive questions concerning the validity of some arguments put forth by the philosophers of science. I hold that Brooke in his analysis succeeds in contending that some assumptions made in case-studies on the neglect of Avogadro's hypothesis should be considered either as oversimplifications or as unnecessary. He further objects to the lack of appreciation (see Frické, note 13) of the conceptual chasm existing between Avogadro's theory of matter and the atomists. With regard to the influence of Berthollet's affinity program, as previously said, I cannot agree with Bonner's view. For Avogadro the quantification of chemistry was a much more important goal than the quantification of affinities. He saw the importance of using two well-defined parameters – densities and volumes – as a means for the correct determination of the weight of chemical elements and their compounds, as well as their combining ratios. Brooke's argument that usefulness represented a strong factor in persuading chemists after 1860 is certainly well grounded. This view confirms my more general interpretation that Avogadro's gas hypothesis could not be accepted until the appropriate time for its appreciation by *both* chemists and

physicists had arrived. And this, contrary to the prevailing opinion, came long after the conclusion of the Karlsruhe Congress.

22. These points are raised by both Bonner and Frické in their works.
23. See, for instance, Frické. *passim*, esp. pp. 288 and 289.
24. As defined by Frické, p. 285.
25. At the same time Cannizzaro seems to ignore Gerhardt's statement that atoms, equivalents, and volumes are synonymous.
26. On this see, for instance, M. Berthelot's 'On systems of chemical notations' (1878).
27. Cannizzaro apparently was unaware of this. In any case, he never mentioned it in the 'Sunto' and in other essays devoted to the atomic-molecular theory.
28. As A. R. Hall points out in the *Scientific Revolution* (Boston: Beacon, 1967), p. 215, chemists of his time distrusted Dalton as a weaver of idle fancies . . . ; quite likely they held a similar view of Avogadro if they knew of his work.
29. The topic of the relationship between science teachers and historians of science has been examined in a penetrating article by S. Brush, 'Should the history of science be rated X?', *Science*, **183** (1974), 1164. I find myself in agreement with Brush, especially when he suggests (p. 1170) that one cause of the widening gap between historians of science and science teachers may be identified when historians de-emphasize the technical content of some scientific theories, by overly relating them to earlier philosophical and cultural movements.

BIBLIOGRAPHY

Chapter One: Amedeo Avogadro, His Country and His Times

Avogadro, A., *J. Phys. Chim. Hist. Nat.* **63** (1806), 450; **65** (1807), 130.

——, *Archives de l'électricite*, **2** (1842), 102.

——, *Bibl. Universelle*, **29** (1840), 142.

——, *Mem. Mat. Fis. Soc. Ital.* **23** (1844), 156, 260.

——, *Fisica dei corpi ponderabili*, vol. I, Turin, Stamperia Reale, 1837.

Briggs, M. S., *Baroque Architecture*, London, 1913.

Brush, S. G., *Brit. J. Hist. Sci.* **5** (1970), 153.

Causey, R. L., *J. Chem. Educat.* **6** (1971), 365.

Cognasso, F., *Novara e il suo territorio*, Novara, 1952.

——, *Vita e cultura in Piemonte*, vol. II, Turin, 1960.

Cossa, A., *Il Conte Amedeo Avogadro di Quaregna*, Milan, 1898.

Court, S., *Ambix*, **19** (1972), 113.

Crosland, M., *The Society of Arcueil*, Cambridge, Mass., Harvard University Press, 1967.

Dionisotti, C., *Memorie Storiche della citta' di Biella*, Biella, 1864.

——, *Storia della Magistratura Piemontese*, vol. I, Turin, 1881.

Fox, R., *The Caloric Theory of Gases*, Oxford, Clarendon Press, 1971.

Frické, M., *Method and Appraisal in the Physical Sciences*, Cambridge University Press, 1976.

Gioberti, V., *Del primato morale e civile degli italiani*, Turin, 1843.

Hay, D., *Italian Renaissance*, Cambridge University Press, 1961.

Kopp, H., *J. Chem. Soc.* **19** (1866), 158.

Lubatti, O., *Rivista Biellese*, **6** (1956), 17.

Manno, A., *Il Patriziato Italiano*, vol. II, Florence, 1895.

Matteucci, C., *Nuovo Cimento*, **5** (1856), 473.

Meyer, L., *Modern Theories of Chemistry*, London, 1888.

Naumann, A., *Ann. Chem. Pharm.* **123** (1867), 341.

Outram, D., *Hist. J.* **19** (1976), 611.

Passerin, E., *L'Italia nell'etá delle riforme*, Vol. III, Storia d'Italia, 1959.

Piria, R., *Nuovo Cimento*, **6** (1857), 25.

Priestley, J., *History and Present State of Electricity*, London, 1775.

Regnault, H. V., *Ann. Chim.* **73** (1840); **3** (1841).

Sismonda, E., *Mem. R. Acc. Sci. Torino*, ser. 2, **16** (1857), LXI.

Tettoni, L., *Notizie genealogiche-storiche intorno alla nobile, antica ed illustre famiglia Avogadro*, Biella, 1845.

Vaccarino, G., *Boll. Storico Bibl. Subalpino*, **1** (1953).

Vinay, G., *L'umanesimo subalpino nel secolo XV*, Turin, 1935.

William, N., *Journals kept in France and Italy from 1848 to 1852*, London, 1871.
Williams, L. P., *Michael Faraday*, New York, Basic Books, 1965.

Chapter Two: Electrical and Electrochemical Studies

Aepinus, F. U., *Tentamen theoriae electricitatis et magnetismi*, St. Petersburg, 1759.
——, *Hist. Acad, Roy. Sciences, Berlin*, 1756.
Ampère, A. M., *Ann. Chim.* **15** (1820), 59.
Avogadro, A., *J. Phys. Chim. Hist. Nat.* **69** (1809), 142; **63** (1806), 450; **65** (1807), 130.
——, *Ann. Chim.* **87** (1813), 286.
——, *Mem. R. Acc. Sci. Torino*, **27** (1823), 43.
Berthollet, C. L., *Chemical Statics*, London, Mawman, 1804 (transl. L. Lambert)
Berzelius, J., *Ann. Chim.* **86** (1813), 146.
——, *Traité de Chimie*, vol. I. Bruxelles, 1845.
——, *Jahresbericht*, **4** (1825), 29; **7** (1828), 15; **20** (1841), 87; **11** (1832), 21.
Barlow, P., *Edin. Phil. J.* **12** (1825), 105.
Botto, G. D., *Bibl. Univ.* **1** (1834), 205.
Caneva, K., *Dict. Sci. Biogr.* vol. 10, New York, Scribners, 1974.
Cappelletti, V., *Diz. Biogr. Ital.* vol. 4, Rome, Treccani, 1962.
Crosland, M., *Dict. Sci. Biogr.* vol. 1, New York, Scribners, 1970.
Daniell, J., *Phil. Trans.* **126** (1836), 106.
Davy, H., *Elements of Chemical Philosophy*, London, 1812.
De La Rive, A., *Ann. Chim.* **37** (1828), 256.
Fechner, G. T., *Elementar Lehrbuch des Elektromagnetismus*, Leipzig, 1830.
Faraday, M., *Experimental Researches on Electricity*, XI series, Nov. 1837; VIII series, June 1834; XII series, February 1838; III series, Jan. 1833.
Gliozzi, M., *Storia della Fisica*, vol. II, Turin, UTET, 1965.
——, *Opere Scelte di Alessandro Volta*, Turin, UTET, 1967.
Harris, W. S., *Phil. Trans.* **16** (1834), 222.
——, *Phil. Mag.* **26** (1863).
Helmholtz, H., Introd. *Principien der Mechanik*, by H. R. Hertz, Leipzig, 1894.
Hoersted, H., *Ann. Chim.* **32** (1823), 358.
Hoppe, F. I., *Geschichte der Elektrizitat*, Leipzig, 1884.
Lauen, von L., *History of Physics*, New York, Academic Press, 1950.
Michelotti, V., *Mem. R. Acc. Sci. Torino,* **26** (1822), 365.
Ohm, G. S., *Repertorium der Physik*, **1** (1833), 392.
——, *Schweigger J.* **60** (1830), 32; **59** (1830), 385.
Partington, J. R., *A History of Chemistry*, vol. IV, London, Macmillan, 1964.
Peltier, G. C., *Ann. Chim.* **2** (1836), 422.
Pfaff, C., *Neu. J. Phys.* **8** (1794), 196.
Poggendorff, J. C., *Pogg. Ann.* **47** (1839), 123.

Richtie, W., *Phil. Trans. London*, **15** (1833), 313.
Schagrin, M. I., *Amer. J. Phys.* **31** (1963), 536.
Volta, A., *Annali Brugnatelli*, **11** (1796), 84.
Wartmann, E., *Arch. Elec.* **1** (1841), 31.
Williams, L. P., *Hist. Sci.* **1** (1962), 9.
Winter, H. J., *Phil. Mag.* **35** (1944), 376.

Chapter Three: The Molecular Hypothesis: the 1811 and 1814 Essays

Achinstein, P., *Law and explanation*, Oxford University Press, 1971.
Ampère, A. M, *Ann. Chim.* **90** (1814), 43.
Avogadro, A., *Bibl. Univ. Genève*, **2** (1842), 102.
——, 'Essai analytique sur l'électricité', ms, Turin, 1803.
——, 'Considérations sur la nature des substances connues', ms, Turin, 1804.
——, *J. Phys. Chim. Hist. Nat.* **73** (1811), 58; **78** (1814), 154.
——, *Mem. Mat. Fis. Soc. Ital. Sci.* **18** (1820), 153.
——, *Mem. R. Acc. Sci. Torino,* **26** (1821), 1.
——, *Fisica dei corpi ponderabili*, Turin, Stampeira Reale, 1837–41.
Baumé, A., *Manuel de chimie*, Paris, 1753.
Berthollet, C. L., *Chemical Statics*, London: Mawman, 1804.
Bonner, J., 'A. Avogadro: a reassessment of his research . . .', Ph. D. dissertation, Johns Hopkins University, 1974.
Buchdahl, G., *Brit. J. Phil. Sci.* **10** (1959), 126.
Cannizzaro, S., *Nuovo Cimento*, **7** (1858), 301.
Cappelletti, V., *Diz. Biogr. Italiano*, vol. IV, Rome, Treccani, 1962.
Cardwell, D. S. (ed.), *John Dalton and the Progress of Science*, New York, Barnes and Noble, 1968.
Coley, N. G., 'Amedeo Avogadro and the molecular hypothesis, dissertation, University of Leicester, 1964.
Crosland, M., *Brit. J. Hist. Sci.* **6** (1973), 308.
Dalton, J., *Ann. Phil.* **3** (1814), 175.
Davy, H., *J. Phys. Chim. Hist. Nat.* **71** (1810), 326.
Debus, H., *Ueber Einige Fundamental Satze der Chemie*, Kassel, 1894.
Fourcroy, A., *Système des connaissances chimiques*, Paris, 1801.
Fullner, J. Z., *Sir H. Davy's Published Works*, Cambridge, Mass., Harvard University Press, 1969.
Gassendi, P., *Sintagma Philosophicum*, Lyon, 1658.
Gay-Lussac, J. L., *Mem. Phys. Chim. Soc. Arcueil*, **2** (1809), 227.
Giua, M., *Storia della chimica*, Turin, Chiantore, 1946.
Graebe, C., *J. Prakt. Chemie,* **87** (1913), 145.
Greenaway, F., *John Dalton and the Atom*, Ithaca, Cornell University Press, 1966.
Guareschi, I., *Discorso storico-critico*, Turin, UTET, 1911.
——, *Opere scelte di Avogadro*, Turin, Reale Accad. Scienze, 1911.
Jaffe, B., *Crucibles*, New York, Schuster, 1930.
Kay-Shuttleworth, U. J., *First Principles of Modern Chemistry*, London, 1868.

Knight, D., *Atoms and Elements*, London, Hutchinson, 1967.
Kopp, H., *Die Entwickelung der Chemie in der Neuren Zeit*, Munich, 1873.
Lowry, T., *Historical Introduction to Chemistry*, London, Macmillan, 1916.
Matteucci, C., *Nuovo Cimento*, **6** (1857), 24.
McKenzie, A. E., *The Major Achievements of Science*, Cambridge University Press, 1960.
Mauskopf, S. H., *Isis*, **60** (1969), 65.
Meldrum, A., *Avogadro and Dalton*, Edinburgh, W. F. Clay, 1904.
Meyer von E., *History of Chemistry*, London, Macmillan, 1906.
Morselli, M. A., *Ambix*, **27** (1980), 147.
Mossotti, O., *Bibl. Univ. Genève*, **6** (1847), 193.
Mundy, B. W., *Chymia*, **12** (1967), 152.
Nash, L. K., *Atomic-Molecular Theory*, Cambridge, Mass., Harvard University Press, 1957.
Partington, J. R., *History of Chemistry*, vol. 4, London, Macmillan, 1964.
Pauling, L., *Science*, **124** (1956), 708.
Smith, D. M., *The Making of Italy*, New York, Harper and Row, 1968.
Williams, L. P., *Hist. Sci.* **1** (1962), 9.
Wurtz, A., *Histoire des doctrines chimiques*, Paris, 1868.
——, *Comptes Rendus*, **84** (1877), 977, 1183, 1262, 1347.
——, *La Théorie Atomique*, Paris, 1879.

Chapter Four: The Molecular Hypothesis: from 1811 to Cannizzaro's 'Sunto' in 1858

Avogadro, A., *J. Phys. Chim. Hist. Nat.* **73** (1811), 60.
——, *Mem. R. Acc. Sci. Torino*, ser. 2, **1** (1839), 179; ser. 2, **8** (1846), 129; **26** (1821), 60.
——, *Ann. Chim.* ser. 3, **14** (1845), 330.
——, *Giorn. Fis. Chim. Storia Naturale*, **31** (1827), 1
——, *Fisica dei corpi ponderabili*, vol. II, Turin, Stamperia Reale, 1838.
Ampère, A. M., *Opere*, M. Bertolini (ed.), Turin, UTET, 1969.
Baudrimont, M. A., *Comptes Rendus*, **20** (1845), 960.
Berthollet, C. L., *Chemical Statics*, London, Mawman, 1804.
Berzelius, J., *Jahresbericht*, **13** (1834), 32, 63; **14** (1835), 84; **16** (1837), 1; **19** (1840), 370; **26** (1847), 42. *Ann. Phil.* **2** (1813), 450; **5** (1815), 10, 123.
Bonner, J., 'A. Avogadro, a reassessment of his research . . .', Ph. D. dissertation, Johns Hopkins University, 1974.
Brock, W. H., *Isis*, **56** (1965), 5.
Brock, W. H., *J. Chem. Education*, **40** (1963), 652.
Brooke, J. H., *Studies Hist. Phil. Sci.* **4** (1973), 90.
Buchdahl, G., *Brit. J. Phil. Sci.* **10** (1959), 120.
Cannizzaro, S., *Nuovo Cimento*, **6** (1857), 428; **7** (1858), 376.
Coley, N. G., "A. Avogadro and the molecular hypothesis", dissertation, Leicester University, 1964.
Crosland, M., *The Society of Arcueil*, Cambridge, Mass., Harvard University Press, 1967.

Dalton, J., *Ann. Phil.* **9** (1817), 189.
Davy, H., *Phil. Mag.* **37** (1811), 411.
Delacre, M., *Histoire de la Chimie*, Paris, 1920.
Donovan, M., *Elements of Chemistry*, London, 1832.
Dumas, J. B., *Comptes Rendus*, **7** (1838), 474; **10** (1840), 149.
——, *Ann. Chim.* **33** (1826), 337; **52** (1833), 400.
Fox, R., *Brit. J. Hist. Sci.* **4** (1968), 18.
Frankland, E., *Phil Tran.* **142** (1852), 417; **145** (1855), 259.
Freund, I., *The Study of Chemical Composition*, repr. New York, Dover, 1968.
Gaudin, A. M., *Ann. Chim.* **52** (1833), 113.
Gerhardt, C., *Comptes Rendus*, **15** (1842), 498.
Giua, M., *Storia della Chimica*, Turin, UTET, 1946.
Gladstone, S., *J. Chem. Educ.* **24** (1947).
Graebe, C., *J. Prakt. Chemie*, **87** (1913), 145.
Greenaway, F., *John Dalton and the Atom*, Ithaca, Cornell University Press, 1966.
Grimaux, L. E., *Théories et notations chimiques*, Paris, 1884.
Herivel, J. W., *Brit. J. Hist. Sci.* **3** (1966), 121.
Hoefer, F., *Histoire de la Physique et Chimie*, Paris, 1863.
Ihde, A., *Great chemists*, E. Farber (ed.), New York, Interscience, 1961.
Knopp, J., *Ann. Chem. Pharm.* **105** (1858), 390.
Laurent, A., *Ann. Chim.* **18** (1846), 266; **66** (1836), 175.
——, *Comptes Rendus*, **19** (1844), 1089.
Liebig, J., *Ann. Pharm.* **31** (1839), 119.
Mauskopf, S., *Isis*, **60** (1969), 60.
Meyer, von E., *Storia della Chimica*, Milan, Hoepli, 1922.
Mitscherlich, E., *Pogg. Ann.* **29** (1833), 193.
Mollet, J., *J. Phys.* **90** (1820), 113.
Partington, J. T., *A History of Chemistry*, vol. IV, London, Macmillan, 1964.
Pebal, L., *Ann. Chemie*, **123** (1862), 199.
Prout, W., *Ann. Phil.* **6** (1815), 323.
——, *Chemistry, Meteorology, and Function of Digestion*, Bridgewater Treatise VIII, Philadelphia, 1834.
Siegfried, R., *Acts of the 10th International Congress of History of Science*, 1962.
Taton, R., *History of Science*, vol. III, New York, 1964.
Than, A. K., *Ann. Chemie*, **131** (1864), 129.
Wurtz, A., *Histoire des doctrines chimiques*, Paris, Hachette, 1868.
——, *Leçons de philosophie chimique*, Paris, 1863.
——, *Comptes Rendus*, **84** (1877), 977.
Williamson, A., *Phil. Mag.* **37** (1851), 350.
——, *J. Chem. Soc.* **4** (1852), 229.

Chapter Five: The Molecular Hypothesis: after the Karlsruhe Congress

Avogadro, A., *Fisica dei corpi ponderabili*, vol. I, Turin, Stamperia Reale, 1837.

Bernoulli, J., *Opera Omnia*, vol III, Lausanne and Genève, Brosquet, 1742.
Boscovich, R., *De Viribus Vivis*, Rome, Monaldini, 1745.
Botto, G. B., *Mem. R. Acc. Sci. Torino*, ser. 2, **2** (1840), 457.
Brock, W. H. and Harris, J., *Ann. Sci.* **31** (1974), 95.
Brodie, B., *Phil. Trans.* **166** (1866), 781.
——, *Chem. News*, **15** (1867), 295.
Cannizzaro, S., *Scritti attorno alla Teoria molecolare e atomica,* Palermo, 1896.
——, *Nuovo Cimento*, **7** (1858), 323.
Clausius, R., *Ann. Phys. Chem.* Band C, 3 (1857), 353; **103** (1858), 644.
Cooke, J. P., *The New Chemistry,* New York, Appleton, 1874.
Davy, H., *The Elements of Chemical Philosophy*, London, Johnson, 1812.
Divers, E., *Report Brit. Assoc. Adv. Sci.* Sept. 1902, p. 557.
Donnan, F. G., *J. Chem. Soc.* **136** (1933), 316.
Duhem, P., *Aim and Structure of Physical Theory*, 2nd ed., Princeton, Princeton University Press, 1954.
Einstein, A., *Ann. Phys.* **17** (1905), 549.
Faraday, M., *Experimental Researches on Electricity*, Series II (1844), p. 290.
——, *Phil. Mag.* **24** (1844), 136; **28** (1846), 345.
Gouy, L. G., *J. Phys.* **7** (1888), 561.
——, *Comptes Rendus*, 109 (1889), 102.
Graham, T., *Phil. Mag.* 4th series (1864), 81.
Greenaway, F., *John Dalton and the Atom,* Ithaca, Cornell University Press, 1966.
Guareschi, I., *Discorso Storico Critico*, Turin, UTET, 1911.
Herapath, J., *Mathematical Physics*, London, 1847.
Hesse, M., *Forces and Fields*, Totowa, N.J., Littlefield, 1965.
Hiebert, E. W., *Studies in the Philosophy of Science*, vol. IV, Boston, Reidel, 1970.
Joule, J., *BAAS,* **2** (1848), 21.
——, *The Scientific Papers*, London, Dawsons, 1963 (reprint).
Kekulé, F., *The Laboratory,* **1** (1867), 304.
——, *Lehrbuch der Organischen Chemie*, Erlangen, 1867.
Knight, D., *Atoms and Elements*, London, 1967.
——, *Isis,* **56** (1965), 12.
Krönig, R., *Ann. Phys.* **99** (1856), 315.
Leibniz, G. W., *Opera Philosophica*, Berlin, Erdmann, ed. 1840.
——, *Journal des Savans*, 1695, p. 294.
——, *La Monadologie*, Hanover, Erdmann, ed. 1840.
Loschmidt, J., *J. Wiener. Berichte*, 2, **52** (1865), 395.
Mach, E., *Popular Scientific Lectures*, Chicago, 1898.
——, *Die Prinzipen der Physikalischen Optik*, Leipzig, 1821.
Maxwell, J. C., *The Scientific Papers*, New York, Dover, 1965 (reprint).
——, *Nature,* **8** (1873), 437.
Maxwell, J. C., 'Atom', *Encyclopaedia Britannica*, 9th ed., vol. III, 1875.
Meldrum, A., *Avogadro and Dalton: The Standing in Chemistry of their Hypotheses*, Edinburgh, 1906.

Meyer, L., *Modern theories of chemistry* (trans. Bedson), London, Longmans, 1888.
Naumann, A., *Berichte*, **12** (1879), 738.
Nye, M. J., *Molecular Reality*, New York, Neal Watson, 1972.
Odling, W., *Chem. News*, **2** (1860), 226.
Ostwald, W., *Studien zur Energetik*, Sachs, Ges. Wills., 1891.
——, *Vorlesungen über Naturalphilosophie*, 10th ed., Leipzig, Van Leitz, 1905.
——, *J. Chem. Soc.* **85** (1904), 508.
Partington, J. R., *A History of Chemistry*, vol. IV, London, Macmillan, 1962.
Perrin, J., *Comptes Rendus*, **146** (1908), 967; **147** (1980), 530; **149** (1909), 447.
——, *Brownian Movement and the Molecular Reality*, (trans. Soddy), London, 1910.
Pattison Muir, M. M., *A History of Chemical Theories and Laws*, London, Wiley, Taylor and Francis, 1907.
Russell, C. A., *The History of Valency*, New York, Humanities Press, 1971.
Scott, W., *The Conflict Between Atomism and Conservatism*, London, McDonald, 1970.
Silliman, R. H., *Isis*, **54** (1963), 472.
Stallo, J., *The Concepts and Theories of Modern Physics*, Cambridge, Mass., Harvard University Press, 1960, reprint.
Thomson, J. J., *Recollections and Reflexions* (reprint), New York, Macmillan, 1937.
——, *Electricity and Matter*, New York, Scribner, 1904.
Thomson, W., *Proc. Roy. Soc.* **6** (1867), 94; **10** (1883), 185.
——, *Brit. Assoc. Report*, (1884), 613.
——, *The Mathematical and Physical Papers*, vol. V, London, Clay, 1890.
Tyndall, J., *Faraday: A Discoverer*, New York, 1868.
Verdet, M., *Ann. Chim.* **50** (1857), 505.
Walden, P., *Berichte*, Band A (1932), p. 101.
Wiener, C. L., *Ann. Phys.* **118** (1863), 79.
Williams, L. P., *Michael Faraday*, New York, Basic Books, 1965.
Williamson, A. W., *J. Chem. Soc.* **22** (1869), 328.
——, *Chemistry for Students*, Oxford, 1868.
——, *Phil. Mag.* **37** (1850), 356.
Wollaston, W. H., *Phil. Trans.* **104** (1814), 1.

Chapter Six: The Major Chemical Essays (1821)

Arago, D. F., *Mémoires de l'Institut*, vol. I, no. 7 (1806), p. 302.
Avogadro, A., *J. Phys. Chim. Hist. Nat.* **73** (1811), 60.
——, *Mem. R. Acc. Sci. Torino*, **26** (1821), 1, 440.
——, *Fisica dei corpi ponderabili*, vol. II, Turin, Stamperia Reale, 1838.
——, *Giorn. Fisica, Chimica, Storia Nat.* **31** (1827), 1, 42.
Berzelius, J., *Ann. Phys.* **7** (1826), 71; 397; **8** (1827), 177.

——, *Ann. Phil.* **2** (1813), 450; **3** (1814), 51, 360; **5** (1815), 265.
——, *Ann. Chim.* **92** (1814), 142; ser. 2, **11** (1819), 120; **94** (1815), 5.
——, *Lehrbuch*, **IV** (1831), 356.
——, *Jahresbericht*, **2** (1823), 39; **1** (1822), 67.
Berzelius, J. and Dulong, P., *Ann. Chim.* **15** (1820), 386.
Cappelletti, V., *Diz. Biogr. Italiano*, vol. IV, Rome, Treccani, 1962.
Coley, N. G., *Ann. Sci.* **20** (1964), 210.
Dalton, J., *Ann. Phil.* **9** (1817), 188.
Davy, H., *Phil. Trans.* **104** (1814), 557.
De Saussure, T., *Ann. Chim.* **89** (1814), 273.
Dulong, P. and Petit, A., *Ann. Chim.* ser 2, **10** (1819), 395.
Fox, R., *Brit. J. Hist. Sci.* **4** (1968), 13.
Gay-Lussac, J. L., *Mem. Soc. Arcueil,* **2** (1809), 227.
Gerhardt, C. F., *Précis de Chimie Organique*, Paris, 1845.
Graebe, C., *J. Prakt. Chemie,* **87** (1912), 153.
Guareschi, I., *A. Avogadro, Opere scelte*, Turin, UTET, 1911.
——, *A. Avogadro e la teoria molecolare*, Turin, UTET, 1911.
Laurent, A., *Ann. Chim.* **61** (1836), 125.
Liebig, J., *Letters on Chemistry*, London, 1844.
Mundy, B. W., *Chymia,* **12** (1967), 151.
Partington, J. R., *A History of Chemistry*, vol. IV, London, 1962.
Prout, W., *Ann. Phil.* **6** (1815), 321; **11** (1818), 352.
Regnault, H., *Ann. Chim.* **73** (1840), 5.
Thenard, L. J., *Ann. Chim.* **9** (1818), 441.
——, *Traité de Chimie*, vol. IV, Paris, 1816.
Thomson, T., *Ann. Phil.* **4** (1814), 84; **5** (1815), 10; **8** (1816), 71; **7** (1818), 12; **16** (1820), 170.
——, *A System of Chemistry*, Edinburgh, 1817.

Chapter Seven: Avogadro's Opus Magnum: Fisica dei Corpi Ponderabili (1837–1841)

Avogadro, A., *Mem. Mat. Fis. Soc. Ital. Sci.* **23** (1844), 273.
——, *Ann. Chim.* **57** (1834), 124.
——, *Mem. Mat. Fis. Soc. Ital. Sci.* **18** (1820), 153.
——, *Bibl. Univ. Genève,* **29** (1840), 142.
——, *Mem. R. Acc. Sci. Torino,* **30** (1826), 80.
——, *Fisica dei corpi ponderabili*, Stamperia Reale (1837–1841).
Ampère, A. M., *Bibl. Univ. Genève,* **49** (1932), 225.
——, *Ann. Chim.* **58** (1835), 432.
Berzelius, J., *Traité de chimie*, vol. 1, Paris, Firmin Didot, 1845.
Brush, S. G., *Brit. J. Hist. Sci.* **5** (1970), 145.
Cannizzaro, S., *Nuovo Cimento,* **20** (1964), 205.
Coley, N. G., *Ann. Sci.* **20** (1964), 205.
Crosland, M., *The Society of Arcueil*, Cambridge, Mass., Harvard University Press, 1967.

——, *Gay-Lussac: Scientist and Bourgeois*, Cambridge University Press, 1978.
Dulong, P. L., *Ann. Chim.* **10** (1819), 395; **41** (1829), 113.
Fox, R., *Hist. Stud. Phys. Sci.* **4** (1975), 89.
——, *Brit. J. Hist. Sci.* **4** (1968), 1.
——, *The Caloric Theory of Gases*, Oxford, Clarendon Press, 1971.
Faraday, M., *Experimental Researches on Electricity*, Series VIII, 1836.
Levere, T. H., *Affinity and Matter*, Oxford, Clarendon Press, 1971.
Mauskopf, S. H., *Isis*, **60** (1969), 61.
Mitscherlich, E., *Ann. Chim.* **55** (1833), 5.
Mundy, B. W., *Chymia*, **12** (1967), 151.
Poisson, S. D., *Traité de Mécanique*, Paris, Bachelier, 1833.
Prevost, P., *Observations sur la Physique*, **38** (1791), 314.

Chapter Eight: Later Works

Avogadro, A., *Mem. R. Acc. Sci. Torino*, **36** (1833), 215.
——, *Ann. Chim.* **49** (1833), 215.
——, *Mem. Mat. Fis. Soc. Ital. Sci.* **23** (1844), 156, 260; **24** (1850), 166.
——, *J. Phys.* **69** (1809), 285.
——, *Bibl. Univ. Genève*, **11** (1849), 285.
——, *Mem. R. Acc. Sci. Torino*, ser. 2, (1835), 171.
Crosland, M., *Historical Studies in the language of Chemistry*, Cambridge, Mass., Harvard University Press, 1962.
Mossotti, O., *Mem. Mat. Fis. Soc. Ital. Sci.* ser. 2, **24** (1846), 49.
Poisson, S. D., *Mem. de l'Institut*, **12** (1811), n. 37.
Regnault, H., *Mem. Acc. Sci. Paris*, **21** (1847), 329.
Williams, L. P., *The Selected Correspondence of Michael Faraday*, Cambridge University Press, 1971.

NAME INDEX

SUBJECT INDEX